Retro-reflective Beamforming Technique for Microwave Power Transmission

The ACES Series on Computational and Numerical Modelling in Electrical Engineering

Andrew F. Peterson, PhD – Series Editor

The volumes in this series will encompass the development and application of numerical techniques to electrical and electronic systems, including the modelling of electromagnetic phenomena over all frequency ranges and closely related techniques for acoustic and optical analysis. The scope includes the use of computation for engineering design and optimization, as well as the application of commercial modelling tools to practical problems. The series will include titles for senior undergraduate and postgraduate education, research monographs for reference, and practitioner guides and handbooks.

Titles in the Series

K. Warnick, **"Numerical Methods for Engineering,"** 2010.

W. Yu, X. Yang and W. Li, **"VALU, AVX and GPU Acceleration Techniques for Parallel FDTD Methods,"** 2014.

A.Z. Elsherbeni, P. Nayeri and C.J. Reddy, **"Antenna Analysis and Design Using FEKO Electromagnetic Simulation Software,"** 2014.

A.Z. Elsherbeni and V. Demir, **"The Finite-Difference Time-Domain Method in Electromagnetics with MATLAB® Simulations, 2nd Edition,"** 2015.

M. Bakr, A.Z. Elsherbeni and V. Demir, **"Adjoint Sensitivity Analysis of High Frequency Structures with MATLAB®,"** 2017.

O. Ergul, **"New Trends in Computational Electromagnetics,"** 2019.

D. Werner, **"Nanoantennas and Plasmonics: Modelling, design and fabrication,"** 2020.

K. Kobayashi and P.D. Smith, **"Advances in Mathematical Methods for Electromagnetics,"** 2020.

V. Lancellotti, **"Advanced Theoretical and Numerical Electromagnetics, Volume 1: Static, stationary and time-varying fields,"** 2021.

V. Lancellotti, **"Advanced Theoretical and Numerical Electromagnetics, Volume 2: Field representations and the method of moments,"** 2021.

S. Roy, **"Uncertainty Quantification of Electromagnetic Devices, Circuits, and Systems,"** 2021.

ACES

Retro-reflective Beamforming Technique for Microwave Power Transmission

Mingyu Lu and Xin Wang

ACES

The Institution of Engineering and Technology

Published by SciTech Publishing, an imprint of The Institution of Engineering and
Technology, London, United Kingdom

The Institution of Engineering and Technology is registered as a Charity in England &
Wales (no. 211014) and Scotland (no. SC038698).

The Institution of Engineering and Technology
Futures Place
Kings Way, Stevenage
Hertfordshire, SG1 2UA, United Kingdom

www.theiet.org

British Library Cataloguing in Publication Data
A catalogue record for this product is available from the British Library

ISBN 978-1-78561-803-1 (hardback)
ISBN 978-1-78561-804-8 (PDF)

Typeset in India by MPS Limited

Cover Image: Futuristic illustration of the propagation of waves in the information
medium: Denis Pobytov/Digital Vision Vectors via Getty Images

Contents

About the authors   ix

1 Overview: wireless power transmission, microwave power transmission, and retro-reflective beamforming   **1**
  1.1 Wireless power transmission to mobile targets   1
  1.2 Microwave as the carrier of wireless power transmission to mobile targets   5
  1.3 Employing phased array for microwave power transmission to mobile targets   9
  1.4 Basic scheme of retro-reflective beamforming for microwave power transmission   17
  References   21

2 Phased array technique for microwave power transmission applications   **25**
  2.1 Basic principles of the phased array as transmitting antenna   25
    2.1.1 One-dimensional linear phased array   33
    2.1.2 Two-dimensional phased array   36
  2.2 Basic principles of the phased array as receiving antenna   38
  2.3 Power transmission efficiency in "electrical far zone"   42
  2.4 Power transmission efficiency in "electrical near zone"   47
  2.5 One numerical example on the performance of a two-dimensional phased array in the "electrical far zone" and "electrical near zone"   53
  2.6 Power transmission efficiency in "geometrical far zone" and "geometrical near zone"   65
  2.7 Mutual coupling among antenna elements in a phased array   73
  References   81

3 Theoretical principles and practical implementation of retro-reflective beamforming technique for microwave power transmission   **83**
  3.1 Retro-reflective beamforming in "electrical far zone"   83
  3.2 Retro-reflective beamforming in "electrical near zone"   86
  3.3 Retro-reflective beamforming in "electrical far zone" and "electrical near zone": comparative studies   90

3.4   Theoretical analysis of retro-reflective beamforming technique
        using a circuit model   105
3.5   Retro-reflective beamforming in "geometrical near zone" and
        "geometrical far zone"   111
3.6   Practical implementation of retro-reflective beamforming
        technique for microwave power transmission   117
References   130

**4 Retro-reflective beamforming technique for microwave power
    transmission in Internet of Things applications   133**
4.1   Microwave power transmission in Internet of Things   133
4.2   Retro-reflective beamforming scheme for microwave power
        transmission in the Internet of Things   135
4.3   Two experimental examples of the retro-reflective beamforming
        scheme in Section 4.2   139
4.4   Theoretical study of retro-reflective beamforming for wireless
        power transmission to multiple targets (with "targets" standing
        for "wireless power receivers")   151
4.5   Experimental study of retro-reflective beamforming for wireless
        power transmission to multiple targets (with "targets" standing
        for "wireless power receivers")   164
References   177

**5 Retro-reflective beamforming technique for microwave power
    transmission in space solar power applications   179**
5.1   Basic concepts of space solar power   179
5.2   Theoretical model of wireless power transmission from a
        geostationary satellite to the Earth based on retro-reflective
        beamforming technique   181
5.3   Theoretical analysis of retro-reflective antenna array in SSPS
        applications using the formulations of Section 2.1 under the
        condition of electrical far zone   188
5.4   Theoretical analysis of retro-reflective antenna array in SSPS
        applications using the formulations of Section 2.4 under the
        condition of electrical near zone   204
5.5   A bench-scale experimental demonstration of wireless power
        transmission from satellite to Earth   213
5.6   Wireless power reception on the Earth   220
References   226

**6 Retro-reflective beamforming technique for microwave power
    transmission in fully-enclosed space   229**
6.1   Technical concept of wireless power transmission in
        fully-enclosed space   229

6.2 Feasibility study of efficient microwave power transmission
in fully-enclosed space: theoretical analysis   232
6.3 Feasibility study of efficient microwave power transmission in
fully-enclosed space: measurement results   241
6.4 Retro-reflective beamforming based on phased arrays versus
retro-reflective beamforming based on parasitic arrays: general
theoretical analysis   246
6.5 Retro-reflective beamforming based on phased arrays versus
retro-reflective beamforming based on parasitic arrays: theoretical
analysis of a special case with the wireless power transmitter
including two antenna elements   252
6.6 Preliminary numerical and experimental results of microwave
power transmission in fully-enclosed space based on
parasitic arrays   260
References   271

Appendices   **275**
Index   **283**

About the authors

Mingyu Lu is a professor at the Department of Electrical and Computer Engineering, West Virginia University Institute of Technology, West Virginia, USA. He is a senior member of the IEEE, a member of the IEEE MTT-25 Technical Committee, and the treasurer of the IEEE West Virginia Section. He has published three book chapters, 49 journal papers, and more than 100 conference papers.

Xin Wang is a professor at the School of Electrical Engineering, Chongqing University, Chongqing, China. He received BS and MS degrees in electronic engineering from Tsinghua University, Beijing, China, and the PhD degree in electrical and computer engineering from Purdue University, West Lafayette, IN, USA. Xin Wang's research areas include wireless power transmission, reconfigurable antennas, and RF circuits. He has co-authored over 50 academic papers, 1 textbook, and 1 book chapter.

Chapter 1

Overview: wireless power transmission, microwave power transmission, and retro-reflective beamforming

Wireless power transmission is a broad topic. This book focuses on *microwave power transmission*, which is a sub-discipline of wireless power transmission. The specific focus/scope of this book is discussed in the first two sections of Chapter 1 (i.e., Sections 1.1 and 1.2). The next two sections of Chapter 1 (i.e., Sections 1.3 and 1.4) present a brief narrative on how the *retro-reflective beamforming technique* could be applied to accomplish efficient microwave power transmission. The retro-reflective beamforming technique is further elaborated in the other chapters of this book.

1.1   Wireless power transmission to mobile targets

This section and the next section (i.e., Section 1.2) intend to define the scope of this book. This section starts with the general concepts of wireless power transmission and then moves to the specific discipline of wireless power transmission to mobile targets (with "targets" standing for "wireless power receivers" throughout this book).

Electricity has two practical meanings in our everyday life: Power and information. These two practical meanings can be readily illustrated by two outlets on the wall in every household. As depicted in Figure 1.1, a television set has cord connections with two outlets: One is a power outlet and the other is a cable TV outlet. From the television set's point of view, the power outlet is a source of power whereas the cable TV outlet is a source of information. The power delivery over a power cord and the information delivery over a cable TV cord can both be characterized by the physical quantity of the Poynting vector, which is the cross product between the electric field vector and magnetic field vector [1]. Thus indeed, the power transmission over a power cord and the information transmission over a cable TV cord share the same physical nature. To be more specific, in Figure 1.1 electrical power with high power level is propagating over the power cord, whereas electrical power with low power level is propagating over the cable TV cord. Obviously, in the meantime, a television signal is attached to the electrical power propagation over the cable TV cord, whereas no information/signal is attached to

Figure 1.1   Illustration of electrical power and electrical information

the electrical power propagation over the power cord. Correspondingly, the television set has an electrical power receiver at its inlet of the power cord, and it has another electrical power receiver at its inlet of the cable TV cord. Of course, the two electrical power receivers do not perform the same task. When electrical power is received by the television set from the power cord, part of the power is converted to optical power such that the television set's display is bright enough for human vision. When electrical power is received by the television set from the cable TV cord, the information/signal attached to the power is detached such that the television set knows what contents should be displayed.

Electrical information/signals can be transmitted by two possible means: Wired and wireless (it must be noted that a DC signal cannot be transmitted wirelessly). For example, landline telephones and cell phones embody wired voice signal transmission and wireless voice signal transmission, respectively. Electrical power can also be transmitted either by wired means or by wireless means (as a note similar to DC signal, DC power cannot be transmitted wirelessly). The electricity distribution network (i.e., electricity grid) that supplies power to society is the best example of wired electrical power transmission. As the practical applications of wireless power transmission, wireless chargers for electric toothbrushes, cell phones, and smartwatches are popular nowadays. While wired propagation and wireless propagation are based on the same physical laws (which are described by Maxwell's equations [1]), they are governed by different boundary conditions. Specifically, wired propagation follows the boundary conditions specified by transmission lines such as a piece of cable TV cord, whereas wireless propagation satisfies the boundary conditions dictated by the environments such as an urban environment.

Based on the discussions on "power versus information" and "wired versus wireless" above, four combinations are tabulated in Table 1.1, including wired power transmission, wired information transmission, wireless power transmission, and wireless information transmission. The wireless power transmission technology is not highly developed today, compared with the other three in Table 1.1. The

Table 1.1   Practical applications of wired power transmission, wired information transmission, wireless power transmission, and wireless information transmission

	Electrical power		Electrical information	
Wired	Example of wired power transmission: Electricity distribution network		Example of wired information transmission: Landline telephone	
Wireless	Example of wireless power transmission: Wireless charger for smartwatch		Example of wireless information transmission: Cell phone	

four items in Table 1.1 rely on the same fundamental physics. Therefore, the wireless power transmission technology is as feasible as the other three in terms of fundamental physics. For instance, if strong power were broadcasted by a cell tower, one might receive sufficient wireless power to charge the battery of his/her cell phone. Nevertheless, the practical implementation of such a brute-force wireless power transmission is prohibitive due to a large number of practical restrictions. Three major practical concerns relevant to wireless power transmission technology are discussed below.

Power transmission efficiency is the top concern pertinent to wireless power transmission. Power transmission efficiency is defined as the ratio between the amount of received power and the amount of transmitted power. When wireless power is delivered from a cell tower to a cell phone over a long distance (say, 200 m), the power transmission efficiency is very poor. A numerical example is provided in Section 1.3, with the power transmission efficiency calculated to be as low as 0.00025. Obviously, poor power transmission efficiency is equivalent to high financial loss. As a matter of fact, power transmission efficiency is an important metric in virtually every Electrical Engineering application, such as electricity grid and cellular communication, albeit wireless power transmission applications are particularly sensitive to the value of power transmission efficiency.

As the second practical concern, wireless power transmission technology will not be accepted by the general public if it is not safe, that is, if wireless power transmission may cause biological hazards to human beings. While the potential hazards of wireless cell phone signals are still under study, it would be simply unacceptable for a cell tower to boost its broadcasting power in order to charge the battery of a remote cell phone. In fact, a range of regulations have been established to safeguard human safety from excessive exposure to wireless technologies [2,3].

Electromagnetic compatibility is the third vital concern the wireless power transmission technology must take into account. As one example of electromagnetic compatibility, there is a National Radio Quiet Zone in the United States, in which cell phone service is strictly limited in order to protect radio astronomical measurements from possible interferences [4]. Apparently, wireless power transmission applications are anticipated to create stronger interferences than wireless signal transmission applications (cell phone communication, for instance). As a result, the development of wireless power transmission technology must comply with laws/regulations enforced by the government, such as those issued by the Federal Communications Commission of the United States [5].

The practical concerns discussed above do not appear highly challenging when the target (i.e., wireless power receiver) is stationary at a fixed location. One of the classic demonstrations of wireless power transmission to a stationary target is shown in Figure 1.2. In 1975, an experiment carried out by NASA JPL at the Goldstone Deep Space Communications Complex, California demonstrated the delivery of 30 kW of wireless power over one mile, i.e., 1.6 kilometers [6,7]. As shown in the photo of Figure 1.2, a narrow power beam was constructed by a large parabolic antenna toward a stationary wireless power receiver one mile away. Since the target is stationary and the path of wireless power transmission is fixed, it is possible to achieve high power transmission efficiency and avoid potential hazards without tremendous technical difficulties.

When the target is not stationary or when its location is not fixed, the practical difficulties associated with accomplishing efficient and safe wireless power trans-mission increase significantly, compared with the scenarios of stationary targets. As a matter of fact, the rapid development of mobile technologies over the past few decades created a vital demand for wireless power transmission to mobile targets. Today, an ordinary person has to manage the rechargeable batteries of multiple devices such as a cell phone, a tablet computer, a Bluetooth headphone, an electric toothbrush, and an electric shaver. With the advent of the Internet of Things and Personal Area Networks, the number of mobile/portable devices is anticipated to keep growing in the near future. If an individual person possesses more than ten

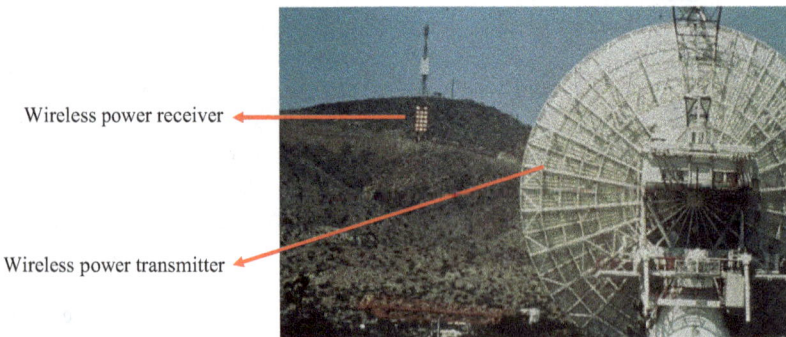

Figure 1.2   A photo of 1975 Goldstone demonstration, in which wireless power was delivered from a stationary transmitter to a stationary receiver. Reproduced from [6], courtesy of NASA.

portable devices, wired charging of these devices will become frustrating. Indeed, the major challenges due to the explosive growth of mobile electronic devices are in the industrial and commercial settings rather than in everyday life. Suppose an electronic tag is attached to each piece of merchandise in a supermarket; the total number of tags the supermarket staff must deal with would be on the order of millions or ten millions. In front of such a large number of mobile/portable devices, wired charging would be practically impossible. Therefore, technologies that could keep track of mobile devices and supply wireless power to them with little human intervention would open the gate to a massive market. This book is motivated by the practical demand for wireless power transmission to mobile targets as well as by the technical difficulties associated with the demand.

The retro-reflective beamforming technique has the potential to address the practical demand for wireless power transmission to mobile targets, as it includes the following two technical elements. First, a directional power beam is generated. Second, the power beam could be steered in real time to aim at a mobile target. The theory and implementation of the retro-reflective beamforming technique are studied in the rest of this book with the aim of accomplishing efficient wireless power delivery to mobile targets subject to the various practical concerns.

1.2   Microwave as the carrier of wireless power transmission to mobile targets

Numerous wireless power transmission technologies have been proposed and are under research currently. This section does not intend to review the existing technologies comprehensively. Instead, with "wireless power transmission to mobile targets" as the goal, the available technologies are assessed and microwave power transmission is identified to be an excellent candidate.

Today, the term "wireless communication" usually refers to transmitting electrical information without using wires/cables (such as in cell phone communication applications), although wireless communication could be fulfilled in non-electrical forms (for instance, the everyday verbal conversation among people is a wireless communication in non-electrical form). Similarly, "wireless power transmission" typically stands for transmitting electrical power without using wires/cables, whereas wireless power transmission does not have to be carried out in the electrical form. For example, many researchers are interested in employing acoustic waves to accomplish wireless power transmission [8]. In acoustic power transmission, electrical power is converted to acoustic power at the transmitter, then the acoustic power propagates to the receiver, and finally, the receiver converts the acoustic power back to electrical power. As far as the theme of this book is concerned (which is wireless power transmission to mobile targets), acoustic power transmission does not appear highly advantageous. For instance, Chapter 5 of this book is pertinent to wireless power transmission in outer space, in which scenario acoustic waves are absent. Acoustic power transmission is particularly appealing in certain media (such as conductive media like human organs and tissues) where wireless transmission in the electrical form may suffer from heavy attenuation. As the applications of delivering wireless power to embedded/implanted devices are not covered by this book, this book places emphasis on "transmitting wireless power in the electrical form" rather than "transmitting wireless power in the acoustic form." It is worth noting that, however, all the basic principles covered by this book (such as those of the phased array technique and retro-reflective beamforming technique) are applicable to not only electromagnetic waves but also acoustic waves.

The technologies for transmitting power in the electrical form wirelessly can be roughly classified into the following three categories.

(i) Technologies based on low-frequency (below 100 MHz, typically) magnetic fields or electric fields.
(ii) Technologies based on high-frequency (above 100 MHz, typically) electromagnetic waves in the microwave, millimeter wave, and Terahertz frequency bands.
(iii) Technologies based on optical waves such as infrared laser and visible laser.

Next, these three categories of technologies are discussed separately.

Inductive coupling is the best-known wireless power transmission technology [9]. Various products based on inductive coupling are commercially available. As an example, wireless charging pads for mobile devices (like cell phones) shown in Figure 1.3(a) are popular nowadays. The inductive coupling technology relies on the low-frequency magnetic flux to achieve the coupling between a wireless power transmitter (a wireless charging pad, for instance) and a wireless power receiver (a cell phone, for instance) [10]. Since low-frequency magnetic fields and low-frequency electric fields are dual to each other in physics, it is unsurprising that capacitive coupling can achieve wireless power transmission as well. For instance, many researchers are endeavoring to charge electric vehicles wirelessly using a low-frequency electric field as demonstrated by Figure 1.3(b) [11]. The coupling efficiency based on low-frequency magnetic field or electric field drops quickly with the increase in distance. Consequently, the mobility of mobile devices (such as cell phones and electric cars) is highly limited when they are charged, as depicted

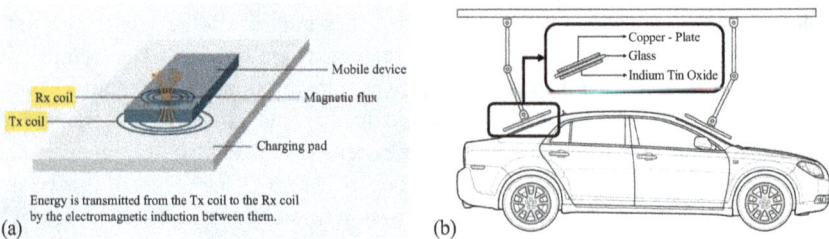

Figure 1.3   Wireless power transmission technologies based on low-frequency magnetic field and electric field. (a) Reproduced from [12], courtesy of JAK Electronics. (b) Reproduced from [11], courtesy of MDPI.

in Figure 1.3. To charge mobile/portable target devices without limiting their mobility in practice, it appears that the wireless charging apparatus has to be bulky and intricate [13–15].

Laser propagation is highly directional and thus could carry wireless power over long distances. As proposed in Figure 1.4(a), a laser beam can be used as the carrier of wireless power to charge cell phones in an indoor environment [16]. A group of German researchers successfully demonstrated delivering wireless power to a rover vehicle through laser in 2004 [17]. In October 2019, the US Naval Research Laboratory demonstrated laser power beaming to the general public at Bethesda, Maryland for 3 days; as shown in Figure 1.4(b), 400 Watts of laser power traveled over 325 m. Although laser power transmission can reach long distances, optical waves suffer from poor penetration capability and poor conversion efficiency between electrical power and optical power. As far as wireless power transmission to mobile targets is concerned, a laser beam, as the carrier of wireless power, is required to be steered in real-time to keep track of mobile targets' location. Though the most straightforward resolution is mounting a laser transmitter over a turn table and then steering the laser beam's direction mechanically, it is not optimal in practice obviously. Numerous techniques of steering laser beams without resorting to mechanical motion (i.e., steering laser beams by electronic means) are under research [18,19], but are not as mature as the phased array technique for steering radio-wave beams.

Compared with the low-frequency regime (i.e., below 100 MHz) and optical regime (i.e., above infrared), the frequency range in between them is an excellent candidate to accomplish efficient wireless power transmission to mobile targets. Enormous research efforts have been reported on employing microwave, millimeter wave, and Terahertz wave frequency bands for wireless power transmission [21–23].

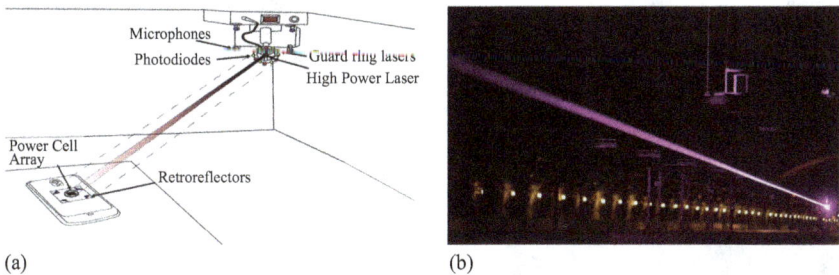

(a)   (b)

Figure 1.4   *Wireless power transmission technologies based on optical wave. (a) A concept of using laser to charge cell phones. Used with permission of ACM (Association for Computing Machinery), from Charging a smartphone across a room using lasers, Vikram et al., 2017, Proceedings of the ACM on Interactive, Mobile, Wearable and Ubiquitous Technologies, vol. 1, no. 4, article 143, permission conveyed through Copyright Clearance Center, Inc. (b) A demonstration in 2019. Reproduced from [20], courtesy of the US Naval Research Laboratory.*

Directional electromagnetic beams can be constructed in these frequency bands, which are analogous to laser beams. More importantly, the phased array technique is well developed to steer the electromagnetic beams via electronic control signals, that is, without any mechanical motion. Beam steering or beamforming by electronic means enables keeping track of mobile targets in real-time, maintaining high power transmission efficiency, and avoiding possible hazards. Particularly, the focus of this book is wireless power transmission based on microwave carriers. Relative to millimeter wave and Terahertz wave, microwave offers better penetration capability and less vulnerability to atmospheric constituents. In addition, a large number of low-cost and mature fabrication processes, components, and circuit schemes are readily available in the microwave frequency band for beam steering or beamforming. Generally speaking, microwave appears more appropriate for wireless power transmission applications than millimeter wave or Terahertz wave, although the situation may evolve with the development of technologies in the future.

Microwave-based wireless power transmission technology, or simply microwave power transmission technology, has a rich history. As a far-from-exhaustive review of the history of microwave power transmission, several historical experiments are highlighted next. In the 1960s, Brown demonstrated supplying microwave power from a ground station to a helicopter (shown in Figure 1.5(a)), which is probably the first impactful and well-documented demonstration of microwave power transmission in history [24,25]. In an experiment carried out at the laboratories of Raytheon in 1975, 54% of power transmission efficiency was measured

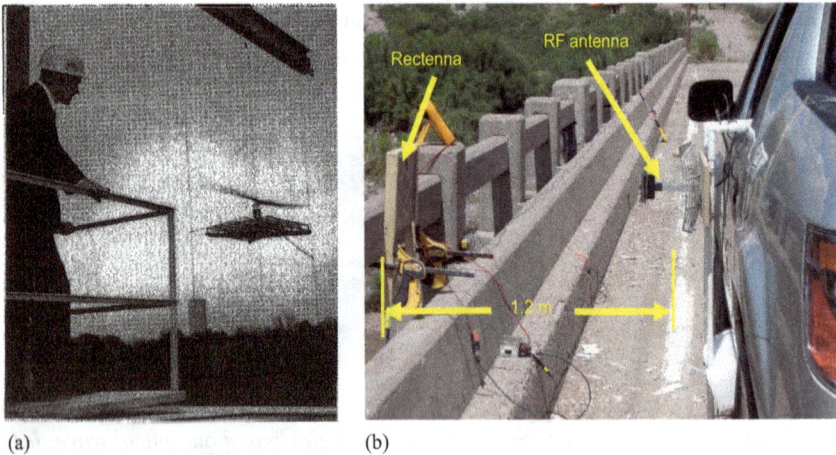

(a)   (b)

Figure 1.5   Two demonstrations of microwave power transmission technology. (a) A demonstration in 1965. Reproduced from [24], courtesy of NASA. (b) A demonstration in 2009. Reprinted from Journal of Sound and Vibration, vol. 329, Mascarenas et al., Experimental studies of using wireless energy transmission for powering embedded sensor nodes, pp 2421–2433, Copyright (2010), with permission from Elsevier.

[26]. The famous Goldstone demonstration in Figure 1.2 was also conducted in 1975 [6,7]. The Stationary High Altitude Relay Program (SHARP) initiated in Canada in the 1980s aimed to provide microwave power to small aircraft [27]. A program similar to SHARP, named MIcrowave Lifted Airplane eXperiment (MILAX), was active in Japan in the 1990s [28]. In 1993, International Space Year - Microwave Energy Transmission (ISY-METS) experiments were conducted in Japan to achieve microwave power transmission between spacecraft [29]. A case study from 1997 to 2004 is reported in [30] to construct a point-to-point wireless electricity transmission to a small isolated village called Grand-Bassin in France. In 2009, the feasibility of using a car-borne power broadcaster to power sensors installed over a bridge was studied in [31], as demonstrated by a photo in Figure 1.5(b). In the 2010s, a range of experiments of microwave power transmission on the ground as well as from ground to a drone were reported [21,32–34]. Quite a few companies are pursuing the commercialization of microwave power transmission technology, although no commercial products based on microwave power transmission have been developed to date [35–38].

While microwave power transmission technology has the potential to accomplish efficient wireless power delivery to mobile targets, its practical implementation involves a large number of technical/engineering problems, which can be roughly classified into the following three groups.

- Technical problems on efficient conversion from DC power to microwave power at the wireless power transmitter.
- Technical problems in generating and reconfiguring a narrow microwave beam from the wireless power transmitter to the wireless power receiver.
- Technical problems on efficient conversion from microwave power to DC power at the wireless power receiver.

This book focuses on the beamforming problems of the second group above. The readers are referred to [39], a book authored by Professor Naoki Shinohara in 2014, for the other technical issues of microwave power transmission.

1.3 Employing phased array for microwave power transmission to mobile targets

As discussed in Section 1.2, *phased array* is the primary enabling technique for microwave power transmission to be an excellent candidate to accomplish wireless power delivery to mobile targets. Specifically, the phased array technique enables generating a narrow microwave beam and then reconfiguring the microwave beam via electronic control. In this section, the underlying theory of the phased array technique is presented. A more systematic narrative of phased array for microwave power transmission applications is provided in Chapter 2. This section can be considered a succinct version of Chapter 2.

As depicted in Figure 1.6, one antenna element resides at the spatial origin and the entire space is composed of free space. The antenna element is excited by a

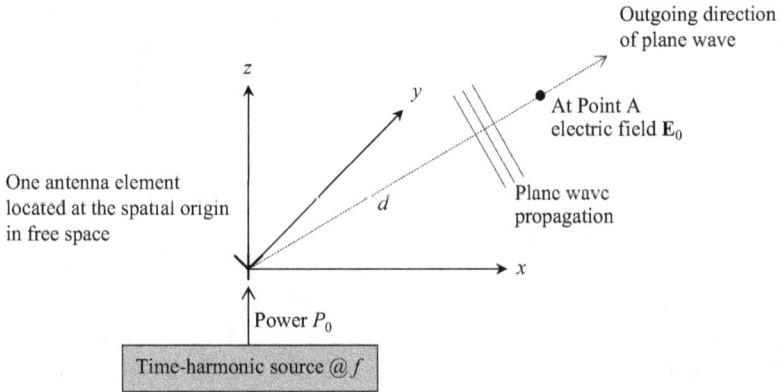

Figure 1.6   Radiation of one antenna element located at the spatial origin

time-harmonic source at a frequency f with a time-average power of P_0. At an observation point A, the electric field radiated by the antenna element is denoted by a phasor vector $\mathbf{E}_0$. It is assumed that Point A is located in the *far zone*; in other words, d, the distance between Point A and the origin, is large enough. "Far zone" will be defined explicitly in Chapter 2. Under the far zone condition, the electromagnetic field in the neighborhood of Point A resembles a plane wave traveling in the outgoing direction. The time-average power density at Point A is

$$\frac{|\mathbf{E}_0|^2}{2\eta_0}, \text{ with unit of } \frac{\text{Watt}}{\text{m}^2} \tag{1.1}$$

In (1.1), the operator "$|\cdot|$" selects the magnitude of the argument, $\eta_0 = \sqrt{\mu_0/\varepsilon_0}$ is the intrinsic impedance of free space, $\varepsilon_0 = 8.85 \times 10^{-12}$ (F/m) is the permittivity of free space, and $\mu_0 = 4\pi \times 10^{-7}$ (H/m) is the permeability of free space. Following [40], the gain value of the antenna element toward Point A is

$$G_0 = \frac{\left(\frac{|\mathbf{E}_0|^2}{2\eta_0}\right)}{\left(\frac{P_0}{4\pi d^2}\right)} \tag{1.2}$$

The denominator on the right-hand side of (1.2) stands for the "isotropic power density," which is the power density when power P_0 is uniformly distributed over a fictitious spherical surface centered at the origin and with d as the radius. Thus, antenna gain value is usually characterized by the unit of "dBi" in the decibel scale, with "i" standing for "isotropic radiator." If a receiving antenna terminated by a matched load is placed at Point A, the power P_r at the receiving antenna's circuit port is

$$P_r = \frac{|\mathbf{E}_0|^2}{2\eta_0} A_e, \tag{1.3}$$

where A_e is the effective aperture of the receiving antenna. The effective aperture and gain value of the receiving antenna are related to each other [40]:

$$\frac{A_e}{G_r} = \frac{(\lambda_0)^2}{4\pi}, \tag{1.4}$$

where G_r is the gain value of the receiving antenna toward the spatial origin, $\lambda_0 = c/f$ is the wavelength in free space, and $c = 1/\sqrt{\varepsilon_0\mu_0}$ is the speed of light in free space. After (1.2) and (1.4) are substituted into (1.3), the Friis transmission equation is arrived at:

$$\frac{P_r}{P_t} = \frac{P_r}{P_0} = \left(\frac{\lambda_0}{4\pi d}\right)^2 G_0 G_r. \tag{1.5}$$

In (1.5), P_r/P_t is the power transmission efficiency between the received power P_r and transmitted power $P_t = P_0$ when there is only one antenna element in the transmitter.

As depicted in Figure 1.7, an array composed of two antenna elements resides around the spatial origin. The two antenna elements are both identical to the antenna element in Figure 1.6. Each antenna element is excited by a time-harmonic source at a frequency f with a power of P_0. The observation point A is located in the far zone of the antenna array. Under the far zone condition, the total electro-magnetic field at Point A is the superposition of two plane waves radiated by the two antenna elements, respectively. Suppose there is no coupling between the two antenna elements. The electric field produced by the first antenna element, $\mathbf{E}_1$, is different from $\mathbf{E}_0$ by a certain phase, and the phase difference is determined by the time-harmonic source attached to the first antenna element. Similarly, $\mathbf{E}_2$, the electric field produced by the second antenna element is different from $\mathbf{E}_0$ by a

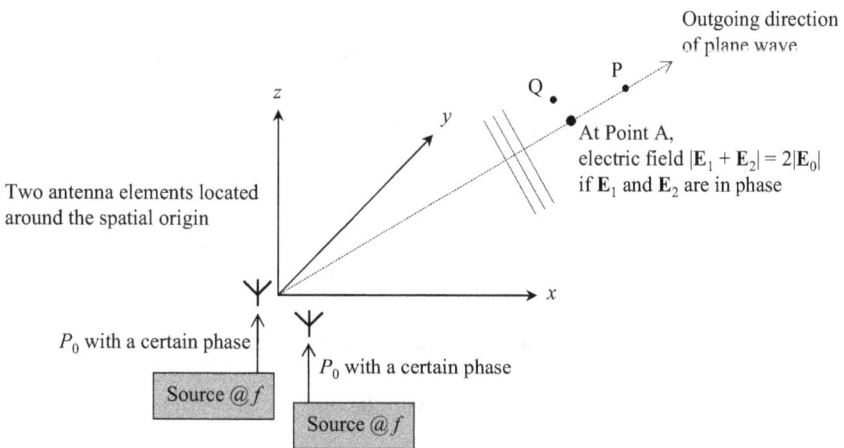

Figure 1.7   Radiation of an array composed of two antenna elements located around the spatial origin

certain phase, and the phase difference is determined by the time-harmonic source attached to the second antenna element. When the two time-harmonic sources are adjusted such that $\mathbf{E}_1$ and $\mathbf{E}_2$ are in phase (i.e., $\mathbf{E}_1$ and $\mathbf{E}_2$ share the same phase), $|\mathbf{E}_1 + \mathbf{E}_2| = 2|\mathbf{E}_0|$. The gain value of the antenna array toward Point A is

$$G_2 = \frac{\left(\frac{|\mathbf{E}_1 + \mathbf{E}_2|^2}{2\eta_0}\right)}{\left(\frac{2P_0}{4\pi d^2}\right)} = \frac{\left(\frac{4|\mathbf{E}_0|^2}{2\eta_0}\right)}{\left(\frac{2P_0}{4\pi d^2}\right)} = 2G_0. \tag{1.6}$$

In (1.6), the total transmitted power is $2P_0$, under the assumption that there is no power coupling between the two antenna elements. As a result of (1.6), the power transmission efficiency increases by a factor of 2, compared with the scenario in Figure 1.6.

$$\frac{P_r}{P_t} = \frac{P_r}{2P_0} = \left(\frac{\lambda_0}{4\pi d}\right)^2 G_2 G_r = \left(\frac{\lambda_0}{4\pi d}\right)^2 (2G_0)G_r. \tag{1.7}$$

Compared with (1.5), the enhancement of power transmission efficiency in (1.7) is because the two antenna elements' radiations are in phase at Point A. As illustrated in Figure 1.7, suppose there are two other observation points, namely Point P and Point Q. Point P is located along the direction from the origin to Point A, i.e., along the direction of the plane wave's propagation direction. It is not difficult to show that, under the far zone condition, the two elements' radiations are in phase at Point P when they are in phase at Point A (as detailed in Chapters 2 and 3 of this book). However, since Point Q deviates from the plane wave's propagation direction, it is unlikely that the two antenna elements' radiations are still in phase. Therefore, the electromagnetic field distribution produced by the two antenna elements behaves like a directional beam toward Point A. The beam width is dictated by the gain value: The larger the gain value is, the narrower the beam is. In addition, the beam can be steered toward another direction if the two time-harmonic sources' phases are adjusted to another set of values.

With more antenna elements included, the antenna array will have greater beam steering or beamforming capabilities. As portrayed in Figure 1.8, a two-dimensional array is constructed with $M \times N$ antenna elements that are identical to one another. In Figure 1.8, the antenna elements are assumed to be planar antennas like patch antennas and slot antennas. As a result, the array has a low profile and can be conformal to the mounting surface, which is highly desirable in practice. Each antenna element is excited by an individual time-harmonic source; in other words, there are $M \times N$ time-harmonic sources in total. The excitation power generated by each time-harmonic source is P_0. When the phase values of the time-harmonic sources are adjusted such that all the elements' radiations are in phase at Point A, the magnitude of the electric field at Point A is $MN|\mathbf{E}_0|$, where $\mathbf{E}_0$ is the electric field when there is only one antenna element active. The gain value associated with the directional beam toward Point A is

$$G_{array} = \frac{\left(\frac{M^2 N^2 |\mathbf{E}_0|^2}{2\eta_0}\right)}{\left(\frac{MNP_0}{4\pi d^2}\right)} = MNG_0. \tag{1.8}$$

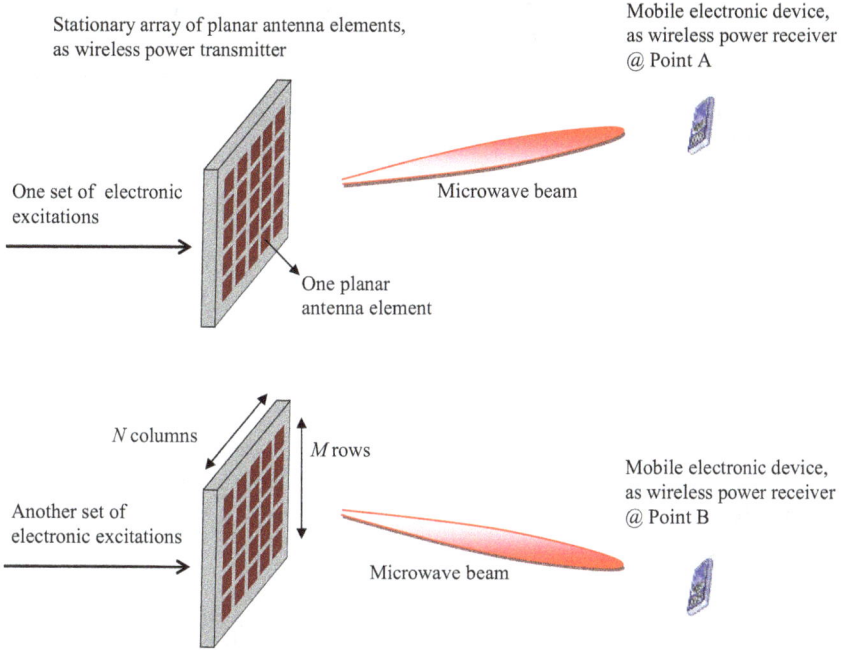

Figure 1.8   Illustration of a two-dimensional phased array with $M \times N$ antenna elements

Apparently, with the increase of $M \times N$ (i.e., the number of antenna elements), G_{array} becomes larger and the beam becomes narrower. The power transmission efficiency between the array and a wireless power receiver located at Point A is

$$\frac{P_r}{P_t} = \frac{P_r}{MNP_0} = \left(\frac{\lambda_0}{4\pi d}\right)^2 G_{array}G_r = \left(\frac{\lambda_0}{4\pi d}\right)^2 (MNG_0)G_r. \tag{1.9}$$

If the beam is required to be steered (toward Point B in Figure 1.8, for instance), adjusting the phase values of the $M \times N$ time-harmonic sources suffices while the antenna array stays stationary. The technique illustrated by Figure 1.8 is therefore termed the *phased array* technique.

The inter-element spacing is an important parameter of a phased array [41]. As derived in Chapter 2 of this book, grating lobes (which are undesired beams as strong as the desired beam) may appear if the inter-element spacing is larger than $\lambda_0/2$. When the inter-element spacing is larger than λ_0, the range of beam steering would be highly limited to avoid grating lobes. Consequently, it is not common for the value of inter-element spacing to be greater than λ_0 in practice. On the other hand, small inter-element spacing may cause various practical complications as well. Specifically, if the inter-element spacing is smaller than $\lambda_0/2$, each antenna element is forced to have a small electrical size and the coupling among antenna elements may become strong [42,43]. Thus in practice, the value of inter-element spacing is not smaller than $\lambda_0/2$ typically. In the several experimental studies presented in Chapters 3, 4, and 5 of this book, the inter-element spacing is chosen between $\lambda_0/2$ and λ_0. In Figure 1.9, the inter-element spacing is assumed to be $0.7 \times \lambda_0$, which is the center of the range $(\lambda_0/2, \lambda_0)$ roughly. The physical dimensions of the phased array are characterized by $L = M \times 0.7 \times \lambda_0$ and $W = N \times 0.7 \times \lambda_0$ approximately. Equation (1.9) can be re-arranged to be

$$
\begin{aligned}
\frac{P_r}{P_t} &= \left(\frac{\lambda_0}{4\pi d}\right)^2 (MNG_0)G_r \\
&= \left(\frac{1}{4\pi d}\right)^2 \frac{(M \times 0.7\lambda_0) \times (N \times 0.7\lambda_0)}{0.7 \times 0.7} G_0 G_r \\
&\cong \left(\frac{1}{4\pi d}\right)^2 (2LW)G_0 G_r
\end{aligned}
\tag{1.10}
$$

When a phased array is applied to wireless power transmission, Equation (1.10) can be used to estimate the power transmission efficiency. Two numerical examples are presented below.

Numerical example 1

Suppose a phased array with physical dimensions of 10 m by 10 m is set up at a cell tower for the purpose of wireless power transmission. Specifically, the power level of the cell phone signal is boosted to charge a cell phone that is 200 m away from the tower and has line-of-sight interaction with the tower. Each antenna element of the phased array is assumed to be a planar antenna with a gain value of $G_0 = 4 = 6$ dBi. The antenna of a cell phone is assumed to have a gain value of $G_r = 2 = 3$ dBi. Following (1.10), the power transmission efficiency is

$$
\begin{aligned}
\frac{P_r}{P_t} &= \left(\frac{1}{4\pi d}\right)^2 (2LW)G_0 G_r \\
&= \left(\frac{1}{4\pi \times 200}\right)^2 (2 \times 10 \times 10) \times 2 \times 4 \\
&\approx 0.00025
\end{aligned}
$$

Array of $M \times N$ planar antenna elements, as wireless power transmitter

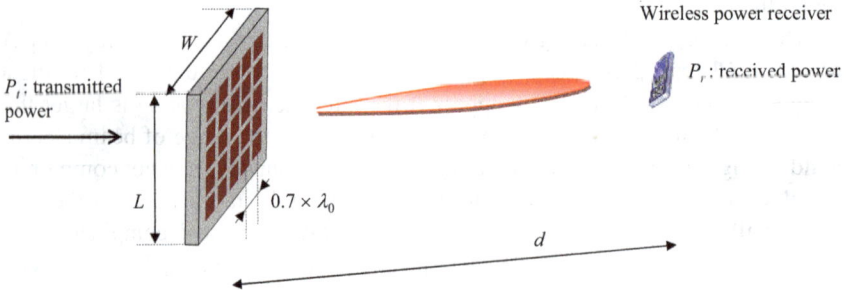

Figure 1.9   Illustration of Equation (1.10)

If the cell phone needs 2.5 Watts of power to charge its rechargeable battery, the power transmitted by the cell tower ought to be 10,000 Watts. The difference between the transmitted power and received power, $10,000 - 2.5 = 9,997.5$ Watts, not only constitutes a tremendous financial loss but also may pollute the environment.

Numerical example 2

In an indoor environment, a phased array with physical dimensions of L by W is employed to deliver wireless power to a mobile electronic device. The phased array and mobile device have line-of-sight interaction. Each antenna element of the phased array is assumed to be a planar antenna with a gain value of $G_0 = 4 = 6$ dBi. The antenna of the mobile device is assumed to have a gain value of $G_r = 2 = 3$ dBi. Following (1.10), the power transmission efficiency is

$$
\begin{aligned}
\frac{P_r}{P_t} &= \left(\frac{1}{4\pi d}\right)^2 (2LW)G_0G_r \\
&= \left(\frac{1}{4\pi d}\right)^2 (2LW) \times 2 \times 4 \\
&= \frac{1}{\pi^2} \times \frac{L}{d} \times \frac{W}{d}
\end{aligned}
$$

If $\frac{L}{d} = \frac{W}{d} = 0.3$, the power transmission efficiency is approximately 1%. This numerical example echoes the conclusion of the previous numerical example: It is extremely difficult, if not impossible, to deliver wireless power to remote power-hungry devices in practice, such as cell phones. If a cell phone demands 2 Watts of wireless power, the transmitted power would be as prohibitive as 200 Watts in an indoor environment. Meanwhile, this numerical example indicates that wireless charging for low-power mobile devices is highly feasible. If 10 mW of power is needed by a low-power device, the transmitted power is about 1 Watt. The power loss of $(1\ \text{W} - 10\ \text{mW} = 990\ \text{mW})$ is tolerable in most of the practical scenarios. At the same time, the possible negative impact associated with a 1-Watt wireless power transmitter is not a serious concern, as it is very common that the power level radiated by a regular cell phone is higher than 1 Watt. Remote delivery of wireless power on the order of milli-Watts, though not sounding extremely exciting, is still valuable in practice. As discussed in Section 1.1, a large number of low-power electronic devices in industrial and commercial settings, such as radio frequency identification tags and wireless sensors, may benefit from wireless power transmission, particularly when wired charging is intractable [44]. In an indoor environment, d would not exceed 10 m typically. With d being 10 m, $L = W = 3$ m if 1% of power transmission efficiency is demanded. As illustrated in Section 1.4, the aperture of $3 \times 3 = 9$ m^2 can be decomposed into multiple smaller apertures distributed in space if one piece of 9-m^2 aperture cannot be accommodated by an indoor environment.

    In fact, the validity of the second numerical example above is questionable. The condition of "far zone" is mentioned several times during the derivation of

(1.10) in this section. In the second numerical example above, however, $\frac{L}{d} = \frac{W}{d} = 0.3$ does not appear to be "far enough." As a matter of fact, the well-known far-field condition of antenna engineering is not satisfied in the first numerical example either (the far-field condition is derived in Chapter 2). Chapter 2 of this book is devoted to investigating the validity of (1.10). As a conclusion of Chapter 2, the well-known far-field condition is not a necessary condition of (1.10). Rather, the validity of (1.10) relies on the following four conditions.

- At the receiving antenna, the electromagnetic field intensity of a phased array with $M \times N$ elements is greater than the electromagnetic field intensity of one single antenna element by a factor of $M \times N$.
- There is no mutual coupling among the antenna elements of the phased array.
- The physical dimension of the wireless power receiver does not exceed a *plane wave region* (the "plane wave region" is defined in Chapter 2).
- The distance between the wireless power transmitter and wireless power receiver (that is, d) is greater (but does not have to be much greater) than the physical size of the wireless power transmitter and wireless power receiver.

As elaborated in Chapter 2, as long as the above four conditions hold true, Equation (1.10) can be used to estimate the power transmission efficiency between a phased array and a mobile target. In Chapter 2, extensive analysis is conducted for the scenario of the second numerical example above, which is "$d = 10$ m, $L = 3$ m, and $W = 3$ m."

Although frequency f does not appear in (1.10), Equation (1.10) provides certain insights toward the selection of operating frequency. The wavelength in free space, λ_0, is determined by the frequency through $\lambda_0 = c/f$. Typically, the physical size of an antenna element in a phased array is close to $\lambda_0/2$, and the inter-element spacing is between $\lambda_0/2$ and λ_0. If the frequency is too low, the physical dimension of the phased array would be unreasonably large in practice. For instance, when the frequency is 100 MHz, the corresponding λ_0 is 3 m, and thus the size of an array with as few as 2 by 2 elements would probably reach 5 m by 5 m. On the other hand, if the frequency is too high, there would be too many antenna elements. As indicated by (1.10), the power transmission efficiency is directly dependent on the area of $L \times W$. Given a certain desired power transmission efficiency, it is possible that L and W must be as large as 3 m in practice. If 30 GHz is selected as the operating frequency, which corresponds to $\lambda_0 = 0.01$ m, the number of antenna elements would be more than 100,000. Controlling such a huge number of antenna elements calls for complicated and expensive circuitries. In this book, the operating frequency is selected between 2 GHz and 6 GHz, which is probably the close-to-optimal frequency range for microwave power transmission after numerous issues/factors are taken into account [45].

In this section, the phased array is assumed to be part of a transmitter. Reciprocal to the beam steering or beamforming when a phased array is used in a transmitter, the receiving pattern of a phased array exhibits a reconfigurable beam when it is used in a receiver. To be specific, a phased array as a receiving antenna is only sensitive to the signal incoming from a certain direction, and the incoming direction to which the phased array is sensitive can be reconfigured via electronic control. Consequently, a phased array is capable of detecting the incoming direction of signals, or the direction of arrival. Reciprocal to the fact that the beam radiated by a phased array becomes narrower when more antenna elements are included in the array, the beamwidth in the receiving pattern decreases when the number of antenna elements in the array increases. In the numerical examples of this section, the value of "$G_r = 2$" is based on two assumptions: The target only has one antenna element, and the receiving pattern of the target's antenna is almost isotropic. Incorporating an antenna array at the target may increase G_r, and in turn, may enhance the power transmission efficiency. Because a mobile target does not have a large physical size typically, incorporating an antenna array over a mobile device does not appear probable in the frequency range of [2 GHz, 6 GHz] but may be possible in higher frequency bands.

1.4   Basic scheme of retro-reflective beamforming for microwave power transmission

This section presents a brief description of how the retro-reflective beamforming technique could enable efficient microwave power transmission to mobile targets. The theory and implementation of retro-reflective beamforming are detailed in Chapter 3 of this book.

Retro-reflectivity has widespread applications in numerous disciplines, whereas its basic concept can be understood readily in optics. In Figure 1.10, the difference between ordinary reflectivity and retro-reflectivity is illustrated using optical waves. When an incident light ray impinges upon an ordinary surface, the reflected light ray's direction is determined by both the incident light ray's direction and the direction normal to the surface. For a retro-reflective surface, however, the direction of the reflected light ray always follows the direction of the incident light ray and does not depend on the direction normal to the retro-reflective surface.

One of the applications of retro-reflectivity in optical engineering is traffic signs, as demonstrated in Figure 1.11. When a car sheds light onto a traffic sign at night, the reflected light is desired to be returned to the driver rather than toward any other direction. Therefore, traffic signs made of retro-reflective surfaces appear more bright/visible at night.

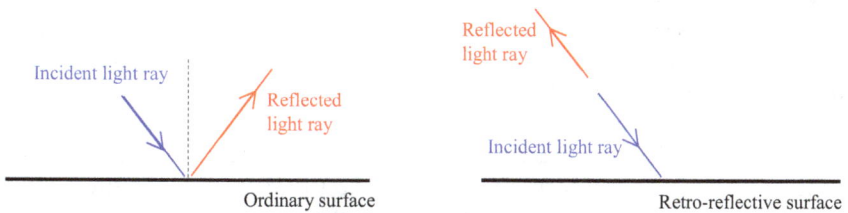

Incident light ray

Reflected light ray

Ordinary surface

Reflected light ray

Incident light ray

Retro-reflective surface

Figure 1.10   An ordinary surface versus a retro-reflective surface in optics

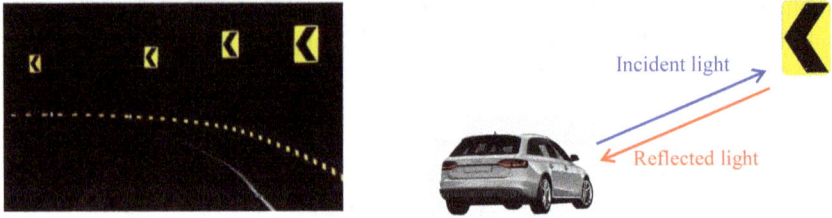

Figure 1.11   Traffic signs made of retro-reflective surfaces appear bright at night

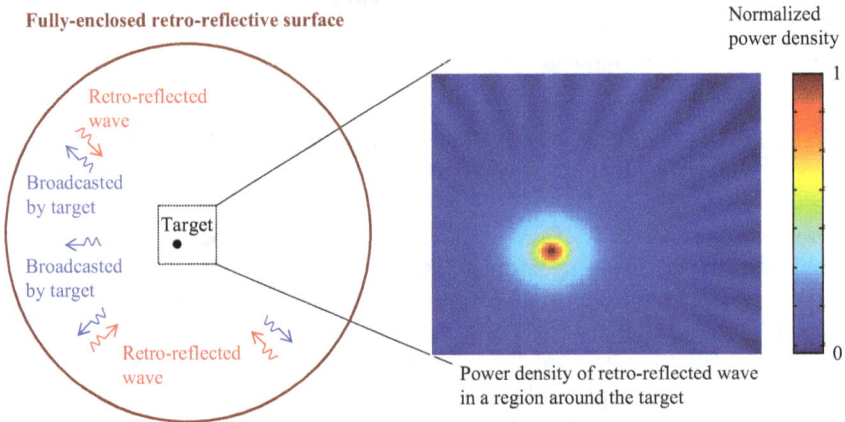

Figure 1.12   Conceptual illustration of retro-reflective beamforming technique

The retro-reflective beamforming technique is inspired by the concept of retro-reflectivity. As illustrated in Figure 1.12, suppose a point target emits waves toward all the directions in the space. If a fully enclosed retro-reflective surface is built around the target, the retro-reflected waves would converge onto the target's location. The technical scheme depicted in Figure 1.12 is sometimes referred to as a "time-reversal sink," that is, it appears that the retro-reflected waves are absorbed by the target. The functionality of the retro-reflective surface can be appreciated by the "rewind" button of a movie player. If the propagation of waves broadcasted by the point target is considered as a movie, the propagation of retro-reflected waves appears like the rewound movie. Specifically, the waves broadcasted by the target diverge in the space whereas the retro-reflected waves converge onto the target. It is worthwhile noting that the retro-reflective beamforming scheme in Figure 1.12 is not limited to optical waves or electromagnetic waves; for instance, an acoustic time-reversal sink is researched in [46].

If the retro-reflective surface in Figure 1.12 is active, i.e., if the retro-reflective surface is attached to a power supply, the retro-reflected waves would be stronger than the waves broadcasted by the target. The resultant retro-reflective beamforming scheme has the potential to achieve efficient wireless power transmission. When a certain target is in need of wireless power, it broadcasts a signal in all directions as a "request for wireless charging"; in this book, the signal that requests wireless charging is termed a *pilot signal*. When the pilot signal hits the retro-reflective surface, the retro-reflected waves are employed as the carrier of wireless power. As shown in Figure 1.12, the spatial distribution of wireless power carried by the retro-reflected waves demonstrates a focal point at the target. In other words, the wireless power transmission resulting from retro-reflective beamforming is dedicated to the target in space. Obviously, the spatially-dedicated wireless power transmission displayed in Figure 1.12 is the key to addressing the efficiency, safety, and electromagnetic compatibility issues raised in Section 1.1. In terms of power level, the retro-reflected waves are much stronger than the pilot signal. Overall, when the retro-reflective beamforming technique is applied to accomplish wireless power transmission, the propagation of wireless power is guided or "ushered" by the propagation of the pilot signal.

An active retro-reflective surface can be implemented by phased arrays in the microwave frequency band. Thus, it is practically viable to achieve efficient microwave power transmission by using the retro-reflective beamforming technique. The retro-reflective beamforming technique for microwave power transmission is illustrated in Figure 1.13. The wireless power transmitter includes one or more than one phased array. A target (i.e., a wireless power receiver) receives wireless power from the wireless power transmitter via the following two steps.

Step (i)   The target broadcasts a pilot signal. The pilot signal is received and analyzed by the phased array(s) of the wireless power transmitter.

Step (ii) Based on the outcome of analyzing the pilot signal, the phased array(s) construct focused microwave power beam(s) onto the location of the target.

The phased arrays' aperture in Figure 1.13 behaves as a retro-reflective surface. A phased array could fulfill the functionality of a retro-reflective surface as it is capable of detecting the incoming direction of the pilot signal and then constructing microwave power beam(s) accordingly. When a phased array is active (that is, when it is attached to a power supply), the power beam(s) in Step (ii) have a higher power level than the pilot signal in Step (i). If the target is not stationary, the microwave power beam(s) would follow the target's location dynamically as long as the target periodically broadcasts the pilot signal. Intuitively, the larger the phased arrays' aperture is, the better the performance of wireless power transmission would be (although building a fully-enclosed retro-reflective surface as depicted in Figure 1.12 using phased arrays is prohibitive in practice).

The underlying principle of retro-reflective beamforming in Figure 1.13 is time-reversal, which takes advantage of channel reciprocity to accomplish a space-time matched filter [47]. Specifically, the propagation of the pilot signal follows the channel from the target to the wireless power transmitter, whereas the propagation of microwave power follows the channel from the wireless power transmitter to the target. If these two channels are reciprocal to each other and if the microwave power excitation is configured to be the time-reversed version of pilot signal reception, the microwave power propagation would be spatially focused onto the

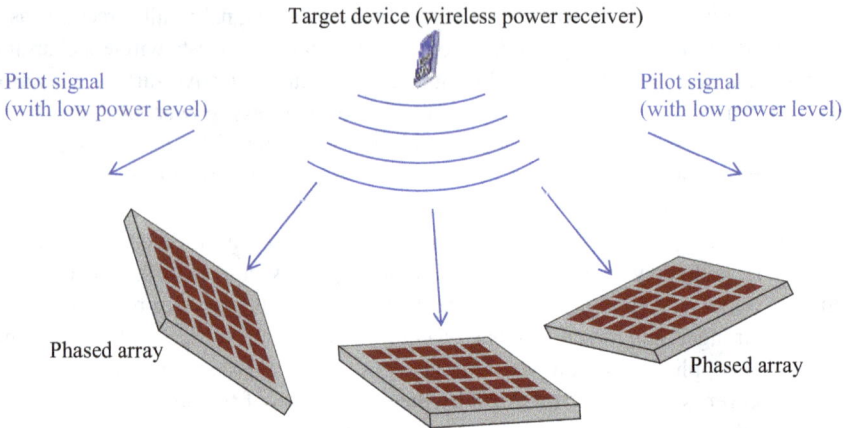

(a)

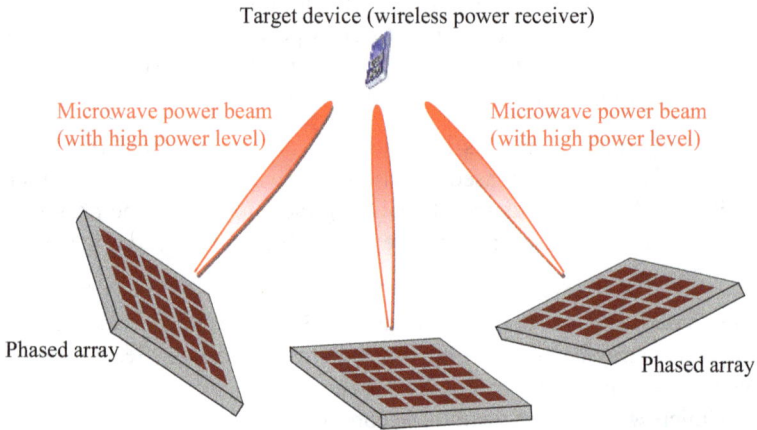

(b)

Figure 1.13   Two-step procedure of retro-reflective beamforming for wireless power transmission (a) Step (i): Pilot signal is broadcasted by target and detected by phased arrays and (b) Step (ii): Microwave power transmitted by phased arrays converges onto target.

location from which the pilot signal stems, that is, the location of the target. Furthermore, spatial focusing due to retro-reflection/time-reversal does not suffer from multi-path in environments [48,49].

The retro-reflective beamforming technique for microwave power transmission applications is elaborated in the rest of this book. It should be noted that retro-reflective beamforming has widespread applications (in wireless communication and radar, for instance), albeit this book focuses on wireless power transmission applications.

References

[1] Balanis C.A. *Advanced Engineering Electromagnetics*. 2nd ed. Hoboken, NJ: John Wiley & Sons; 2012.

[2] IEEE C95.1-1991: Safety levels with respect to human exposure to radio frequency electromagnetic fields, 3 kHz to 300 GHz. (2005).

[3] International Commission On Non-Ionizing Radiation Protection. [Online]; [cited 2023]; Available from: https://www.icnirp.org/.

[4] United States National Radio Quiet Zone. [Online]; [cited 2023]; Available from: https://en.wikipedia.org/wiki/United_States_National_Radio_Quiet_Zone.

[5] Federal Communications Commission. [Online]; [cited 2023]; Available from: https://www.fcc.gov/.

[6] Dickinson R.M. *Evaluation of a microwave high-power reception-conversion array for wireless power transmission*. Jet Propulsion Lab., Report No.: Tech. Memo 33–741, 1975.

[7] Dickinson R.M., Brown W.C. *Radiated microwave power transmission system efficiency measurements*. Jet Propulsion Lab., Report No.: Tech. Memo 33–727, 1975.

[8] Zaid T., Saat S., Jamal N., Yusop Y., Huzaimah Husin S., Hindustan I. 'A study on performance of the acoustic energy transfer system through air medium using ceramic disk ultrasonic transducer'. *Journal of Applied Sciences*. 2016;6:580–7.

[9] Degen C. 'Inductive coupling for wireless power transfer and near-field communication'. *EURASIP Journal on Wireless Communications and Networking*. 2021:121.

[10] Garnica J., Chinga R.A., Lin J. 'Wireless power transmission: From far field to near field'. *Proceedings of the IEEE*. 2013;101(6):1321–31.

[11] Yi K. 'Capacitive coupling wireless power transfer with quasi-LLC resonant converter using electric vehicles' windows'. *Electronics*. 2020;9(4):676.

[12] JAK Electronics. [Online]; [cited 2023]; Available from: https://www.jak-electronics.com/blog/wireless-charging-explained-working-and-standards.

[13] Chabalko M.J., Shahmohammadi M., Sample A.P. 'Quasistatic cavity resonance for ubiquitous wireless power transfer'. *PLoS ONE*. 2017;12(2):e0169045.

[14] Sasatani T., Yang J., Chabalko M.J., Kawahara Y., Sample A.P. 'Room-wide wireless charging and load-modulation communication via quasistatic cavity resonance'. *Proceedings of the ACM on Interactive, Mobile, Wearable and Ubiquitous Technologies*. 2018;2(4):1–23.

[15] Sasatani T., Sample A.P., Kawahara Y. 'Room-scale magnetoquasistatic wireless power transfer using a cavity-based multimode resonator'. *Nature Electronics*. 2021;4:689–97.

[16] Iyer V., Bayati E., Nandakumar R., Majumdar A., Gollakota S. 'Charging a smartphone across a room using lasers'. *Proceedings of the ACM on Interactive Mobile Wearable and Ubiquitous Technologies* 2017;1(4): Article 143.

[17] Steinsiek F., Weber K.H., Foth W.P., Foth H.J., Schafer C. 'Wireless power transmission experiment using an airship as relay system and a moveable rover as ground target for later planetary exploration missions'. *Presented at the 8th ESA Workshop on Advanced Space Technologies for Robotics and Automation*; Noordwijk, The Netherlands. 2004.

[18] Kim S.M., Kim S.M. 'Wireless optical energy transmission using optical beamforming'. *Optical Engineering*. 2013;52(4):043205.

[19] Chung S., Abediasl H., Hashemi H. 'A monolithically integrated large-scale optical phased array in silicon-on-insulator CMOS'. *IEEE Journal of Solid-State Circuits*. 2018;53(1):275–96.

[20] US Naval Research Laboratory Image Gallery. [Online]; [cited 2023]; Available from: https://media.defense.gov/2021/Apr/20/2002624015/-1/-1/0/180212-N-NO204-442.JPG.

[21] Rodenbeck C.T., Jaffe P.I., Strassner II B.H., Hausgen P.E., McSpadden J. O., Kazemi H., *et al.* 'Microwave and millimeter wave power beaming'. *IEEE Journal of Microwaves*. 2021;1(1):229–59.

[22] Mizojiri S., Shimamura K. 'Wireless power transfer via subterahertz-wave'. *Applied Sciences*. 2018;8(12):2653.

[23] Rizzo L., Federici J.F., Gatley S., Gatley I., Zunino J.L., Duncan K.J. 'Comparison of Terahertz, microwave, and laser power beaming under clear and adverse weather conditions'. *Journal of Infrared, Millimeter, and Terahertz Waves*. 2020;41(8):979–96.

[24] Brown W.C. 'The history of the development of the rectenna'. *Presented at Solar Power Satellite Microwave Transmission and Reception*; Houston, Texas. 1980.

[25] Brown W.C. *Experimental airborne microwave supported platform*. Raytheon CO Burlington MA Spencer Lab Report, 1965.

[26] Brown W.C. *Free-space microwave power transmission study, combined phase III and final report*. Raytheon Company, Report No.: Tech. Report No. PT-4601, 1975.

[27] Schlesak J., Alden A., Ohno T. 'SHARP rectenna and low altitude flight trials'. *Presented at Global Telecommunications Conference*; New Orleans, LA. 1985.

[28] Fujino Y., Ito T., Fujita M., Kaya N., Matsumoto H., Kawabata K., *et al.* 'A rectenna for MILAX'. *Presented at the 1st Annual Wireless Power Transmission Conference*; San Antonio, TX. 1993.

[29] Kaya N., Kojima H., Matsumoto H., Hinada M., Akiba R. 'ISY-METS rocket experiment for microwave energy transmission'. *Acta Astronautica*. 1994;34;43–6.

[30] Celeste A., Jeantya P., Pignolet G. 'Case study in reunion island'. *Acta Astronautica*. 2004;54(4):253–8.

[31] Mascarenas D.L., Flynn E.B., Todd M.D., *et al.* 'Experimental studies of using wireless energy transmission for powering embedded sensor nodes'. *Journal of Sound and Vibration*. 2010;329(12):2421–33.

[32] Sasaki S., Tanaka K., Maki K.-I. 'Microwave power transmission technologies for solar power satellites'. *Proceedings of the IEEE*. 2013;101(6):1438–47.

[33] Mihara S., Maekawa K., Nakamura S., *et al.* 'The plan of microwave power transmission development for SSPS and its industry application'. *Presented at 2018 Asia-Pacific Microwave Conference*; Kyoto, Japan. 2018.

[34] Takahashi T., Sasaki T., Homma Y., Mihara S., Sasaki K., Nakamura S., *et al.* 'Phased array system for high efficiency and high accuracy microwave power transmission'. *Presented at IEEE International Symposium on Phased Array Systems and Technology*; Waltham, MA. 2016.

[35] GuRu. [Online]; [cited 2023]; Available from: https://guru.inc/.

[36] PowerCast Corporation. [Online]; [cited 2023]; Available from: https://www.powercastco.com/.

[37] Ossia Inc. [Online]; [cited 2023]; Available from: https://www.ossia.com/.

[38] Energous. [Online]; [cited 2023]; Available from: https://energous.com/.

[39] Shinohara N. *Wireless Power Transfer via Radiowaves*. ISTE Ltd and John Wiley & Sons, Inc.; 2014.

[40] Balanis C.A. *Antenna Theory: Analysis and Design*. 3rd ed. New York: Wiley-Interscience; 2005.

[41] Adnan N.H.M., Rafiqul I.M., Alam A.Z. 'Effects of inter-element spacing and number of elements on planar array antenna characteristics'. *Indonesian Journal of Electrical Engineering and Computer Science*. 2018;10(1):230–40.

[42] Lui H.-S., Hui H.T., Leong M.S. 'A note on the mutual-coupling problems in transmitting and receiving antenna arrays'. *IEEE Antennas and Propagation Magazine*. 2009;51(5):171–6.

[43] Singh H., Sneha H.L., Jha R.M. 'Mutual coupling in phased arrays: A review'. *International Journal of Antennas and Propagation*. 2013:348123.

[44] Visser H.J., Vullers R.J.M. 'RF energy harvesting and transport for wireless sensor network applications: principles and requirements'. *Proceedings of the IEEE*. 2013;101(6):1410–23.

[45] Strassner B., Chang K. 'Microwave power transmission: historical milestones and system components'. *Proceedings of the IEEE*. 2013;101(6):1379–96.

[46] de Rosny J., Fink M. 'Overcoming the diffraction limit in wave physics using a time-reversal mirror and a novel acoustic sink'. *Physical Review Letters*. 2002;89(12):124301.

[47] Henty B.E., Stancil D.D. 'Multipath-enabled super-resolution for rf and microwave communication using phase-conjugate arrays'. *Physical Review Letters*. 2004;93:243904.

[48] de Rosny J., Fink M. 'Focusing properties of near-field time reversal'. *Physical Review A - Atomic, Molecular, and Optical Physics*. 2007;76(6):065801.

[49] Lerosey G., de Rosny J., Tourin A., Fink M. 'Focusing beyond the diffraction limit with far-field time reversal'. *Science*. 2007;315:1120–2.

Chapter 2

Phased array technique for microwave power transmission applications

The retro-reflective beamforming technique constitutes the theme of this book, and the theory of phased array is the foundation of the retro-reflective beamforming technique. In Sections 2.1 and 2.2, the classic theory of phased array is reviewed. In Sections 2.3–2.7, the power transmission efficiency is analyzed when a phased array is employed to transmit microwave power to a mobile target. Based on this chapter, the retro-reflective beamforming technique is discussed in Chapter 3 for microwave power transmission applications.

2.1 Basic principles of the phased array as transmitting antenna

The classic theory of phased array is reviewed in this section and the next section. The contents of this section and the next section follow [1] to a large extent.

As depicted in Figure 2.1, one antenna element is radiating in free space. The antenna element is excited by a time-harmonic source with a frequency of f. The electromagnetic fields radiated by the antenna element are also time-harmonic with the frequency of f. In this book, a time-harmonic physical quantity is represented by a phasor with $e^{j\omega t}$ as the time-dependence factor, where $j = \sqrt{-1}$ is the imaginary unit and $\omega = 2\pi f$ is the angular frequency. In addition, a time-harmonic power quantity is always characterized by its time-average power value in this book.

In Figure 2.1, the excitation voltage at the antenna element's circuit port is assumed to be 1 Volt. The location of the antenna element is denoted by a position vector $\mathbf{r}_s$. The electromagnetic fields radiated by the antenna element are observed at an observation point denoted by a position vector $\mathbf{r}_o$. Assume that $\mathbf{r}_o$ is located in the *far zone* of the antenna element ("far zone" will be defined in Section 2.3 explicitly). Under the far zone assumption, the electromagnetic fields around $\mathbf{r}_o$ behave as a plane wave [2]. Specifically, the electric field $\mathbf{E}_0$ and magnetic field $\mathbf{H}_0$ observed at $\mathbf{r}_o$ are

$$
\begin{aligned}
\mathbf{E}_0 &= \mathbf{U}_0 \frac{e^{-jk_0 r_{os}}}{r_{os}} \\
\mathbf{H}_0 &= \frac{\hat{\mathbf{r}}_{os} \times \mathbf{U}_0}{\eta_0} \frac{e^{-jk_0 r_{os}}}{r_{os}}
\end{aligned}
\tag{2.1}
$$

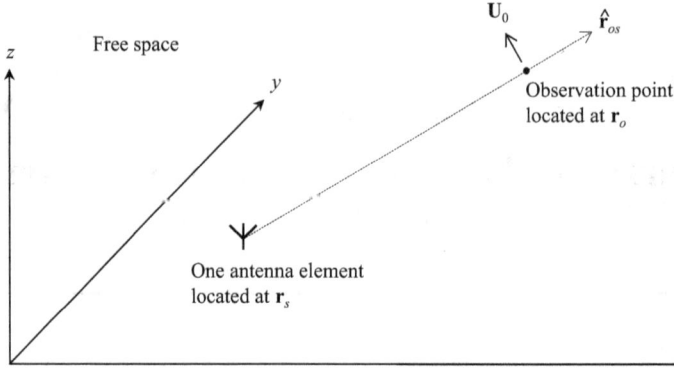

Figure 2.1   Configuration with one antenna element and one observation point

In (2.1), $r_{os} = |\mathbf{r}_{os}|$ is the distance between $\mathbf{r}_s$ and $\mathbf{r}_o$, $\mathbf{r}_{os} = \mathbf{r}_o - \mathbf{r}_s$ is the spatial vector starting from $\mathbf{r}_s$ and ending at $\mathbf{r}_o$, $\hat{\mathbf{r}}_{os} = \mathbf{r}_{os}/r_{os}$ is the unit vector representing the propagation direction of the plane wave, the operator "$|\cdot|$" selects the magnitude of the argument, $k_0 = \omega\sqrt{\varepsilon_0\mu_0}$ is the wavenumber in free space, $\eta_0 = \sqrt{\mu_0/\varepsilon_0}$ is the intrinsic impedance of free space, $\varepsilon_0 = 8.85 \times 10^{-12}$ (F/m) is the permittivity of free space, and $\mu_0 = 4\pi \times 10^{-7}$ (H/m) is the permeability of free space. The direction of the electric field (which is along $\mathbf{U}_0$), the direction of the magnetic field (which is along $\hat{\mathbf{r}}_{os} \times \mathbf{U}_0$), and $\hat{\mathbf{r}}_{os}$ are perpendicular to each other. The Poynting vector at $\mathbf{r}_o$, which embodies the "time-average power flow," is

$$\frac{1}{2}\mathrm{Re}\{\mathbf{E}_0 \times (\mathbf{H}_0)^*\} = \hat{\mathbf{r}}_{os}\frac{|\mathbf{U}_0|^2}{2\eta_0}\frac{1}{(r_{os})^2}, \tag{2.2}$$

where the operator "$\mathrm{Re}\{\cdot\}$" selects the real part of the argument and the superscript "*" stands for the complex conjugation operation.

If the antenna element's input impedance at its circuit port is Z_0, the time-average power transmitted by the antenna element associated with "excitation voltage being 1 Volt" is

$$P_t = \frac{1}{2}\mathrm{Re}\left\{\frac{1}{Z_0}\right\}. \tag{2.3}$$

If the antenna element were a lossless isotropic radiator, it would distribute the transmitted power P_t uniformly toward all directions, resulting in the isotropic Poynting vector at $\mathbf{r}_o$ with magnitude of

$$\frac{P_t}{4\pi(r_{os})^2}. \tag{2.4}$$

The gain value of the antenna element at $\mathbf{r}_o$ is defined by comparing the Poynting vector in (2.2) with the isotropic Poynting vector in (2.4):

$$G_0 = \frac{\frac{|\mathbf{U}_0|^2}{2\eta_0}\frac{1}{(r_{os})^2}}{\frac{P_t}{4\pi(r_{os})^2}} = \frac{4\pi \, |\mathbf{U}_0|^2}{\eta_0 \mathrm{Re}\left\{\frac{1}{Z_0}\right\}}. \tag{2.5}$$

The definition in (2.5) assumes that the antenna element in Figure 2.1 is lossless. It should be noted that antenna loss is inevitable in practice. In this book, the antenna elements are not electrically small; to be specific, all the antenna elements in the experimental demonstrations of this book have a physical size close to half wavelength. In addition, this book is concerned with the frequency range of [2 GHz, 6 GHz]. It is well known that the antenna loss in the frequency range of [2 GHz, 6 GHz] is not serious typically when the antenna is not electrically small. The antenna elements are therefore assumed to be lossless throughout this book, which makes the narratives/derivations more succinct. In practice, nevertheless, it is possible that the antenna loss must be characterized and taken into account rigorously.

Suppose a receiving antenna is placed at $\mathbf{r}_o$. Also, suppose the receiving antenna has an effective aperture of A_e and it is terminated by a matched load. The time-average power received by the matched load is

$$P_r = \frac{1}{2}\mathrm{Re}\{\mathbf{E}_0 \times (\mathbf{H}_0)^*\} \cdot \hat{\mathbf{r}}_{os} A_e = \frac{|\mathbf{U}_0|^2}{2\eta_0}\frac{1}{(r_{os})^2}A_e. \tag{2.6}$$

The effective aperture and gain value of the receiving antenna are related to each other by

$$\frac{A_e}{G_r} = \frac{(\lambda_0)^2}{4\pi}, \tag{2.7}$$

where G_r is the gain value of the receiving antenna along the direction of $-\hat{\mathbf{r}}_{os}$ and $\lambda_0 = 2\pi/k_0$ is the wavelength in free space. After the substitution of (2.5) and (2.7), Equation (2.6) becomes

$$P_r = \frac{|\mathbf{U}_0|^2}{2\eta_0}\frac{1}{(r_{os})^2}A_e = \frac{P_t}{4\pi}G_0\frac{1}{(r_{os})^2}A_e = \frac{P_t}{4\pi}G_0\frac{1}{(r_{os})^2}\frac{(\lambda_0)^2}{4\pi}G_r. \tag{2.8}$$

Re-arrangement of (2.8) leads to the Friis transmission equation:

$$\text{Power transmission efficiency} = \frac{P_r}{P_t} = G_0 G_r \left(\frac{\lambda_0}{4\pi r_{os}}\right)^2. \tag{2.9}$$

Based on the configuration of Figure 2.1, consider the radiation of an array of antenna elements in Figure 2.2. As depicted in Figure 2.2, the radiator or transmitter includes N antenna elements, each of which is identical to the antenna element in Figure 2.1. The location of the antenna elements is denoted by position

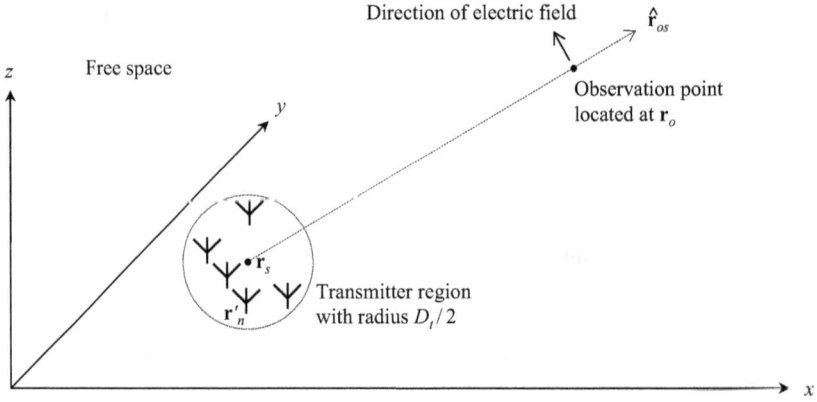

Figure 2.2   Configuration with one antenna array and one observation point

vector $\mathbf{r}'_n$, $n = 1, 2, 3, \ldots, N$. The antenna elements reside within a spherical region centered at $\mathbf{r}_s$ and with a diameter of D_t. Each of the antenna elements is excited individually by voltage $X_n e^{j\psi_n}$, $n = 1, 2, 3, \ldots, N$. If the mutual coupling among the antenna elements is neglected, the electric field $\mathbf{E}$ observed at an observation point $\mathbf{r}_o$ is the linear superposition of electric fields radiated by the individual antenna elements:

$$\mathbf{E}(\mathbf{r}_o) = \sum_{n=1}^{N} X_n e^{j\psi_n} \mathbf{U}_n \frac{e^{-jk_0 r_{on}}}{r_{on}}. \tag{2.10}$$

In (2.10), the direction of $\mathbf{U}_n$ represents the polarization direction radiated by the n-th element and $r_{on} = |\mathbf{r}_o - \mathbf{r}'_n|$ is the distance between $\mathbf{r}_o$ and the n-th antenna element. If $r_{os} \gg D_t$, the following two relationships are approximately valid.

$$\frac{1}{r_{o1}} = \frac{1}{r_{o2}} = \frac{1}{r_{o3}} = \cdots = \frac{1}{r_{oN}} = \frac{1}{r_{os}} \quad \mathbf{U}_1 = \mathbf{U}_2 = \mathbf{U}_3 = \cdots = \mathbf{U}_N = \mathbf{U}_0$$
$$\tag{2.11}$$

In addition, assume that

$$X_1 = X_2 = X_3 = \cdots = X_N = X_0, \tag{2.12}$$

that is, the antenna elements are excited with a uniform amplitude of X_0. Substituting (2.11) and (2.12) into (2.10) yields

$$\mathbf{E}(\mathbf{r}_o) = \sum_{n=1}^{N} X_n e^{j\psi_n} \mathbf{U}_n \frac{e^{-jk_0 r_{on}}}{r_{on}} = \frac{X_0 \mathbf{U}_0}{r_{os}} \sum_{n=1}^{N} e^{j\psi_n} e^{-jk_0 r_{on}}. \tag{2.13}$$

Delicate derivations are required for the phase term $e^{-jk_0 r_{on}}$ in (2.13). Specifically, the distance r_{on} in (2.13) can be expressed as

$$
\begin{aligned}
r_{on} &= |\mathbf{r}_o - \mathbf{r}'_n| \\
&= \sqrt{[(\mathbf{r}_o - \mathbf{r}_s) - (\mathbf{r}'_n - \mathbf{r}_s)] \cdot [(\mathbf{r}_o - \mathbf{r}_s) - (\mathbf{r}'_n - \mathbf{r}_s)]} \\
&= \sqrt{|\mathbf{r}_o - \mathbf{r}_s|^2 - 2(\mathbf{r}_o - \mathbf{r}_s) \cdot (\mathbf{r}'_n - \mathbf{r}_s) + |\mathbf{r}'_n - \mathbf{r}_s|^2} \\
&= r_{os}\sqrt{1 - \frac{2(\mathbf{r}_o - \mathbf{r}_s) \cdot (\mathbf{r}'_n - \mathbf{r}_s)}{(r_{os})^2} + \frac{|\mathbf{r}'_n - \mathbf{r}_s|^2}{(r_{os})^2}}
\end{aligned}
\tag{2.14}
$$

By making sure of the following Taylor series

$$
\sqrt{1+x} = 1 + \frac{x}{2} - \frac{x^2}{8} + \frac{x^3}{16} + o(x^3),
\tag{2.15}
$$

Equation (2.14) becomes

$$
\begin{aligned}
r_{on} &= r_{os}\sqrt{1 - \frac{2(\mathbf{r}_o - \mathbf{r}_s) \cdot (\mathbf{r}'_n - \mathbf{r}_s)}{(r_{os})^2} + \frac{|\mathbf{r}'_n - \mathbf{r}_s|^2}{(r_{os})^2}} \\
&= r_{os}\left\{
\begin{aligned}
&1 + \frac{1}{2}\left[-\frac{2(\mathbf{r}_o - \mathbf{r}_s) \cdot (\mathbf{r}'_n - \mathbf{r}_s)}{(r_{os})^2} + \frac{|\mathbf{r}'_n - \mathbf{r}_s|^2}{(r_{os})^2}\right] \\
&- \frac{1}{8}\left[-\frac{2(\mathbf{r}_o - \mathbf{r}_s) \cdot (\mathbf{r}'_n - \mathbf{r}_s)}{(r_{os})^2} + \frac{|\mathbf{r}'_n - \mathbf{r}_s|^2}{(r_{os})^2}\right]^2 \\
&+ \frac{1}{16}\left[-\frac{2(\mathbf{r}_o - \mathbf{r}_s) \cdot (\mathbf{r}'_n - \mathbf{r}_s)}{(r_{os})^2} + \frac{|\mathbf{r}'_n - \mathbf{r}_s|^2}{(r_{os})^2}\right]^3 + \cdots
\end{aligned}
\right\}
\end{aligned}
\tag{2.16}
$$

Straightforwardly, Equation (2.16) can be rewritten as

$$
\begin{aligned}
r_{on} &= r_{os} - \hat{\mathbf{r}}_{os} \cdot (\mathbf{r}'_n - \mathbf{r}_s) + \frac{|\mathbf{r}'_n - \mathbf{r}_s|^2}{2r_{os}} \\
&\quad - \frac{r_{os}}{8}\left[-\frac{2(\mathbf{r}_o - \mathbf{r}_s) \cdot (\mathbf{r}'_n - \mathbf{r}_s)}{(r_{os})^2} + \frac{|\mathbf{r}'_n - \mathbf{r}_s|^2}{(r_{os})^2}\right]^2 \\
&\quad + \frac{r_{os}}{16}\left[-\frac{2(\mathbf{r}_o - \mathbf{r}_s) \cdot (\mathbf{r}'_n - \mathbf{r}_s)}{(r_{os})^2} + \frac{|\mathbf{r}'_n - \mathbf{r}_s|^2}{(r_{os})^2}\right]^3 + \cdots
\end{aligned}
\tag{2.17}
$$

On the basis of Figure 2.2, consider the configuration shown in Figure 2.3. There is only one difference between Figures 2.2 and 2.3: The observation point is fixed at $\mathbf{r}_o$ in Figure 2.2 whereas the observation point $\mathbf{r}$ resides in a certain "receiver region" in Figure 2.3. To be more specific, the observation point $\mathbf{r}$ resides within a spherical region centered at $\mathbf{r}_o$ and with a diameter of D_r in Figure 2.3. It is

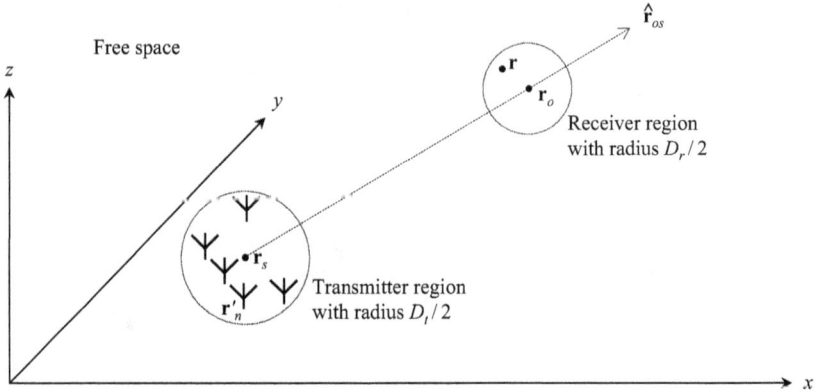

Figure 2.3   Configuration with one antenna array and one receiver region

obvious that under the conditions of $r_{os} \gg D_t$ and $r_{os} \gg D_r$, the expression of (2.13) can be extended to Figure 2.3:

$$\mathbf{E}(\mathbf{r}) = \frac{X_0 U_0}{r_{os}} \sum_{n=1}^{N} e^{j\psi_n} e^{-jk_0|\mathbf{r} - \mathbf{r}'_n|}. \tag{2.18}$$

Next, derivations similar to (2.14) are conducted for the term of $|\mathbf{r} - \mathbf{r}'_n|$ in (2.18)

$$
\begin{aligned}
&|\mathbf{r} - \mathbf{r}'_n| \\
&= \sqrt{(\mathbf{r} - \mathbf{r}'_n) \cdot (\mathbf{r} - \mathbf{r}'_n)} \\
&= \sqrt{[(\mathbf{r}_o - \mathbf{r}_s) + (\mathbf{r} - \mathbf{r}_o) - (\mathbf{r}'_n - \mathbf{r}_s)] \cdot [(\mathbf{r}_o - \mathbf{r}_s) + (\mathbf{r} - \mathbf{r}_o) - (\mathbf{r}'_n - \mathbf{r}_s)]} \\
&= \sqrt{|\mathbf{r}_o - \mathbf{r}_s|^2 + 2(\mathbf{r}_o - \mathbf{r}_s) \cdot [(\mathbf{r} - \mathbf{r}_o) - (\mathbf{r}'_n - \mathbf{r}_s)] + |(\mathbf{r} - \mathbf{r}_o) - (\mathbf{r}'_n - \mathbf{r}_s)|^2} \\
&= r_{os} \sqrt{1 + \frac{2(\mathbf{r}_o - \mathbf{r}_s) \cdot [(\mathbf{r} - \mathbf{r}_o) - (\mathbf{r}'_n - \mathbf{r}_s)]}{(r_{os})^2} + \frac{|(\mathbf{r} - \mathbf{r}_o) - (\mathbf{r}'_n - \mathbf{r}_s)|^2}{(r_{os})^2}}
\end{aligned} \tag{2.19}
$$

By making use of the Taylor series in (2.15), Equation (2.19) becomes

$$
\begin{aligned}
&|\mathbf{r} - \mathbf{r}'_n| \\
&= r_{os} \sqrt{1 + \frac{2(\mathbf{r}_o - \mathbf{r}_s) \cdot [(\mathbf{r} - \mathbf{r}_o) - (\mathbf{r}'_n - \mathbf{r}_s)]}{(r_{os})^2} + \frac{|(\mathbf{r} - \mathbf{r}_o) - (\mathbf{r}'_n - \mathbf{r}_s)|^2}{(r_{os})^2}} \\
&= r_{os} \left\{
\begin{aligned}
&1 + \frac{1}{2}\left[\frac{2(\mathbf{r}_o - \mathbf{r}_s) \cdot [(\mathbf{r} - \mathbf{r}_o) - (\mathbf{r}'_n - \mathbf{r}_s)]}{(r_{os})^2} + \frac{|(\mathbf{r} - \mathbf{r}_o) - (\mathbf{r}'_n - \mathbf{r}_s)|^2}{(r_{os})^2}\right] \\
&- \frac{1}{8}\left[\frac{2(\mathbf{r}_o - \mathbf{r}_s) \cdot [(\mathbf{r} - \mathbf{r}_o) - (\mathbf{r}'_n - \mathbf{r}_s)]}{(r_{os})^2} + \frac{|(\mathbf{r} - \mathbf{r}_o) - (\mathbf{r}'_n - \mathbf{r}_s)|^2}{(r_{os})^2}\right]^2 \\
&+ \frac{1}{16}\left[\frac{2(\mathbf{r}_o - \mathbf{r}_s) \cdot [(\mathbf{r} - \mathbf{r}_o) - (\mathbf{r}'_n - \mathbf{r}_s)]}{(r_{os})^2} + \frac{|(\mathbf{r} - \mathbf{r}_o) - (\mathbf{r}'_n - \mathbf{r}_s)|^2}{(r_{os})^2}\right]^3 + \cdots
\end{aligned}
\right\}
\end{aligned} \tag{2.20}
$$

Straightforwardly, Equation (2.20) can be rewritten as

$$
|\mathbf{r} - \mathbf{r}'_n| = r_{os} + \hat{\mathbf{r}}_{os} \cdot (\mathbf{r} - \mathbf{r}_o) - \hat{\mathbf{r}}_{os} \cdot (\mathbf{r}'_n - \mathbf{r}_s) + \frac{|(\mathbf{r} - \mathbf{r}_o) - (\mathbf{r}'_n - \mathbf{r}_s)|^2}{2r_{os}}
$$

$$
- \frac{r_{os}}{8} \left[\frac{2(\mathbf{r}_o - \mathbf{r}_s) \cdot [(\mathbf{r} - \mathbf{r}_o) - (\mathbf{r}'_n - \mathbf{r}_s)]}{(r_{os})^2} + \frac{|(\mathbf{r} - \mathbf{r}_o) - (\mathbf{r}'_n - \mathbf{r}_s)|^2}{(r_{os})^2} \right]^2
$$

$$
+ \frac{r_{os}}{16} \left[\frac{2(\mathbf{r}_o - \mathbf{r}_s) \cdot [(\mathbf{r} - \mathbf{r}_o) - (\mathbf{r}'_n - \mathbf{r}_s)]}{(r_{os})^2} + \frac{|(\mathbf{r} - \mathbf{r}_o) - (\mathbf{r}'_n - \mathbf{r}_s)|^2}{(r_{os})^2} \right]^3 + \cdots
$$

$$(2.21)$$

When the *far zone condition* is satisfied between the transmitter region and receiver region, the first three terms on the right-hand side of (2.21) are retained and the other terms are neglected. According to the classic theory of antenna engineering, the far zone condition is determined by the fourth term on the right-hand side of (2.21) [1]. Specifically, the transmitter region and receiver region are in each other's far zone if

$$
k_0 \frac{|(\mathbf{r} - \mathbf{r}_o) - (\mathbf{r}'_n - \mathbf{r}_s)|^2}{2r_{os}} < \frac{\pi}{8}.
$$

$$(2.22)$$

The value of $|\mathbf{r} - \mathbf{r}_o|$ does not exceed $D_r/2$ and the value of $|\mathbf{r}'_n - \mathbf{r}_s|$ does not exceed $D_t/2$, which are obvious from Figure 2.3. Moreover, the maximum value of $|(\mathbf{r} - \mathbf{r}_o) - (\mathbf{r}'_n - \mathbf{r}_s)|^2$ is $(D_t + D_r)^2/4$. Therefore, the far zone condition in (2.22) can be rewritten as

$$
\frac{2\pi}{\lambda_0} \frac{(D_t + D_r)^2}{4} \frac{1}{2r_{os}} < \frac{\pi}{8}.
$$

$$(2.23)$$

Since "r_{os}" stands for the distance between the transmitter and receiver, it is also termed as "d" in this book. Straightforwardly, (2.23) can be re-arranged to be

$$
d > \frac{2(D_t + D_r)^2}{\lambda_0}.
$$

$$(2.24)$$

Equation (2.24) is usually referred to as the far-field condition in antenna engineering. When the far-field condition in (2.24) is satisfied, only the first three terms on the right-hand side of (2.21) are retained, and thus (2.18) becomes

$$
\mathbf{E}(\mathbf{r}) = \frac{X_0 \mathbf{U}_0}{r_{os}} \sum_{n=1}^{N} e^{j\psi_n} e^{-jk_0 |\mathbf{r} - \mathbf{r}'_n|}
$$

$$
= \frac{X_0 \mathbf{U}_0}{r_{os}} \sum_{n=1}^{N} e^{j\psi_n} e^{-jk_0 r_{os}} e^{-jk_0 \hat{\mathbf{r}}_{os} \cdot (\mathbf{r} - \mathbf{r}_o)} e^{jk_0 \hat{\mathbf{r}}_{os} \cdot (\mathbf{r}'_n - \mathbf{r}_s)}
$$

$$
= \frac{X_0 \mathbf{U}_0 e^{-jk_0 r_{os}}}{r_{os}} e^{-jk_0 \hat{\mathbf{r}}_{os} \cdot (\mathbf{r} - \mathbf{r}_o)} \sum_{n=1}^{N} e^{j\psi_n} e^{jk_0 \hat{\mathbf{r}}_{os} \cdot (\mathbf{r}'_n - \mathbf{r}_s)}
$$

$$(2.25)$$

If the antenna array is desired to generate a radiation beam toward $\mathbf{r}_b$ that is located in the far zone of the antenna array, the excitation phases should be chosen as

$$\psi_n = -k_0 \hat{\mathbf{r}}_{bs} \cdot (\mathbf{r}'_n - \mathbf{r}_s), \quad n = 1, 2, 3, \ldots, N, \tag{2.26}$$

where $\hat{\mathbf{r}}_{bs} = (\mathbf{r}_b - \mathbf{r}_s)/r_{bs}$ and $r_{bs} = |\mathbf{r}_b - \mathbf{r}_s|$. When the electric field is observed at $\mathbf{r}_o = \mathbf{r}_b$, electric fields contributed by the antenna elements are in phase (that is, share the same phase), and thus

$$\mathbf{E}(\mathbf{r}_b) = N X_0 \mathbf{U}_0 \frac{e^{-jk_0 r_{bs}}}{r_{bs}}. \tag{2.27}$$

Meanwhile, the magnetic fields contributed by the antenna elements are in phase at $\mathbf{r}_o = \mathbf{r}_b$ as well. As a result, the Poynting vector at $\mathbf{r}_b$ is

$$\frac{1}{2} \mathrm{Re}\{\mathbf{E} \times \mathbf{H}^*\} = \hat{\mathbf{r}}_{bs} N^2 (X_0)^2 \frac{|\mathbf{U}_0|^2}{2\eta_0} \frac{1}{(r_{bs})^2}. \tag{2.28}$$

When the observation point $\mathbf{r}$ is in the neighborhood of $\mathbf{r}_b$, the electromagnetic fields behave as a plane wave traveling toward the direction of $\hat{\mathbf{r}}_{bs}$, as evidenced by the term $e^{-jk_0 \hat{\mathbf{r}}_{os} \cdot (\mathbf{r} - \mathbf{r}_o)}$ in (2.25) with $\mathbf{r}_o$ replaced by $\mathbf{r}_b$. When $\mathbf{r}$ is not in the neighborhood of $\mathbf{r}_b$, the terms ignored in (2.21) may become not negligible, which would invalidate (2.25).

When there is only one antenna element in free space, the input impedance at its circuit port was assumed to be Z_0 in this section. If an array of antenna elements reside within a compact region, the mutual coupling among the antenna elements is unavoidable and the input impedance at each antenna element's circuit port deviates from Z_0. In this section, it is assumed that there is no mutual coupling among the antenna elements and the input impedance at each antenna element's circuit port remains to be Z_0, which facilitates the narrative of this section. Section 2.7 is devoted to investigating the impact of mutual coupling among antenna elements. Because the antenna elements are excited by a uniform voltage amplitude of X_0, the total amount of power transmitted by the antenna array is

$$P_t = N \frac{(X_0)^2}{2} \mathrm{Re}\left\{\frac{1}{Z_0}\right\}. \tag{2.29}$$

The gain value of the antenna array toward $\mathbf{r}_b$, G_{array}, is obtained by comparing the Poynting vector in (2.28) with the isotropic Poynting vector:

$$G_{array} = \frac{\left(N^2 (X_0)^2 \frac{|\mathbf{U}_0|^2}{2\eta_0 (r_{bs})^2}\right)}{\left(\frac{P_t}{4\pi (r_{bs})^2}\right)} = N \frac{4\pi |\mathbf{U}_0|^2}{\eta_0 \mathrm{Re}\left\{\frac{1}{Z_0}\right\}} = N G_0, \tag{2.30}$$

where G_0 is the gain value of one antenna element in (2.5) with $\mathbf{r}_o = \mathbf{r}_b$. Equation (2.30) indicates that the gain value of an N-element antenna array is N times larger than the gain value of one antenna element toward $\mathbf{r}_b$. The enhancement of gain value is because the N antenna elements' contributions are in phase at $\mathbf{r}_b$, resulting from the choice of excitation phases in (2.26). At an observation point other than $\mathbf{r}_b$, nevertheless, it is very likely that the fields radiated by the N antenna elements are no longer in phase and the gain value is smaller than NG_0. Thus, the radiation pattern of the antenna array exhibits a beam toward $\mathbf{r}_b$. Since the beam can be steered by adjusting the values of the excitation phase, the technique presented above is usually termed as the "phased array technique."

When a receiving antenna is placed at $\mathbf{r}_b$ and terminated by a matched load, the received power P_r is

$$P_r = P_t G_{array} G_r \left(\frac{\lambda_0}{4\pi r_{bs}}\right)^2 = P_t (NG_0) G_r \left(\frac{\lambda_0}{4\pi r_{bs}}\right)^2, \tag{2.31}$$

where G_r is the gain value of the receiving antenna toward $-\hat{\mathbf{r}}_{bs}$ direction. Because $G_{array} = NG_0$ at $\mathbf{r}_b$, the power transmission efficiency (that is, P_r/P_t) associated with the antenna array is N times greater than that associated with one single antenna element with $\mathbf{r}_o = \mathbf{r}_b$ in (2.9).

To illustrate the basic principles of phased array, two specific scenarios are discussed in Sections 2.1.1 and 2.1.2, respectively. The first scenario is a one-dimensional linear array, and the second scenario is a two-dimensional array.

2.1.1   One-dimensional linear phased array

Suppose an array of N antenna elements is deployed along the x-axis, as illustrated in Figure 2.4. The N antenna elements are identical to one another. The spacing between any two adjacent antenna elements is s. The N antenna elements are located at

$$x'_n = \left(n - \frac{N+1}{2}\right)s, \ n = 1, 2, 3, \ldots, N \tag{2.32}$$

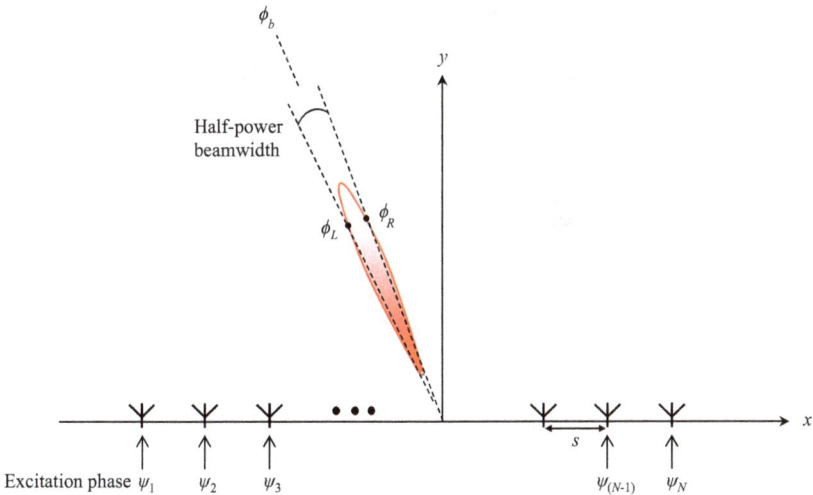

Figure 2.4   Illustration of a one-dimensional phased array

The electric field radiated by the antenna array is observed at an observation point $\mathbf{r}_o$ in the x-y plane. The cylindrical coordinates of $\mathbf{r}_o$ are $(\rho_o, \phi_o, 0)$; equivalently, the Cartesian coordinates of $\mathbf{r}_o$ are $x_o = \rho_o \cos(\phi_o)$, $y_o = \rho_o \sin(\phi_o)$, and $z_o = 0$. It is assumed that $\rho_o \to \infty$.

When there is only one antenna element located at the spatial origin and it is excited by a voltage with an amplitude of 1 Volt and phase of zero, the electric field radiated by the antenna element is

$$\mathbf{E}_0(\mathbf{r}_o) = \mathbf{U}_0(\phi_o)\frac{e^{-jk_0\rho_o}}{\rho_o}. \tag{2.33}$$

Following (2.25), when the N antenna elements are excited by $X_0 e^{j\psi_n}$, $n = 1, 2, 3, \ldots, N$, the electric field radiated by the antenna array is

$$\mathbf{E}(\mathbf{r}_o) = \frac{X_0 \mathbf{U}_0(\phi_o)e^{-jk_0\rho_o}}{\rho_o}\sum_{n=1}^{N} e^{j\psi_n} e^{jk_0\widehat{\mathbf{r}_o}\cdot\mathbf{r}'_n}$$

$$= X_0\frac{\mathbf{U}_0(\phi_o)e^{-jk_0\rho_o}}{\rho_o}\sum_{n=1}^{N} e^{j\psi_n} e^{jk_0 x'_n \cos(\phi_o)} \tag{2.34}$$

where $\mathbf{r}'_n = x'_n \widehat{\mathbf{x}}$ is the position vector of the n-th antenna element and $\widehat{\mathbf{r}}_o = \cos(\phi_o)\widehat{\mathbf{x}} + \sin(\phi_o)\widehat{\mathbf{y}}$ embodies the direction from the array's center to the observation point. Usually, the term

$$\sum_{n=1}^{N} e^{j\psi_n} e^{jk_0 x'_n \cos(\phi_o)} \tag{2.35}$$

is defined as the "array factor." It is interesting to note that the electric field radiated by the antenna array in (2.34) is the product between $\mathbf{E}_0$ and the array factor mathematically.

The excitation phases are chosen as formulated in (2.26):

$$\psi_n = -k_0 \widehat{\mathbf{p}}_b \cdot \mathbf{r}'_n = -k_0 x'_n \cos(\phi_b), \quad n = 1, 2, 3, \ldots, N \tag{2.36}$$

where $\widehat{\mathbf{p}}_b = \cos(\phi_b)\widehat{\mathbf{x}} + \sin(\phi_b)\widehat{\mathbf{y}}$. Consequently, the array's radiation pattern exhibits a beam along ϕ_b. Specifically with the excitation phases in (2.36), the array factor is

$$\sum_{n=1}^{N} e^{jk_0 x'_n[\cos(\phi_o) - \cos(\phi_b)]} = \sum_{n=1}^{N} e^{jk_0\left(n-\frac{N+1}{2}\right)s[\cos(\phi_o) - \cos(\phi_b)]}. \tag{2.37}$$

Due to the following mathematical identity

$$\sum_{n=1}^{N} e^{jn\xi} = \frac{\sin\left\{\frac{N\xi}{2}\right\}}{\sin\left\{\frac{\xi}{2}\right\}}e^{j\frac{N+1}{2}\xi}, \tag{2.38}$$

the array factor in (2.37) is a Dirichlet function:

$$\sum_{n=1}^{N} e^{jk_0 x'_n [\cos(\phi_o) - \cos(\phi_b)]} = \sum_{n=1}^{N} e^{jk_0 \left(n - \frac{N+1}{2}\right) s [\cos(\phi_o) - \cos(\phi_b)]}$$

$$= \frac{\sin\left\{\dfrac{Nk_0 s [\cos(\phi_o) - \cos(\phi_b)]}{2}\right\}}{\sin\left\{\dfrac{k_0 s [\cos(\phi_o) - \cos(\phi_b)]}{2}\right\}} \tag{2.39}$$

When $\phi_o = \phi_b$, the value of the array factor is N. When ϕ_o deviates from ϕ_b, the array factor's value drops. As a result, the array factor exhibits a beam along ϕ_b. If the array factor's value drops to $N/\sqrt{2}$ at $\phi_o = \phi_L$ and $\phi_o = \phi_R$ as depicted in Figure 2.4, the half-power beamwidth is defined as $\phi_L - \phi_R$. Obviously, the beamwidth decreases with the increase of N.

With the choice of excitation phases in (2.36), the following relationship is satisfied.

$$\psi_1 + k_0 x'_1 \cos(\phi_o) = \psi_2 + k_0 x'_2 \cos(\phi_o) = \cdots$$

$$= \psi_N + k_0 x'_N \cos(\phi_o), \text{ when } \phi_o = \phi_b. \tag{2.40}$$

In other words, the excitation phases in (2.36) ensure that the N antenna elements' contributions are in phase at $\phi_o = \phi_b$. Equation (2.40) is equivalent to

$$k_0 x'_n \cos(\phi_b) - k_0 x'_{(n-1)} \cos(\phi_b) = -(\psi_n - \psi_{n-1}), \quad n = 2, 3, 4, \ldots, N. \tag{2.41}$$

Because $\cos(-\phi_b) = \cos(\phi_b)$, the N antenna elements' contributions are in phase at $\phi_o = -\phi_b$ as well. Consequently, the array factor exhibits a beam along $-\phi_b$ when a beam is desired along ϕ_b. In practice, the beam along $-\phi_b$ is usually eliminated by minimizing $U_0(-\phi_b)$, that is, by minimizing the radiation of individual antenna elements toward $-\phi_b$. If the following relationship is satisfied at ϕ_g,

$$k_0 x'_n \cos(\phi_g) - k_0 x'_{(n-1)} \cos(\phi_g) = -(\psi_n - \psi_{n-1}) \pm 2\pi,$$

$$n = 2, 3, 4, \ldots, N \tag{2.42}$$

the N antenna elements' contributions are in phase, and as a result, a *grating lobe* appears along ϕ_g. A grating lobe is as strong as the beam desired along ϕ_b, and thus should be avoided typically. Evaluating the difference between (2.41) and (2.42) leads to

$$k_0 s \left[\cos(\phi_b) - \cos(\phi_g)\right] = \pm 2\pi. \tag{2.43}$$

If $s < \lambda_0/2 = \pi/k_0$, the relationship in (2.43) would never be satisfied in practice, because $|\cos(\phi_b) - \cos(\phi_g)| \leq 2$ with any real values of ϕ_b and ϕ_g. As a result, grating lobes can be avoided when the inter-element spacing s is smaller than half wavelength. However, $s < \lambda_0/2$ is not very common in practice as small inter-element spacing causes various complications. For instance, each antenna element is forced to have a small electrical size when $s < \lambda_0/2$, which would lead to lower antenna efficiency (or higher antenna loss, equivalently). As another harmful consequence of small s, the coupling among antenna elements would become strong (the impacts of strong mutual coupling among antenna elements are studied in Section 2.7). The specific technical issues pertinent to $s < \lambda_0/2$ are not elaborated in this book. When the inter-element spacing is greater than half wavelength, the occurrence of a grating lobe becomes possible. It is well known that the beam-steering capability of a phased array would be highly restricted in order to avoid the occurrence of grating lobes when $s > \lambda_0$ (the issues pertinent to $s > \lambda_0$ are not ela-borated in this book, either). In this book, the value of s is always chosen between $\lambda_0/2$ and λ_0.

2.1.2   Two-dimensional phased array

The configuration of the one-dimensional phased array in Section 2.1.1 can be extended to the configuration of the two-dimensional phased array in Figure 2.5. As shown in Figure 2.5, antenna elements are deployed into N columns along the x direction and M rows along the z direction. The $N \times M$ antenna elements are identical to one another; in Figure 2.5, each of them is assumed to be a microstrip patch printed over a printed circuit board. The inter-element spacing values are s_x along the x direction and s_z along the z direction, respectively. Specifically, the antenna elements are located at

$$\begin{aligned} \mathbf{r}'_{nm} &= \hat{\mathbf{x}}x'_n + \hat{\mathbf{z}}z'_m \\ &= \hat{\mathbf{x}}\left(n - \frac{N+1}{2}\right)s_x + \hat{\mathbf{z}}\left(m - \frac{M+1}{2}\right)s_z, \quad \begin{array}{l} n = 1,2,3,\cdots,N \\ m = 1,2,3,\cdots,M \end{array} \end{aligned} \quad (2.44)$$

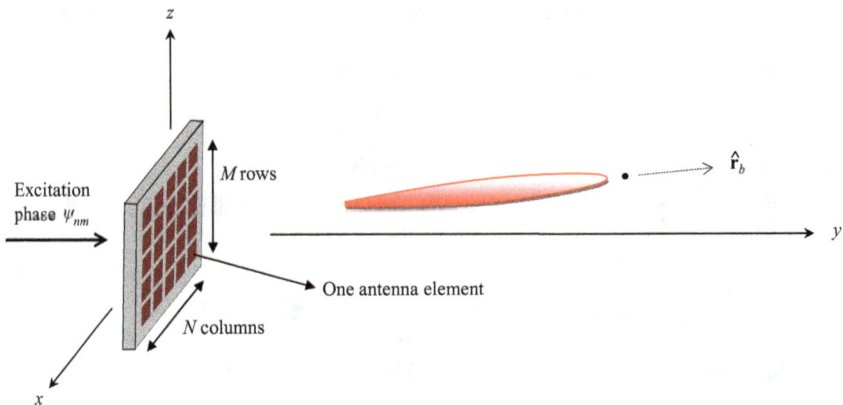

Figure 2.5   Illustration of a two-dimensional phased array

The electric field radiated by the antenna array is observed at an observation point $\mathbf{r}_o$ with coordinates (r_o, θ_o, ϕ_o) in the spherical coordinate system. It is assumed that $r_o \to \infty$.

When there is only one antenna element located at the spatial origin and it is excited by a voltage with an amplitude of 1 Volt and phase of zero, the electric field radiated by the antenna element is

$$\mathbf{E}_0(\mathbf{r}_o) = \mathbf{U}_0(\theta_o, \phi_o) \frac{e^{-jk_0 r_o}}{r_o}. \tag{2.45}$$

When the $N \times M$ antenna elements are excited by $X_0 e^{j\psi_{nm}}$, $n = 1, 2, 3, \ldots, N$, $m = 1, 2, 3, \ldots, M$, the electric field is

$$\mathbf{E}(\mathbf{r}_o) = \frac{X_0 \mathbf{U}_0 e^{-jk_0 r_o}}{r_o} \sum_{n=1}^{N} \sum_{m=1}^{M} e^{j\psi_{nm}} e^{jk_0 \widehat{\mathbf{r}}_o \cdot \mathbf{r}'_{nm}}, \tag{2.46}$$

following the derivation of (2.25). In (2.46),

$$\widehat{\mathbf{r}}_o = \widehat{\mathbf{x}} \sin(\theta_o)\cos(\phi_o) + \widehat{\mathbf{y}} \sin(\theta_o)\sin(\phi_o) + \widehat{\mathbf{z}} \cos(\theta_o), \tag{2.47}$$

$$\widehat{\mathbf{r}}_o \cdot \mathbf{r}'_{nm} = x'_n \sin(\theta_o)\cos(\phi_o) + z'_m \cos(\theta_o), \tag{2.48}$$

and the term

$$\sum_{n=1}^{N} \sum_{m=1}^{M} e^{j\psi_{nm}} e^{jk_0 \widehat{\mathbf{r}}_o \cdot \mathbf{r}'_{nm}} \tag{2.49}$$

is the array factor.

Following (2.26), the excitation phases are chosen as

$$\psi_{nm} = -k_0 \widehat{\mathbf{r}}_b \cdot \mathbf{r}'_{nm}$$

$$= -k_0 x'_n \sin(\theta_b)\cos(\phi_b) - k_0 z'_m \cos(\theta_b), \quad \begin{array}{l} n = 1, 2, 3, \cdots, N \\ m = 1, 2, 3, \cdots, M \end{array} \tag{2.50}$$

where

$$\widehat{\mathbf{r}}_b = \widehat{\mathbf{x}} \sin(\theta_b)\cos(\phi_b) + \widehat{\mathbf{y}} \sin(\theta_b)\sin(\phi_b) + \widehat{\mathbf{z}} \cos(\theta_b) \tag{2.51}$$

is the direction of a beam desired in the radiation pattern. With the excitation phases in (2.50), the array factor is

$$\sum_{n=1}^{N} \sum_{m=1}^{M} e^{j\psi_{nm}} e^{jk_0 \widehat{\mathbf{r}}_o \cdot \mathbf{r}'_{nm}} = \sum_{n=1}^{N} e^{jk_0 x'_n [\sin(\theta_o)\cos(\phi_o) - \sin(\theta_b)\cos(\phi_b)]} \sum_{m=1}^{M} e^{jk_0 z'_m [\cos(\theta_o) - \cos(\theta_b)]}.$$

$$\tag{2.52}$$

Due to (2.38), the array factor in (2.52) is the product of two Dirichlet functions:

$$
\sum_{n=1}^{N}\sum_{m=1}^{M} e^{j\psi_{nm}} e^{jk_0\widehat{\mathbf{r}}_o\cdot\mathbf{r}'_{nm}}
$$

$$
= \frac{\sin\left\{\dfrac{Nk_0s_x[\sin(\theta_o)\cos(\phi_o) - \sin(\theta_b)\cos(\phi_b)]}{2}\right\}}{\sin\left\{\dfrac{k_0s_x[\sin(\theta_o)\cos(\phi_o) - \sin(\theta_b)\cos(\phi_b)]}{2}\right\}} \times \frac{\sin\left\{\dfrac{Mk_0s_z[\cos(\theta_o) - \cos(\theta_b)]}{2}\right\}}{\sin\left\{\dfrac{k_0s_z[\cos(\theta_o) - \cos(\theta_b)]}{2}\right\}}
$$

$$(2.53)$$

The array factor exhibits a beam along (θ_b, ϕ_b), as the $N \times M$ antenna elements' contributions are in phase when $\theta_o = \theta_b$ and $\phi_o = \phi_b$. The beam radiated by the phased array can be steered via adjusting the excitation phase values ψ_{nm}.

If the radiation pattern of a phased array only includes one narrow beam, the value of beamwidth can be estimated by the formulation of $\sqrt{41,253/G_{array}}$ (in degree), where G_{array} is the peak gain value of the beam [1]. When $M = N = 5$ and $G_0 = 4$, $G_{array} = NMG_0 = 100$ and the beamwidth is estimated to be $\sqrt{41,253/100} \cong 20$ degrees. If a fictitious planar screen is placed 10 m away from the phased array and is geometrically perpendicular to the beam's propagation direction, the beam's footprint over the screen is a circular region with a radius of approximately $10 \times \tan(10°) = 1.8$ m. Thus, if a wireless power receiver is located 10 m away from the phased array and it is desired to collect most of the power carried by the beam, the wireless power receiver's physical dimension should be on the order of meters. If a wireless power receiver with a small physical size resides 10 m away from the phased array, a beamwidth narrower than $20°$ is necessary in order to yield high power transmission efficiency, which requires the phased array (as a wireless power transmitter) to include more than 5×5 antenna elements.

2.2   Basic principles of the phased array as receiving antenna

When a phased array is employed by a wireless transmitter, its radiation pattern can be reconfigured by adjusting the excitation phase to the array's elements, as described in Section 2.1. In this section, a phased array is employed by a wireless receiver with its reception pattern reconfigured via phase adjustment.

As depicted in Figure 2.6, there is one antenna element residing in free space as a receiving antenna. Suppose a transmitting antenna is in the far zone of the receiving antenna. As a result, the electromagnetic wave radiated by the transmitting antenna behaves as a plane wave when it reaches the receiving antenna. The Poynting vector of the incident plane wave is assumed to be $W^{inc}\widehat{\mathbf{k}}^{inc}$, where $\widehat{\mathbf{k}}^{inc}$ is the propagation direction of the incident plane wave and W^{inc} is the power density carried by the incident plane wave with unit of Watt per m². The incident plane wave has planar wavefronts, as illustrated in Figure 2.6. To be specific, the incident plane wave's phase is unchanged over each planar wavefront. When the receiving

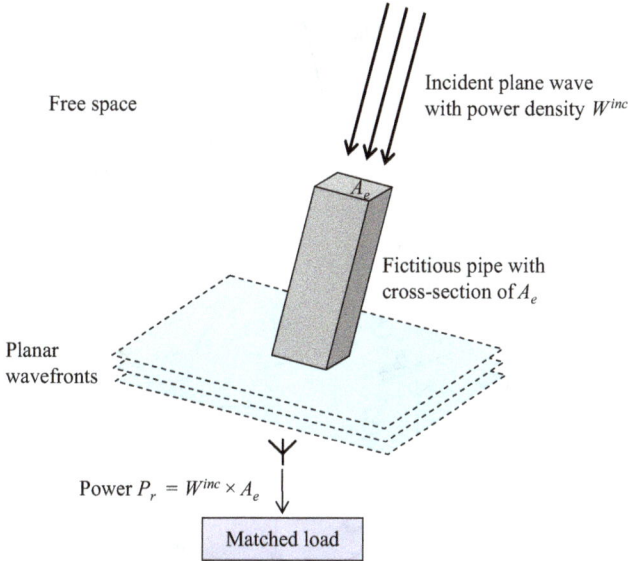

Figure 2.6   Illustration of one antenna element as a receiving antenna

antenna element is terminated by a matched load, the amount of power delivered to the load is termed P_r. The effective aperture A_e of the receiving antenna element is defined as

$$A_e = \frac{P_r}{W^{inc}}, \text{ with a unit of m}^2. \tag{2.54}$$

As illustrated in Figure 2.6, it appears that the power flowing along a fictitious "pipe" with a cross-section area of A_e is absorbed by the receiving antenna. Because the value of P_r depends on the incident plane wave's propagation direction, A_e is also dependent on the incident plane wave's propagation direction.

The effective aperture is related to the antenna gain through the following relationship,

$$A_e = \frac{(\lambda_0)^2}{4\pi} G_r, \tag{2.55}$$

where G_r is the gain value toward the $-\widehat{\mathbf{k}}^{inc}$ direction when the antenna element is used as a transmitting antenna.

Consider an array of N antenna elements deployed along a straight line, as illustrated in Figure 2.7. Each of the N antenna elements is identical to that in Figure 2.6. The spacing between any two adjacent antenna elements is s. If the value of spacing s is small, there might be strong mutual coupling among the antenna elements in the array, which is harmful in practice typically. The magnitude of mutual coupling among antenna elements can be estimated by A_e roughly. Suppose each antenna element is a microstrip patch with a broadside gain value of

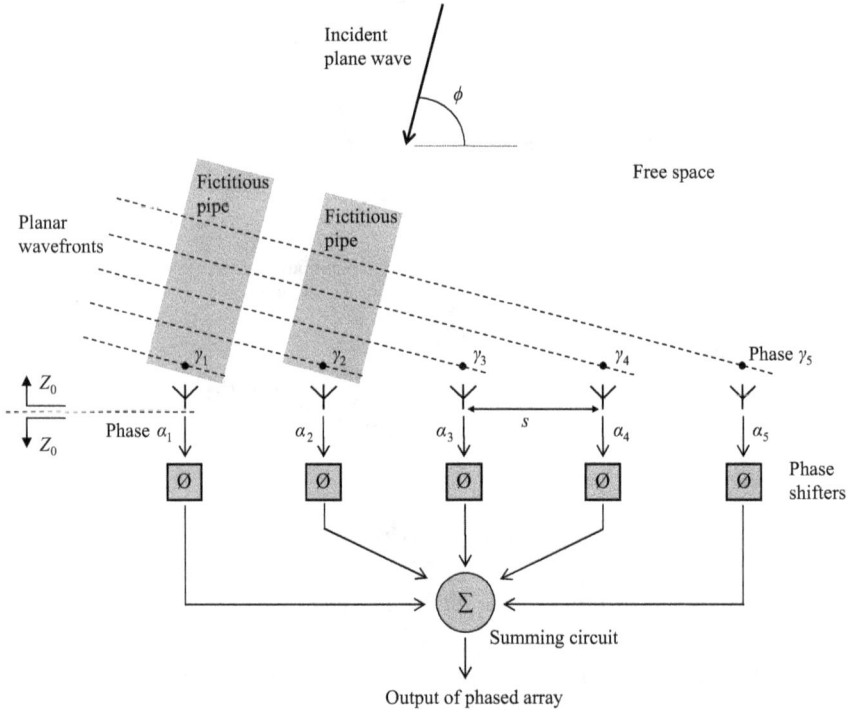

Figure 2.7   Illustration of a phased array as a receiving antenna

$G_r = 4$. According to (2.55), the effective aperture of each antenna element is $A_e = (0.56\lambda_0) \times (0.56\lambda_0)$ when $G_r = 4$ if it has a square shape. Obviously, when the inter-element spacing is smaller than $0.56\lambda_0$, the "pipes" that deliver power to the antenna elements would overlap with each other in the space and thus the coupling among antenna elements seems unavoidable. On the other hand, increasing the value of inter-element spacing beyond $0.56\lambda_0$ appears an effective means to mitigate the mutual coupling among the antenna elements.

The phase values of the incident plane wave in front of the antenna elements are defined as $\gamma_1, \gamma_2, \ldots, \gamma_N$, as shown in Figure 2.7. Apparently,

$$\gamma_n - \gamma_{n-1} = k_0 s \cos(\phi), \quad n = 2, 3, \ldots, N. \tag{2.56}$$

The phase value developed at each antenna element's circuit port is defined as $\alpha_1, \alpha_2, \ldots, \alpha_N$. If each antenna element interacts with the incident plane wave independently,

$$\alpha_1 - \gamma_1 = \alpha_2 - \gamma_2 = \cdots = \alpha_N - \gamma_N = \text{constant}. \tag{2.57}$$

Consequently, the linear pattern among $\{\gamma_1, \gamma_2, \cdots, \gamma_N\}$ is transferred to $\{\alpha_1, \alpha_2, \cdots, \alpha_N\}$:

$$\alpha_n - \alpha_{n-1} = k_0 s \cos(\phi), n = 2, 3, \ldots, N. \tag{2.58}$$

Once the linear pattern among $\{a_1, a_2, \cdots, a_N\}$ is detected, the direction of the incident plane wave, i.e., the direction of arrival, can be obtained. Of course, if the mutual coupling among antenna elements is not negligible, the relationship in (2.58) does not hold true and certain compensation techniques must be applied to resolve the mutual coupling [3,4].

In Figure 2.7, one phase shifter is applied to each antenna element before the antenna elements' outputs are synthesized. Under the condition that there is no mutual coupling among the antenna elements, the input impedance of each antenna element is Z_0, as specified in Figure 2.7. Suppose the circuit design in Figure 2.7 ensures that each antenna element's termination at its circuit port is equivalent to $(Z_0)^*$ for the sake of conjugate matching. To facilitate the narrative below, Z_0 is assumed to be real-valued. At the n-th antenna element's circuit port, suppose the voltage is $A_0 e^{j\alpha_n}$, the current is $\frac{A_0}{Z_0} e^{j\alpha_n}$, and the time-average power is $P_0 = \frac{1}{2} \left\{ A_0 e^{j\alpha_n} \frac{1}{Z_0} (A_0 e^{j\alpha_n})^* \right\} = \frac{(A_0)^2}{2Z_0}$.

If the phase shifters are adjusted to yield phase shift values of $-\alpha_1, -\alpha_2, \ldots,$ $-\alpha_N$, respectively, the n-th phase shifter's output has the voltage of $A_0 e^{j\alpha_n} e^{-j\alpha_n} = A_0$ and the current of $\frac{A_0}{Z_0} e^{j\alpha_n} e^{-j\alpha_n} = \frac{A_0}{Z_0}$.

Suppose the N currents are added up by the summing circuit. At the output of the summing circuit, the voltage is A_0, the current is $N \frac{A_0}{Z_0}$, and the time-average power is $\frac{1}{2} \left\{ A_0 N \frac{A_0}{Z_0} \right\} = N \frac{(A_0)^2}{2Z_0} = N P_0$.

As a result, the output power of the phased array is enhanced by a factor of N compared with the output power of each individual antenna element. In turn, the effective aperture associated with the phased array is $N A_e$ with respect to the incident plane wave. When the incident plane wave's propagation direction changes, it is very likely that the N antenna elements' outputs are no longer in phase. Therefore, the reception pattern of the phased array exhibits a beam and it is reconfigurable via adjusting the N phase shifters.

A special case is studied below with the assumption of $s = \lambda_0$. Also, suppose the mutual coupling among the antenna elements is weak such that (2.58) holds true. When $\phi = 60°$,

$$a_n - a_{n-1} = k_0 s \cos(\phi) = \frac{2\pi}{\lambda_0} \times \lambda_0 \times \cos(60°) = \pi. \tag{2.59}$$

Obviously, $a_n - a_{n-1} = \pi$ remains true when $\phi = -60°$. Thus, $\phi = 60°$ and $\phi = -60°$ cannot be distinguished from each other as far as $\{a_1, a_2, \ldots, a_N\}$ are concerned. The ambiguity between $\phi = 60°$ and $\phi = -60°$ can be eliminated in practice by ensuring that the individual antenna elements are insensitive to the incident plane wave along $\phi = -60°$. However, there is also an ambiguity between $\phi = 60°$ and $\phi = 120°$, because

$$a_n - a_{n-1} = \frac{2\pi}{\lambda_0} \times \lambda_0 \times \cos(120°) = -\pi. \tag{2.60}$$

The ambiguity between $\phi = 60°$ and $\phi = 120°$ is reciprocal to the grating lobe phenomena when a phased array is used as a transmitting antenna. It is not difficult to show that the ambiguity reciprocal to the grating lobe phenomena can be avoided if s is smaller than half wavelength.

2.3   Power transmission efficiency in "electrical far zone"

Two *far zone conditions* appear in Section 2.1, which are articulated below.

(i)   The condition of "$d \gg D_t$ and $d \gg D_r$" appears above (2.18). This condition is termed as *the condition of geometrical far zone* in this book.

(ii) The far-field condition $d > 2(D_t + D_r)^2/\lambda_0$ is derived in (2.24). This condition is termed as *the condition of electrical far zone* in this book.

Although the two conditions above both specify that the receiver is far away from the transmitter, they measure "farness" by the geometrical means and the electrical means, respectively. This section aims to thoroughly define and explain these two far zone conditions. Moreover, it is shown in this section that the power transmission efficiency would be poor when the condition of electrical far zone is satisfied.

Consider the scenario in Figure 2.2, in which the total electric field at $\mathbf{r}_o$ is the sum of N electric fields, each contributed by one antenna element of the phased array:

$$\mathbf{E}(\mathbf{r}_o) = \sum_{n=1}^{N} X_n e^{j\psi_n} \mathbf{U}_n \frac{e^{-jk_0|\mathbf{r}_o-\mathbf{r}'_n|}}{|\mathbf{r}_o - \mathbf{r}'_n|}. \tag{2.61}$$

Suppose $X_1 = X_2 = X_3 = \cdots = X_N = X_0$; in other words, suppose the N antenna elements are excited with a uniform amplitude. The geometrical far zone condition leads to $\frac{1}{|\mathbf{r}_o-\mathbf{r}'_1|} = \frac{1}{|\mathbf{r}_o-\mathbf{r}'_2|} = \frac{1}{|\mathbf{r}_o-\mathbf{r}'_3|} = \cdots = \frac{1}{|\mathbf{r}_o-\mathbf{r}'_N|} = \frac{1}{|\mathbf{r}_o-\mathbf{r}_s|}$ and $\mathbf{U}_1 = \mathbf{U}_2 = \mathbf{U}_3 = \cdots = \mathbf{U}_N = \mathbf{U}_0$. Thus under the condition of geometrical far zone, Equation (2.61) becomes

$$\mathbf{E}(\mathbf{r}_o) = \frac{X_0 \mathbf{U}_0}{|\mathbf{r}_o - \mathbf{r}_s|} \sum_{n=1}^{N} e^{j\psi_n} e^{-jk_0|\mathbf{r}_o-\mathbf{r}'_n|}. \tag{2.62}$$

In other words, under the condition of geometrical far zone, the N electric fields share the same magnitude. When the condition of electrical far zone is satisfied, the choice of excitation phases in (2.26) results in $\sum_{n=1}^{N} e^{j\psi_n} e^{-jk_0|\mathbf{r}_o-\mathbf{r}'_n|} = N$ at $\mathbf{r}_o = \mathbf{r}_b$. In other words, the condition of electrical far zone along with (2.26) ensures that the N electric fields share the same phase at $\mathbf{r}_o = \mathbf{r}_b$. In summary, the two far zone conditions correlate with the N electric fields' magnitude and phase, respectively.

Consider the electromagnetic fields radiated by one antenna element. Because the spatial distribution of electric current over the antenna element's body can be modeled by an array of Hertzian dipoles, it is valid to describe the interaction between one antenna element and one observation region using Figure 2.3. The two far zone conditions are therefore applicable when a wireless transmitter only includes one antenna element with D_t representing the physical dimension of the antenna element. In this book, it is assumed that the two far zone conditions always hold true as far as one antenna element is concerned. In other words, the electromagnetic fields radiated by one antenna element always behave as a plane wave.

As a matter of fact, the condition of geometrical far zone and the condition of electrical far zone can be applied to assess every wireless system once the values of D_t (the physical dimension of transmitter), D_r (the physical dimension of receiver), d (the distance between transmitter and receiver), and λ_0 (the wavelength in free space) are available. There are four combinations among "electrical far zone," "electrical near zone," "geometrical far zone," and "geometrical near zone." In Table 2.1, one example of practical application is identified for each of the four combinations. Both the geometrical far zone condition and electrical far zone condition are satisfied in cell phone communication (and in most of the practical wireless communication applications as well). In wireless power transmission applications based on the inductive coupling technology [5], the wireless power transmitter and wireless power receiver are close to each other geometrically, but the condition of $d > 2(D_t + D_r)^2/\lambda_0$ is satisfied. Obviously, the statement of "the transmitter and receiver of an inductive coupling system are in each other's electrical far zone" sounds weird. Therefore, the criterion of $d > 2(D_t + D_r)^2/\lambda_0$ is not sufficient to judge the "electrical farness." The inductive coupling technology takes advantage of low frequency band whereas the other three applications in Table 2.1

Table 2.1 Four combinations among "electrical far zone," "electrical near zone," "geometrical far zone," and "geometrical near zone"

	Geometrical far zone (d is far greater than D_t and D_r)	Geometrical near zone (d is not far greater than D_t or D_r)
Electrical far zone $d > \dfrac{2(D_t + D_r)^2}{\lambda_0}$	Example: Wireless signal transmission from a cell tower to a cell phone • $f = 2.4$ GHz ($\lambda_0 \cong 0.1$ meter) • $D_t = 1$ meter; $D_r = 0.1$ meter • $d = 1,000$ meters	Example: Wireless power transmission based on inductive coupling technology • $f = 1$ MHz ($\lambda_0 = 300$ meters) • $D_t = 0.1$ meter; $D_r = 0.1$ meter • $d = 0.1$ meter
Electrical near zone $d < \dfrac{2(D_t + D_r)^2}{\lambda_0}$	Example: Microwave power transmission from a satellite over the earth's geostationary orbit to the earth (discussed in Chapter 5) • $f = 5.8$ GHz ($\lambda_0 \cong 0.05$ meter) • $D_t = 1,000$ meters; $D_r = 4,000$ meters • $d = 36,000,000$ meters	Example: Microwave power transmission to mobile devices in an indoor environment (discussed in Chapter 4) • $f = 5.8$ GHz ($\lambda_0 \cong 0.05$ meter) • $D_t = 3$ meters; $D_r = 0.1$ meter • $d = 10$ meters

belong to microwave engineering. When the operating frequency is as low as 1 MHz, "electrical farness" should be judged by $d \gg \lambda_0$ rather than $d > 2(D_t + D_r)^2/\lambda_0$. In microwave engineering, nevertheless, d is much greater than λ_0 typically, and thus it is better to measure the "electrical farness" by $d > 2(D_t + D_r)^2/\lambda_0$ (which is in terms of phase). It is unlikely that the condition of electrical far zone would be satisfied in microwave power transmission applications. The two specific microwave power transmission applications at the bottom row of Table 2.1 are discussed in Chapter 4 and Chapter 5 of this book, respectively.

When the phased array technique is applied to wireless power transmission, power transmission efficiency is always the top technical concern. In contrast, certain technical metrics (such as bit error rate) are more important than the power transmission efficiency in wireless communication and radar applications. Suppose a phased array composed of $N \times N$ planar antenna elements with an inter-element spacing of $0.7\lambda_0$ is employed by a wireless power transmitter, as illustrated in Figure 2.5. When N is a large integer, D_t is approximately $\sqrt{2}N \times 0.7\lambda_0$. In addition, suppose the wireless power receiver's physical dimension is negligible, i.e., $D_r = 0$. Then, the condition of electrical far zone becomes $d > 2(D_t)^2/\lambda_0$. A rearrangement leads to the following inequality:

$$\frac{d}{D_t} > \frac{2D_t}{\lambda_0} = \frac{2\sqrt{2}N \times 0.7\lambda_0}{\lambda_0} = 2N. \tag{2.63}$$

As discussed in Section 2.1.2, a phased array is desired to generate a narrow radiation beam when it is employed by a wireless power transmitter to transmit microwave power to a mobile wireless power receiver. A large value of N is required to generate a narrow beam, as a conclusion of Section 2.1.2. As derived at the end of Section 2.1.2, the beamwidth associated with "$N = 5$" is about $20°$, which is not narrow enough for wireless power transmission over a distance of 10 m. If $N > 5$, the inequality in (2.63) becomes $(d/D_t) > 2N > 10$, which is equivalent to $d \gg D_t$ in engineering practice. Therefore, when the condition of electrical far zone is satisfied, it is very likely that the condition of geometrical far zone would be satisfied as well when a phased array is employed by a wireless power transmitter.

Before the end of this section, it is proved that the power transmission efficiency is poor when the condition of electrical far zone is met.

Given a pair of wireless power transmitter and wireless power receiver, define $d_{far} = 2(D_t + D_r)^2/\lambda_0$ as the smallest distance between them that satisfies the condition of electrical far zone. Next, define $(P_r/P_t)_{far}$ as the power transmission efficiency when $d = d_{far}$. When d is greater than d_{far}, the power transmission efficiency is expected to be poorer than $(P_r/P_t)_{far}$. Thus, $(P_r/P_t)_{far}$ stands for the upper limit of power transmission efficiency in the electrical far zone.

Suppose a phased array composed of $N \times N$ planar antenna elements with an inter-element spacing of $0.7\lambda_0$ is employed by a wireless power transmitter, as illustrated in Figure 2.8. Each antenna element is assumed to have a gain value of $G_0 = 4 = 6$ dBi. The wireless power receiver is assumed to be a small mobile electronic device with one antenna for wireless power reception. The gain value of the wireless power receiver's antenna is assumed to be $G_r = 2 = 3$ dBi. It is also assumed that the $N \times N$ antenna elements' contributions are in phase at the wireless power receiver such that the gain value of the antenna array is $G_{array} = N^2 G_0$ at the wireless power receiver. With $D_t = \sqrt{2}N \times 0.7\lambda_0$ and $D_r = 0$, the power transmission efficiency $(P_r/P_t)_{far}$ is evaluated through the Friis transmission equation as

$$
\begin{aligned}
\left(\frac{P_r}{P_t}\right)_{far} &= \left(\frac{\lambda_0}{4\pi d_{far}}\right)^2 G_{array} G_r \\
&= \left(\frac{(\lambda_0)^2}{4\pi \times 2(D_t)^2}\right)^2 (N^2 G_0) \times 2 \\
&= \left(\frac{(\lambda_0)^2}{4\pi \times 2(\sqrt{2}N \times 0.7\lambda_0)^2}\right)^2 (N^2 \times 4) \times 2 \\
&= \frac{1}{8\pi^2 N^2}
\end{aligned}
\tag{2.64}
$$

When $N = 5$, $(P_r/P_t)_{far}$ is as poor as 0.05%. When $N > 5$, $(P_r/P_t)_{far}$ would be poorer than 0.05%.

Based on the numerical example in Figure 2.8, the antenna of the wireless power receiver is assumed to be identical to the phased array in the wireless power transmitter, as displayed in Figure 2.9. Moreover, it is assumed that the gain value of the receiving phased array is identical to the gain value of the

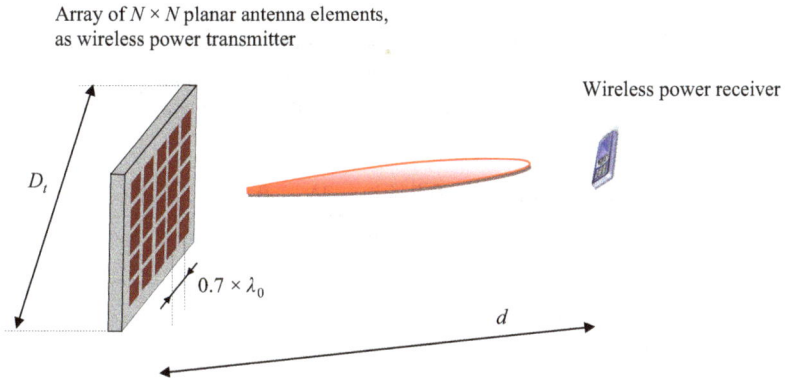

Figure 2.8   Evaluation of power transmission efficiency with a phased array and a small wireless power receiver in the electrical far zone

Array of $N \times N$ planar antenna elements, as wireless power transmitter

Array of $N \times N$ planar antenna elements, as wireless power receiver

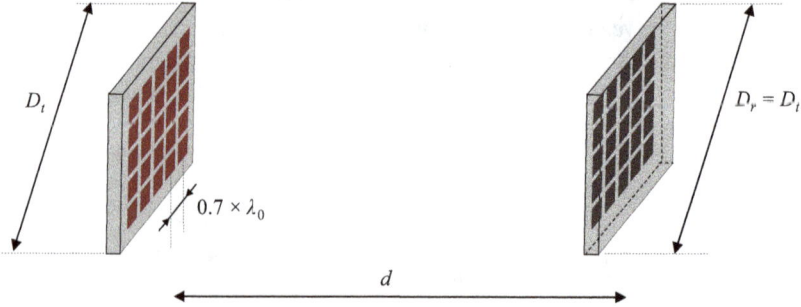

Figure 2.9   Evaluation of power transmission efficiency with two phased arrays in each other's electrical far zone

transmitting phased array. The power transmission efficiency $(P_r/P_t)_{far}$ is as poor as 0.15%:

$$
\begin{aligned}
\left(\frac{P_r}{P_t}\right)_{far} &= \left(\frac{\lambda_0}{4\pi d_{far}}\right)^2 G_{array} G_{array} \\
&= \left(\frac{(\lambda_0)^2}{4\pi \times 2(D_t + D_r)^2}\right)^2 (N^2 G_0) \times (N^2 G_0) \\
&= \left(\frac{(\lambda_0)^2}{4\pi \times 2(2\sqrt{2}N \times 0.7\lambda_0)^2}\right)^2 (N^2 \times 4) \times (N^2 \times 4) \\
&= \left(\frac{1}{8\pi}\right)^2 = 0.15\%
\end{aligned}
\tag{2.65}
$$

In Figure 2.9, the surface area of the wireless power transmitter's aperture A_t and the surface area of the wireless power receiver's aperture A_r are both approximately $(0.7N\lambda_0)^2$. The condition of electrical far zone $d > 2(D_t + D_r)^2/\lambda_0$ can be re-arranged to be

$$
d > \frac{2(D_t + D_r)^2}{\lambda_0} = \frac{2\left(\sqrt{2}N \times 0.7\lambda_0 + \sqrt{2}N \times 0.7\lambda_0\right)^2}{\lambda_0} = \frac{16\sqrt{A_t A_r}}{\lambda_0}. \tag{2.66}
$$

Equation (2.66) is equivalent to $\sqrt{A_t A_r}/(d\lambda_0) < 0.0625$. In [6,7], the dependence of power transmission efficiency on the parameter of $\tau = \sqrt{A_t A_r}/(d\lambda_0)$ is analyzed. Based on the analytical results of [6,7], the power transmission efficiency would approach 100% when $\tau > 2$, and the power transmission efficiency is less than 1% with $\tau < 0.0625$. Thus, the power transmission efficiency obtained in (2.65) is in agreement with the theoretical analysis of [6,7].

From the two numerical examples above, it seems fairly safe to conclude that the power transmission efficiency would be low in practice under the condition of electrical far zone when the phased array technique is applied to microwave power transmission.

2.4   Power transmission efficiency in "electrical near zone"

As discussed in Section 2.3, when a phased array is employed to transmit microwave power, it is very probable that a wireless power receiver must stay in the electrical near zone in order to achieve a reasonably high power transmission efficiency in practice. This section aims to show that the Friis transmission equation remains applicable to estimate the power transmission efficiency when the condition of electrical far zone is not satisfied.

Assume a two-dimensional phased array is employed by a wireless power transmitter, as illustrated in Figure 2.10. The phased array includes $N \times M$ planar antenna elements, with N columns along the x direction and M rows along the z direction. The values of inter-element spacing are s_x and s_z along x and z directions, respectively. Approximately, the physical dimensions of the phased array are $L = N \times s_x$ by $W = M \times s_z$. The geometrical center of the phased array is at the spatial origin. Specifically, the $N \times M$ antenna elements are located at

$$
\begin{aligned}
\mathbf{r}'_{nm} &= \widehat{\mathbf{x}} x'_n + \widehat{\mathbf{z}} z'_m \\
&= \widehat{\mathbf{x}} \left(n - \frac{N+1}{2} \right) s_x + \widehat{\mathbf{z}} \left(m - \frac{M+1}{2} \right) s_z, \quad
\begin{matrix} n = 1,2,3,\ldots,N \\ m = 1,2,3,\ldots,M \end{matrix}
\end{aligned} \tag{2.67}
$$

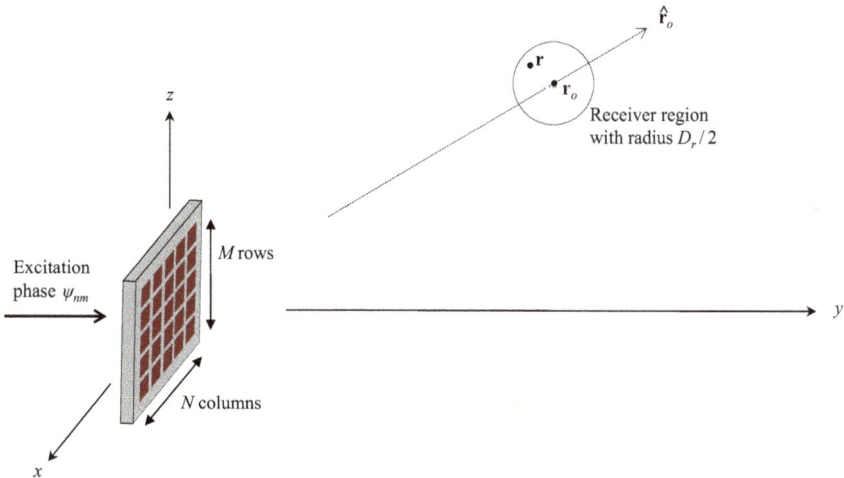

Figure 2.10   Interaction between a two-dimensional phased array and a receiver region in the electrical near zone

The wireless power receiver covers a certain region centered at $\mathbf{r}_o$ and with a diameter of D_r, as depicted in Figure 2.10.

In this section, it is assumed that the condition of electrical far zone (i.e., the far-field condition in Equation (2.24)) is not satisfied. Meanwhile, in this section, it is assumed that the condition of geometrical far zone holds true.

When the antenna elements of the phased array are excited by $X_0 e^{j\psi_{nm}}$, $n = 1, 2, 3, \ldots, N$, $m - 1, 2, 3, \ldots, M$, the electric field observed at an observation point $\mathbf{r}$ is

$$\mathbf{E}(\mathbf{r}) = \sum_{n=1}^{N}\sum_{m=1}^{M} X_0 e^{j\psi_{nm}} \mathbf{U}_{nm} \frac{e^{-jk_0|\mathbf{r}-\mathbf{r}'_{nm}|}}{|\mathbf{r}-\mathbf{r}'_{nm}|}. \tag{2.68}$$

Under the condition of geometrical far zone, (2.68) becomes

$$\mathbf{E}(\mathbf{r}) = \frac{X_0 \mathbf{U}_0}{r_o} \sum_{n=1}^{N}\sum_{m=1}^{M} e^{j\psi_{nm}} e^{-jk_0|\mathbf{r}-\mathbf{r}'_{nm}|}, \tag{2.69}$$

where $r_o = |\mathbf{r}_o|$ is the distance from $\mathbf{r}_o$ to the spatial origin and $\mathbf{U}_0$ is the polarization vector associated with one single antenna element at the spatial origin.

Following the derivation of (2.21), the distance between the observation point and an individual antenna element is

$$|\mathbf{r}-\mathbf{r}'_{nm}| = r_o + \hat{\mathbf{r}}_o\cdot(\mathbf{r}-\mathbf{r}_o) - \hat{\mathbf{r}}_o\cdot\mathbf{r}'_{nm} + \frac{|(\mathbf{r}-\mathbf{r}_o)-\mathbf{r}'_{nm}|^2}{2r_o}$$

$$-\frac{r_o}{8}\left[\frac{2\mathbf{r}_o\cdot[(\mathbf{r}-\mathbf{r}_o)-\mathbf{r}'_{nm}]}{(r_o)^2}+\frac{|(\mathbf{r}-\mathbf{r}_o)-\mathbf{r}'_{nm}|^2}{(r_o)^2}\right]^2$$

$$+\frac{r_o}{16}\left[\frac{2\mathbf{r}_o\cdot[(\mathbf{r}-\mathbf{r}_o)-\mathbf{r}'_{nm}]}{(r_o)^2}+\frac{|(\mathbf{r}-\mathbf{r}_o)-\mathbf{r}'_{nm}|^2}{(r_o)^2}\right]^3+\cdots \tag{2.70}$$

where $\hat{\mathbf{r}}_o = \mathbf{r}_o/r_o$. In addition, following the derivation of (2.17), the distance between the center of the wireless power receiver region and an individual antenna element is

$$|\mathbf{r}_o-\mathbf{r}'_{nm}| = r_o - \hat{\mathbf{r}}_o\cdot\mathbf{r}'_{nm} + \frac{|\mathbf{r}'_{nm}|^2}{2r_o}$$

$$-\frac{r_o}{8}\left[-\frac{2\mathbf{r}_o\cdot\mathbf{r}'_{nm}}{(r_o)^2}+\frac{|\mathbf{r}'_{nm}|^2}{(r_o)^2}\right]^2$$

$$+\frac{r_o}{16}\left[-\frac{2\mathbf{r}_o\cdot\mathbf{r}'_{nm}}{(r_o)^2}+\frac{|\mathbf{r}'_{nm}|^2}{(r_o)^2}\right]^3+\cdots \tag{2.71}$$

The difference between (2.70) and (2.71) is

$$
|\mathbf{r} - \mathbf{r}'_{nm}| - |\mathbf{r}_o - \mathbf{r}'_{nm}| = \hat{\mathbf{r}}_o \cdot (\mathbf{r} - \mathbf{r}_o) + \frac{|(\mathbf{r} - \mathbf{r}_o) - \mathbf{r}'_{nm}|^2}{2r_o} - \frac{|\mathbf{r}'_{nm}|^2}{2r_o}
$$

$$
+ \left\{ -\frac{r_o}{8} \left[\frac{2\mathbf{r}_o \cdot [(\mathbf{r} - \mathbf{r}_o) - \mathbf{r}'_{nm}]}{(r_o)^2} + \frac{|(\mathbf{r} - \mathbf{r}_o) - \mathbf{r}'_{nm}|^2}{(r_o)^2} \right]^2 + \frac{r_o}{8} \left[-\frac{2\mathbf{r}_o \cdot \mathbf{r}'_{nm}}{(r_o)^2} + \frac{|\mathbf{r}'_{nm}|^2}{(r_o)^2} \right]^2 \right\}
$$

$$
+ \left\{ \frac{r_o}{16} \left[\frac{2\mathbf{r}_o \cdot [(\mathbf{r} - \mathbf{r}_o) - \mathbf{r}'_{nm}]}{(r_o)^2} + \frac{|(\mathbf{r} - \mathbf{r}_o) - \mathbf{r}'_{nm}|^2}{(r_o)^2} \right]^3 - \frac{r_o}{16} \left[-\frac{2\mathbf{r}_o \cdot \mathbf{r}'_{nm}}{(r_o)^2} + \frac{|\mathbf{r}'_{nm}|^2}{(r_o)^2} \right]^3 \right\}
$$

$$
+ \cdots
$$

$$(2.72)$$

In this section, the first three terms on the right-hand side of (2.72) are retained, and the high-order terms beyond the first line of (2.72) are neglected. The magnitude of these high-order terms is examined in Section 2.5 with the aid of one numerical example. After the high-order terms are ignored, (2.72) becomes

$$
|\mathbf{r} - \mathbf{r}'_{nm}| = |\mathbf{r}_o - \mathbf{r}'_{nm}| + \hat{\mathbf{r}}_o \cdot (\mathbf{r} - \mathbf{r}_o) + \frac{|(\mathbf{r} - \mathbf{r}_o) - \mathbf{r}'_{nm}|^2}{2r_o} - \frac{|\mathbf{r}'_{nm}|^2}{2r_o}.
$$

$$(2.73)$$

Since

$$
|(\mathbf{r} - \mathbf{r}_o) - \mathbf{r}'_{nm}|^2 = [(\mathbf{r} - \mathbf{r}_o) - \mathbf{r}'_{nm}] \cdot [(\mathbf{r} - \mathbf{r}_o) - \mathbf{r}'_{nm}]
$$
$$
= |\mathbf{r} - \mathbf{r}_o|^2 - 2(\mathbf{r} - \mathbf{r}_o) \cdot \mathbf{r}'_{nm} + |\mathbf{r}'_{nm}|^2,
$$

$$(2.74)$$

Equation (2.73) can be re-arranged to be

$$
|\mathbf{r} - \mathbf{r}'_{nm}| = |\mathbf{r}_o - \mathbf{r}'_{nm}| + \hat{\mathbf{r}}_o \cdot (\mathbf{r} - \mathbf{r}_o) + \frac{|\mathbf{r} - \mathbf{r}_o|^2}{2r_o} - \frac{(\mathbf{r} - \mathbf{r}_o) \cdot \mathbf{r}'_{nm}}{r_o}.
$$

$$(2.75)$$

Substituting (2.75) into (2.69) yields

$$
\mathbf{E}(\mathbf{r}) = \frac{X_0 U_0}{r_o} \sum_{n=1}^{N} \sum_{m=1}^{M} e^{j\psi_{nm}} e^{-jk_0|\mathbf{r} - \mathbf{r}'_{nm}|}
$$

$$
= \frac{X_0 U_0}{r_o} \sum_{n=1}^{N} \sum_{m=1}^{M} e^{j\psi_{nm}} e^{-jk_0|\mathbf{r}_o - \mathbf{r}'_{nm}|} e^{-jk_0 \hat{\mathbf{r}}_o \cdot (\mathbf{r} - \mathbf{r}_o)} e^{-jk_0 \frac{|\mathbf{r} - \mathbf{r}_o|^2}{2r_o}} e^{jk_0 \frac{(\mathbf{r} - \mathbf{r}_o) \cdot \mathbf{r}'_{nm}}{r_o}}
$$

$$(2.76)$$

If the condition of electrical far zone is not satisfied, the choice of excitation phases in (2.50) does not offer optimal performance. Under the condition of electrical near zone, the excitation phases should be chosen as

$$
\psi_{nm} = k_0 |\mathbf{r}_b - \mathbf{r}'_{nm}|.
$$

$$(2.77)$$

With the excitation phases in (2.77), (2.76) becomes

$$\mathbf{E}(\mathbf{r}) = \frac{X_0 U_0}{r_o} \sum_{n=1}^{N} \sum_{m=1}^{M} e^{jk_0|\mathbf{r}_b - \mathbf{r}'_{nm}|} e^{-jk_0|\mathbf{r}_o - \mathbf{r}'_{nm}|} e^{-jk_0\widehat{\mathbf{r}}_o \cdot (\mathbf{r} - \mathbf{r}_o)} e^{-jk_0\frac{|\mathbf{r} - \mathbf{r}_o|^2}{2r_o}} e^{jk_0\frac{(\mathbf{r} - \mathbf{r}_o)\cdot \mathbf{r}'_{nm}}{r_o}} \qquad (2.78)$$

When $\mathbf{r}_o = \mathbf{r}_b$

$$\mathbf{E}(\mathbf{r}) = \frac{X_0 U_0}{r_b} \sum_{n=1}^{N} \sum_{m=1}^{M} e^{-jk_0\widehat{\mathbf{r}}_b \cdot (\mathbf{r} - \mathbf{r}_b)} e^{jk_0\frac{|\mathbf{r} - \mathbf{r}_b|^2}{2r_b}} e^{jk_0\frac{(\mathbf{r} - \mathbf{r}_b)\cdot \mathbf{r}'_{nm}}{r_b}}$$

$$= \frac{X_0 U_0}{r_b} e^{-jk_0\widehat{\mathbf{r}}_b \cdot (\mathbf{r} - \mathbf{r}_b)} e^{-jk_0\frac{|\mathbf{r} - \mathbf{r}_b|^2}{2r_b}} \sum_{n=1}^{N} \sum_{m=1}^{M} e^{jk_0\frac{(\mathbf{r} - \mathbf{r}_b)\cdot \mathbf{r}'_{nm}}{r_b}} \qquad (2.79)$$

where $r_b = |\mathbf{r}_b|$ and $\widehat{\mathbf{r}}_b = \mathbf{r}_b/r_b$. By defining

$$Q_{phase} = e^{-jk_0\frac{|\mathbf{r} - \mathbf{r}_b|^2}{2r_b}} \qquad (2.80)$$

and

$$Q_{amplitude} = \sum_{n=1}^{N} \sum_{m=1}^{M} e^{jk_0\frac{(\mathbf{r} - \mathbf{r}_b)\cdot \mathbf{r}'_{nm}}{r_b}}, \qquad (2.81)$$

Equation (2.79) can be rewritten as

$$\mathbf{E}(\mathbf{r}) = \frac{X_0 U_0}{r_b} e^{-jk_0\widehat{\mathbf{r}}_b \cdot (\mathbf{r} - \mathbf{r}_b)} Q_{phase} Q_{amplitude} \qquad (2.82)$$

The dot product between $\mathbf{r} - \mathbf{r}_b$ and $\mathbf{r}'_{nm}$ in (2.81) is

$$(\mathbf{r} - \mathbf{r}_b) \cdot \mathbf{r}'_{nm} = [(\mathbf{r} - \mathbf{r}_b) \cdot \widehat{\mathbf{x}}]\left(n - \frac{N+1}{2}\right)s_x + [(\mathbf{r} - \mathbf{r}_b) \cdot \widehat{\mathbf{z}}]\left(m - \frac{M+1}{2}\right)s_z. \qquad (2.83)$$

As a result, $Q_{amplitude}$ can be evaluated as

$$Q_{amplitude} = \sum_{n=1}^{N} \sum_{m=1}^{M} e^{jk_0\frac{(\mathbf{r} - \mathbf{r}_b)\cdot \mathbf{r}'_{nm}}{r_b}}$$

$$= \sum_{n=1}^{N} e^{j\frac{k_0}{r_b}[(\mathbf{r} - \mathbf{r}_b)\cdot \widehat{\mathbf{x}}]\left(n - \frac{N+1}{2}\right)s_x} \sum_{m=1}^{M} e^{j\frac{k_0}{r_b}[(\mathbf{r} - \mathbf{r}_b)\cdot \widehat{\mathbf{z}}]\left(m - \frac{M+1}{2}\right)s_z}$$

$$= \sum_{n=1}^{N} e^{j\left(n - \frac{N+1}{2}\right)\xi_x} \sum_{m=1}^{M} e^{j\left(m - \frac{M+1}{2}\right)\xi_z} \qquad (2.84)$$

$$= \frac{\sin\left\{\frac{N\xi_x}{2}\right\}}{\sin\left\{\frac{\xi_x}{2}\right\}} \times \frac{\sin\left\{\frac{M\xi_z}{2}\right\}}{\sin\left\{\frac{\xi_z}{2}\right\}}$$

where

$$\xi_x = \frac{k_0}{r_b}[(\mathbf{r} - \mathbf{r}_b) \cdot \widehat{\mathbf{x}}]s_x \qquad (2.85)$$

and

$$\xi_z = \frac{k_0}{r_b}[(\mathbf{r} - \mathbf{r}_b) \cdot \hat{\mathbf{z}}]s_z. \tag{2.86}$$

Equation (2.38) is used during the derivation of (2.84).

When $\mathbf{r} = \mathbf{r}_b$, $Q_{phase} = 1$ and $Q_{amplitude} = NM$. As a result, the electric field at $\mathbf{r} = \mathbf{r}_b$ is

$$\mathbf{E}(\mathbf{r}_b) = \frac{X_0 \mathbf{U}_0}{r_b} NM, \tag{2.87}$$

meaning that the electric fields contributed by the $N \times M$ antenna elements are in phase (that is, fully constructive among one another) at $\mathbf{r} = \mathbf{r}_b$. When $\mathbf{r} \neq \mathbf{r}_b$, the value of Q_{phase} deviates from 1 and the value of $Q_{amplitude}$ deviates from NM, and as a result, the $N \times M$ antenna elements' contributions are no longer completely constructive among one another. If Q_{phase} does not deviate much from 1 and $Q_{amplitude}$ does not deviate much from NM when $\mathbf{r}$ is in the close neighborhood of $\mathbf{r}_b$, the term $e^{-jk_0\mathbf{r}_b \cdot (\mathbf{r} - \mathbf{r}_b)}$ in (2.82) indicates that the total electric field contributed by the $N \times M$ antenna elements behaves as a plane wave around $\mathbf{r}_b$.

As discussed in Section 2.1, the choice of excitation phases in (2.50) has two direct consequences when the condition of electrical far zone is satisfied. First, the electromagnetic fields contributed by all the antenna elements are in phase at $\mathbf{r}_b$. Second, the total electromagnetic fields behave as a plane wave in the neighborhood of $\mathbf{r}_b$. When the condition of electrical far zone is not satisfied, the two phenomena above would no longer hold true with the excitation phases in (2.50). Under the condition of electrical near zone, however, the two phenomena would be restored around $\mathbf{r}_b$ by the choice of excitation phases in (2.77). In the electrical near zone, the excitation phases in (2.77) guarantee that the electromagnetic fields contributed by all the antenna elements are in phase at $\mathbf{r}_b$. The "plane wave behavior" associated with (2.77) is determined by Q_{phase} and $Q_{amplitude}$. The term Q_{phase}, with a constant amplitude of 1, determines the deviation of (2.82) from a plane wave in terms of phase. The term $Q_{amplitude}$, which is a real number (in other words, with phase being zero), determines the deviation of (2.82) from a plane wave in terms of amplitude. The impacts of Q_{phase} and $Q_{amplitude}$ are assessed in Section 2.5 with the aid of one numerical example. As observed in Section 2.5, a certain "plane wave region" always exists around $\mathbf{r}_b$, and the size of "plane wave region" depends on Q_{phase} and $Q_{amplitude}$.

Suppose Q_{phase} and $Q_{amplitude}$ have weak variations around $\mathbf{r}_b$, that is, suppose the value of Q_{phase} stays close to 1 and the value of $Q_{amplitude}$ stays close to NM around $\mathbf{r}_b$. The choice of excitation phases in (2.77) ensures that the contributions from all the antenna elements of the phased array are in phase at $\mathbf{r}_b$. Specifically, the electric field at $\mathbf{r}_b$ is $MN\mathbf{E}_0$ and the magnetic field at $\mathbf{r}_b$ is $MN\mathbf{H}_0$, where $\mathbf{E}_0$ is the electric field and $\mathbf{H}_0$ is the magnetic field radiated by one single antenna element when the antenna element is located at the spatial origin. Moreover, the electromagnetic fields around $\mathbf{r}_b$ behave as a plane wave traveling toward $\hat{\mathbf{r}}_b$.

Therefore, the Poynting vector around $\mathbf{r}_b$ is

$$\frac{|MN\mathbf{E}_0|^2}{2\eta_0}\hat{\mathbf{r}}_b. \tag{2.88}$$

It then appears reasonable to extend the definition of antenna gain to the electrical near zone as

$$G_{array} = \frac{\frac{|MN\mathbf{E}_0|^2}{2\eta_0}}{\frac{P_t}{4\pi(r_b)^2}} = \frac{\frac{|MN\mathbf{E}_0|^2}{2\eta_0}}{\frac{P_t}{4\pi d^2}}. \tag{2.89}$$

If there is no mutual coupling among antenna elements, the total amount of power transmitted by the phased array is $P_t = MNP_0$, with $P_0 = \frac{(X_0)^2}{2}\mathrm{Re}\{\frac{1}{Z_0}\}$ as the power supplied to each antenna element. Straightforwardly,

$$G_{array} = \frac{\frac{|MN\mathbf{E}_0|^2}{2\eta_0}}{\frac{MNP_0}{4\pi d^2}} = MN\frac{\frac{|\mathbf{E}_0|^2}{2\eta_0}}{\frac{P_0}{4\pi d^2}} = MNG_0, \tag{2.90}$$

where G_0 is the antenna gain of one single antenna element toward $\hat{\mathbf{r}}_b$ direction when the antenna element is located at the spatial origin. If one receiving antenna is located at $\mathbf{r}_b$ and terminated by a matched load, the power received by the matched load is

$$P_r = \frac{|MN\mathbf{E}_0|^2}{2\eta_0}A_e, \tag{2.91}$$

where A_e is the effective aperture of the receiving antenna in reaction to an incoming plane wave traveling toward $\hat{\mathbf{r}}_b$. For (2.91) to be valid, the effective aperture and physical aperture of the receiving antenna ought to be smaller than the size of the "plane wave region." The effective aperture and gain value of the receiving antenna are related to each other by

$$\frac{A_e}{G_r} = \frac{(\lambda_0)^2}{4\pi}, \tag{2.92}$$

where G_r is the antenna gain toward $-\hat{\mathbf{r}}_b$ direction when the receiving antenna is used as a transmitting antenna. Equations (2.90), (2.91), and (2.92) jointly lead to

$$\frac{P_r}{P_t} = \left(\frac{\lambda_0}{4\pi d}\right)^2 G_{array}G_r = \left(\frac{\lambda_0}{4\pi d}\right)^2 (MNG_0)G_r. \tag{2.93}$$

Equation (2.93) is essentially identical to the Friis transmission equation in (2.31) that was derived under the condition of electrical far zone. In summary, the Friis transmission equation can be applied in the electrical near zone under the following four conditions.

Condition (i) Excitation phases are chosen as in (2.77), with a wireless power receiver at $\mathbf{r}_b$.

Condition (ii) There is no mutual coupling among the antenna elements of the phased array.

Condition (iii) The wireless power receiver's size is smaller than the size of the "plane wave region."

Condition (iv) The condition of geometrical far zone is satisfied.

Equation (2.93) is the same as Equation (1.9) in Chapter 1. Based on Equation (1.9), Equation (1.10) was derived in Chapter 1 to estimate the power transmission efficiency from a two-dimensional phased array to a mobile target. As found in Section 1.3, the power transmission efficiency could reach 1% when the phased array has the physical dimensions of 3 m by 3 m and when the target resides 10 m away from the phased array (although 10 m is not much greater than 3 m, the geometrical nearness does not invalidate (2.93), as shown in Sections 2.5 and 2.6). As discussed in Section 1.3, the power transmission efficiency of 1% may be adequate to enable certain practical applications of microwave power transmission.

2.5 One numerical example on the performance of a two-dimensional phased array in the "electrical far zone" and "electrical near zone"

As analyzed in Section 2.4, the excitation phases of (2.77) ensure that the contributions from all the antenna elements of a phased array are in phase at $\mathbf{r}_b$ when $\mathbf{r}_b$ resides in the electrical near zone of the phased array. As another property associated with (2.77), the electromagnetic fields behave as a plane wave traveling toward $\hat{\mathbf{r}}_b$ in a certain "plane wave region" around $\mathbf{r}_b$. It is noted that the above two properties are demonstrated by the excitation phases of (2.50) when the condition of electrical far zone is satisfied. As a matter of fact, (2.50) is a special case of (2.77). The excitation phases in (2.77) are determined by $|\mathbf{r}_b - \mathbf{r}'_{nm}|$, and the excitation phases in (2.50) are determined by $-\hat{\mathbf{r}}_b \cdot \mathbf{r}'_{nm}$. Following the derivations of (2.14) to (2.16), it is easy to prove that

$$|\mathbf{r}_b - \mathbf{r}'_{nm}| = r_b \sqrt{1 - \frac{2\mathbf{r}_b \cdot \mathbf{r}'_{nm}}{(r_b)^2} + \frac{|\mathbf{r}'_{nm}|^2}{(r_b)^2}} \xrightarrow{r_b \to \infty} r_b - \hat{\mathbf{r}}_b \cdot \mathbf{r}'_{nm}. \qquad (2.94)$$

Therefore, (2.77) and (2.50) are equivalent to each other as long as $r_b = d$ is large enough. It must be noted that "relative difference" or "percent difference" is an incorrect measurement for assessing the difference between $|\mathbf{r}_b - \mathbf{r}'_{nm}|$ and $r_b - \hat{\mathbf{r}}_b \cdot \mathbf{r}'_{nm}$ in (2.94). This is because these two terms are used to obtain phase values and two phase values are close to each other only if the difference between them is much smaller than 360°. Specifically, $|\mathbf{r}_b - \mathbf{r}'_{nm}|$ and $r_b - \hat{\mathbf{r}}_b \cdot \mathbf{r}'_{nm}$ in (2.94) are close to each other if $k_0 |\mathbf{r}_b - \mathbf{r}'_{nm}| - k_0(r_b - \hat{\mathbf{r}}_b \cdot \mathbf{r}'_{nm})$ is much smaller than 360°, which leads to the condition of electrical far zone.

A numerical example is presented in this section to demonstrate the performance associated with the excitation phases in (2.77) as well as the performance associated with the excitation phases in (2.50).

The setup of the numerical example is illustrated in Figure 2.11. A two-dimensional phased array resides at the spatial origin. The phased array includes $N \times M$ antenna elements with spacing s_x along the x direction and spacing s_z along the z direction. The physical dimensions of the phased array are approximately $L = N \times s_x$ and $W = M \times s_z$. The location of antenna elements $\mathbf{r}'_{nm}$ is

$$
\begin{aligned}
\mathbf{r}'_{nm} &= \widehat{\mathbf{x}} x'_n + \widehat{\mathbf{z}} z'_m \\
&= \widehat{\mathbf{x}}\left(n - \frac{N+1}{2}\right)s_x + \widehat{\mathbf{z}}\left(m - \frac{M+1}{2}\right)s_z, \quad \begin{array}{l} n = 1, 2, 3, \ldots, N \\ m = 1, 2, 3, \ldots, M \end{array}
\end{aligned}
$$

The antenna elements are excited by a uniform amplitude of 1 and by phase values of ψ_{nm}, $n = 1, 2, 3, \ldots, N$, $m = 1, 2, 3, \ldots M$. The observation point $\mathbf{r}$ is located in a planar region centered at $\mathbf{r}_o$, which is characterized by spherical coordinates ($r_o = 10$ m, $\theta_o = 60°$, $\phi_o = 135°$). The following three unit vectors are defined at $\mathbf{r}_o$:

$$
\begin{aligned}
\widehat{\mathbf{r}}_o &= \sin(\theta_o)\cos(\phi_o)\widehat{\mathbf{x}} + \sin(\theta_o)\sin(\phi_o)\widehat{\mathbf{y}} + \cos(\theta_o)\widehat{\mathbf{z}} \\
\widehat{\boldsymbol{\theta}}_o &= \cos(\theta_o)\cos(\phi_o)\widehat{\mathbf{x}} + \cos(\theta_o)\sin(\phi_o)\widehat{\mathbf{y}} - \sin(\theta_o)\widehat{\mathbf{z}} \\
\widehat{\boldsymbol{\phi}}_o &= -\sin(\phi_o)\widehat{\mathbf{x}} + \cos(\phi_o)\widehat{\mathbf{y}}
\end{aligned} \tag{2.95}
$$

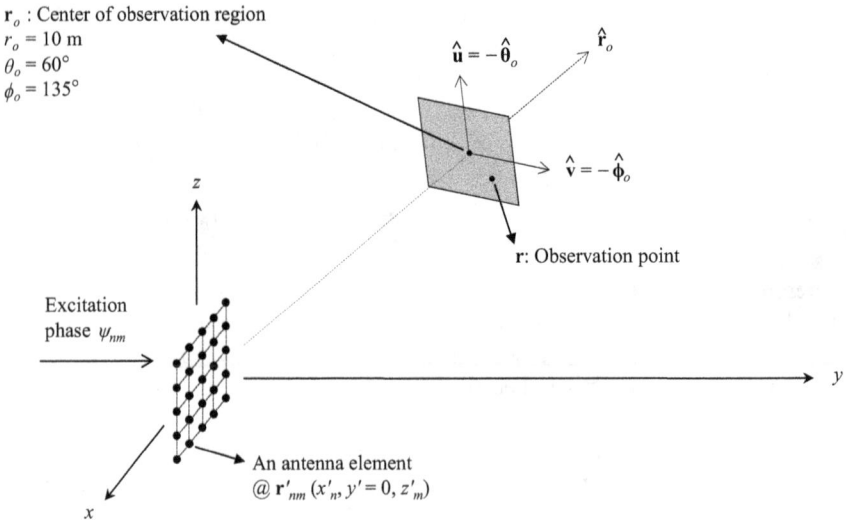

Figure 2.11   Setup of a numerical example to study the radiation of a two-dimensional phased array

A local Cartesian coordinate system is established at $\mathbf{r}_o$ with $\hat{\mathbf{u}} = -\hat{\boldsymbol{\theta}}_o$, $\hat{\mathbf{v}} = -\hat{\boldsymbol{\phi}}_o$, and $\hat{\mathbf{r}}_o$ as the three axes. The observation point $\mathbf{r}$ is characterized by coordinates (u, v) in the local Cartesian coordinate system. The size of the observation region is 30 cm by 30 cm; in other words, the range of u is $[-15 \text{ cm}, +15 \text{ cm}]$ and the range of v is $[-15 \text{ cm}, +15 \text{ cm}]$. In this numerical example, the operating frequency is assumed to be 6 GHz, and $s_x = s_z = 0.7\lambda_0$.

The two-fold sum in (2.69) is defined as the *constructive coefficient*:

$$\Phi(u, v) = \frac{1}{NM} \sum_{n=1}^{N} \sum_{m=1}^{M} e^{j\psi_{nm}} e^{-jk_0 |\mathbf{r} - \mathbf{r}'_{nm}|} \tag{2.96}$$

If the contributions from the $N \times M$ antenna elements are in phase (i.e., fully constructive among one another) at a certain observation point (u, v), the amplitude of constructive coefficient would have the value of 1. Of course, if the contributions from the antenna elements are not completely constructive among one another, the amplitude of the constructive coefficient would drop from 1. When the excitation phase values are chosen as $\psi_{nm} = k_0 | \mathbf{r}_o - \mathbf{r}'_{nm} |$ by using $\mathbf{r}_o = \mathbf{r}_b$ in (2.77), the constructive coefficient can be evaluated analytically as

$$\Phi(u, v) = \frac{1}{NM} Q_{phase} Q_{amplitude} \tag{2.97}$$

according to (2.79) and (2.82) with $\mathbf{r}_o = \mathbf{r}_b$.

Some numerical results of the constructive coefficient Φ evaluated via (2.96) and (2.97) are plotted in Figures 2.12–2.14. In each of Figures 2.12–2.14, there are four subplots. Specifically, Subplot (a) and Subplot (b) are obtained by evaluating the two-fold summation in (2.96), and Subplot (c) and Subplot (d) are obtained via the closed-form expression in (2.97). The physical dimensions of the phased array are approximately (0.1 m × 0.1 m), (1 m × 1 m), and (3 m × 3 m) in Figures 2.12–2.14, respectively. In Figures 2.12–2.14, $|\Phi|$ (the amplitude of Φ) is plotted in Subplot (a) and Subplot (c), and $|\angle\Phi - \angle\Phi_0|$ is plotted in Subplot (b) and Subplot (d) where $\angle\Phi$ is the phase of Φ and Φ_0 is the constructive coefficient at $\mathbf{r}_o$ (that is, with $u = 0$ and $v = 0$).

In Figures 2.12–2.14, the subplots obtained via (2.96) and the subplots obtained via (2.97) are almost the same as each other, indicating that the closed-form formulation in (2.97) is very accurate. The difference between the results evaluated by (2.96) and the results evaluated by (2.97) is studied at the end of this section.

Both the amplitude distribution and phase distribution in Figure 2.12 are fairly uniform. Specifically, the amplitude values are close to 1 throughout the observation region, and the largest phase value (which is at the corners of the observation region) is as small as 16.2°. When the size of the phased array is 0.1 m by 0.1 m, the condition of electrical far zone is satisfied. Thus, the field distribution over the observation region resembles a plane wave, as expected.

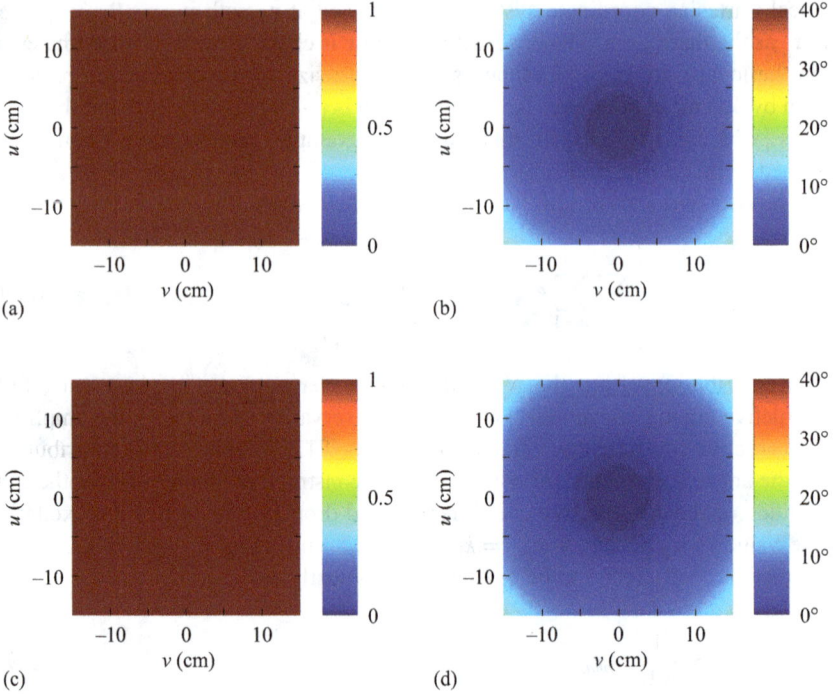

Figure 2.12 Numerical results of constructive coefficient Φ with $\psi_{nm} = k_0 \mid \mathbf{r}_o - \mathbf{r}'_{nm} \mid$, when the size of a phased array is about 0.1 m by 0.1 m. (a) Amplitude of Φ evaluated via (2.96); (b) phase of Φ evaluated via (2.96); (c) amplitude of Φ evaluated via (2.97); and (d) phase of Φ evaluated via (2.97).

When the size of the phased array increases to 1 m by 1 m, the condition of electrical far zone is not satisfied. In Figure 2.13, the distribution of amplitude is slightly nonuniform, while the value at $\mathbf{r}_o$ ($u = 0$, $v = 0$) remains to be 1. The phase distribution in Figure 2.13 stays unchanged in comparison with that in Figure 2.12. The phase distribution is governed by Q_{phase}. It is noticed that Q_{phase} does not depend on the size of the phased array. As a consequence, increasing the size of the phased array from 0.1 m to 1 m does not impact the phase distribution at all. When the observation point $\mathbf{r}$ is located at one of the four corners of the observation region, $\mid \mathbf{r} - \mathbf{r}_o \mid = \sqrt{2} \times 0.15$ m. The phase of Q_{phase} in (2.80) with $\mathbf{r}_o = \mathbf{r}_b$ is

$$k_0 \frac{\mid \mathbf{r} - \mathbf{r}_o \mid^2}{2r_o} = \frac{360°}{\lambda_0} \times \frac{(\sqrt{2} \times 0.15)^2}{2r_o} = \frac{360°}{0.05} \times \frac{(\sqrt{2} \times 0.15)^2}{2 \times 10} = 16°, \quad (2.98)$$

which agrees with the phase plots of Figures 2.12 and 2.13 excellently.

When the size of the phased array increases to 3 m by 3 m, the observation region located 10 m away resides in the electrical near zone of the phased array undoubtedly. A focal point is clearly visible in the amplitude plots of Figure 2.14. The peak amplitude of 1 is reached at the focal point's center ($u = 0$, $v = 0$). The

(a)   (b)

(c)   (d)

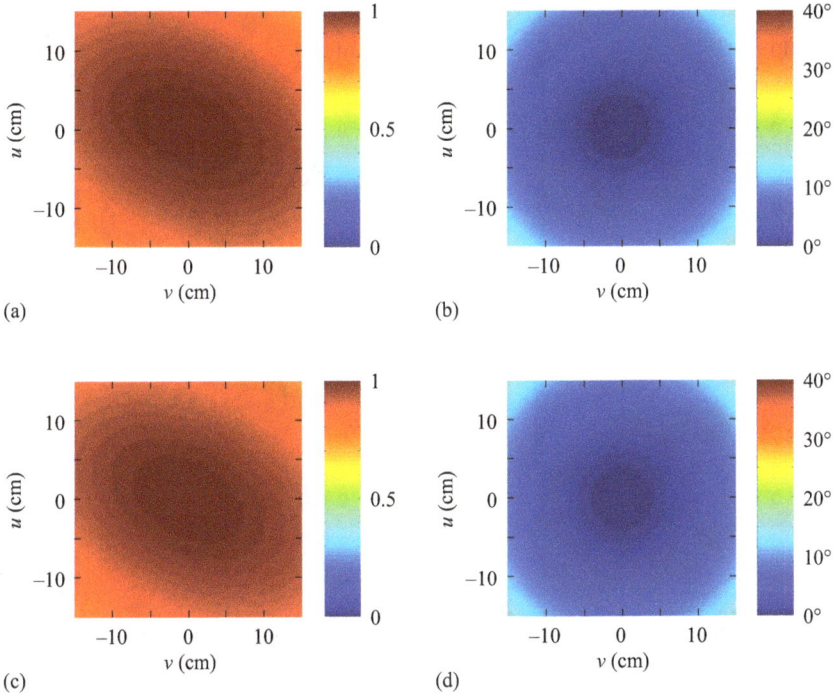

Figure 2.13   *Numerical results of constructive coefficient Φ with*
$\psi_{nm} = k_0 \mid \mathbf{r}_o - \mathbf{r}'_{nm} \mid$, *when the size of a phased array is about 1 m*
by 1 m. (a) Amplitude of Φ evaluated via (2.96); (b) phase of Φ
evaluated via (2.96); (c) amplitude of Φ evaluated via (2.97); and (d)
phase of Φ evaluated via (2.97).

spatial size of the focal point is determined by $Q_{amplitude}$. At the top-right corner of
the observation region, $u = 0.15$ m and $v = 0.15$ m. Thus,

$$\mathbf{r} - \mathbf{r}_o = 0.15\widehat{\mathbf{u}} + 0.15\widehat{\mathbf{v}} = -0.15\widehat{\boldsymbol{\theta}}_o - 0.15\widehat{\boldsymbol{\phi}}_o$$
$$= -0.15 \times \{\cos(60°)\cos(135°)\widehat{\mathbf{x}} + \cos(60°)\sin(135°)\widehat{\mathbf{y}} - \sin(60°)\widehat{\mathbf{z}}\}$$
$$-0.15 \times \{-\sin(135°)\widehat{\mathbf{x}} + \cos(135°)\widehat{\mathbf{y}}\} \text{ (m)}$$
$$\tag{2.99}$$

The x component of (2.99) is

$$(\mathbf{r} - \mathbf{r}_o) \cdot \widehat{\mathbf{x}} = -0.15 \times \cos(60°)\cos(135°) + 0.15 \times \sin(135°)$$
$$= -0.15 \times \frac{1}{2} \times \left(-\frac{\sqrt{2}}{2}\right) + 0.15 \times \frac{\sqrt{2}}{2} \tag{2.100}$$
$$\approx 0.16 \text{ m}$$

After (2.100) is substituted into (2.85) with $\mathbf{r}_o = \mathbf{r}_b$,

$$\xi_x = \frac{k_0}{r_o}[(\mathbf{r} - \mathbf{r}_o) \cdot \widehat{\mathbf{x}}]s_x = \frac{2\pi}{\lambda_0 r_o} \times 0.16 \times s_x \tag{2.101}$$

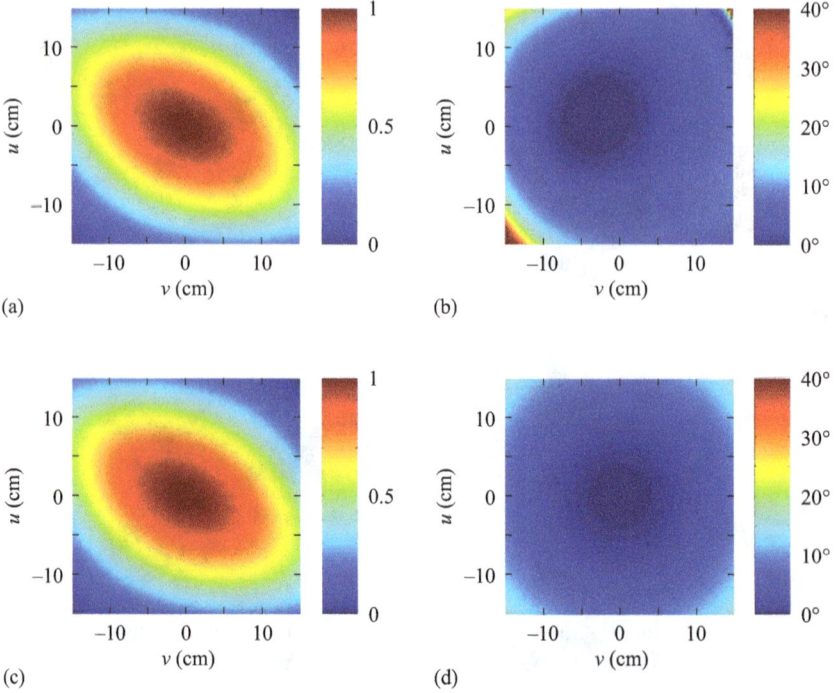

(a)   (b)   (c)   (d)

Figure 2.14   Numerical results of constructive coefficient Φ with $\psi_{nm} = k_0 \mid \mathbf{r}_o - \mathbf{r}'_{nm} \mid$, when the size of the phased array is about 3 m by 3 m. (a) Amplitude of Φ evaluated via (2.96); (b) phase of Φ evaluated via (2.96); (c) amplitude of Φ evaluated via (2.97); and (d) phase of Φ evaluated via (2.97).

and

$$\frac{N\xi_x}{2} = \frac{N}{2} \times \frac{2\pi}{\lambda_0 r_o} \times 0.16 \times s_x = (N \times s_x) \times \frac{\pi}{\lambda_0 r_o} \times 0.16$$

$$= 3 \times \frac{\pi}{0.05 \times 10} \times 0.16 = 0.96\pi \tag{2.102}$$

Because $N\xi_x/2$ is close to π, the value of $\sin(N\xi_x/2)$ is almost zero, and as a result, the value of $Q_{amplitude}$ in (2.84) is almost zero. Correspondingly, a null appears at the top-right corner of the amplitude plots in Figure 2.14. The phase distribution of Figure 2.14 remains quite uniform as those in Figures 2.12 and 2.13. The large phase values at two corners of the observation region in Figure 2.14(b) are due to numerical errors when the amplitude values approach zero at these corners. In the region of $u \in [-5 \text{ cm}, +5 \text{ cm}]$ and $v \in [-5 \text{ cm}, +5 \text{ cm}]$, the amplitude is greater than 0.7 and the phase distribution has little non-uniformity. Therefore, the electromagnetic fields in Figure 2.14 resemble a plane wave quite well in the region of $u \in [-5 \text{ cm}, +5 \text{ cm}]$ and $v \in [-5 \text{ cm}, +5 \text{ cm}]$. This means that the electromagnetic fields generated by a (3 m × 3 m) phased array behave as a plane wave over a distance of 10 m as long as the wireless power receiver's physical dimension is smaller than 10 cm (which is equal to $2\lambda_0$).

When the excitation phases are chosen as $\psi_{nm} = -k_0\widehat{\mathbf{r}}_o\cdot\mathbf{r}'_{nm}$ following (2.50) with $\mathbf{r}_o = \mathbf{r}_b$, numerical results evaluated by the two-fold summation in (2.96) are plotted in Figures 2.15–2.17.

When the size of the phased array is 0.1 m by 0.1 m, the condition of electrical far zone is satisfied. As discussed at the beginning of this section, Equations (2.50) and (2.77) are equivalent to each other under the condition of electrical far zone. Consequently, the plots in Figures 2.15 and 2.12 appear almost identical to each other.

When the size of the phased array increases to (1 m × 1 m) or (3 m × 3 m), the condition of electrical far zone is not satisfied. In the electrical near zone, the excitation phases of $\psi_{nm} = -k_0\widehat{\mathbf{r}}_o\cdot\mathbf{r}'_{nm}$ no longer ensure that the contributions from the $N \times M$ antenna elements are in phase at $\mathbf{r}_o$. As expected, the amplitude values in Figure 2.16 are smaller than 1 considerably, and the amplitude values in Figure 2.17 are smaller than 1 significantly.

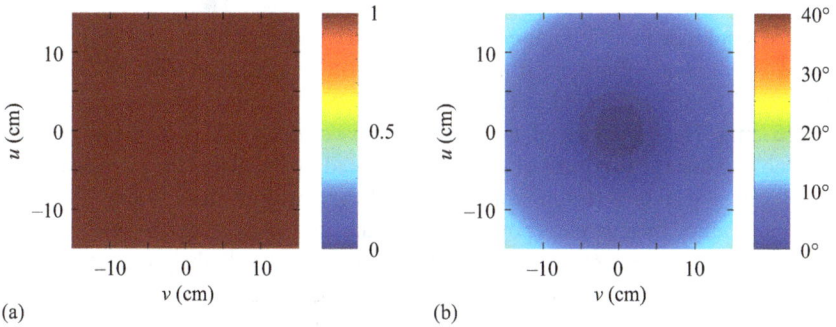

(a)   (b)

Figure 2.15   Numerical results of constructive coefficient Φ with $\psi_{nm} = -k_0\widehat{\mathbf{r}}_o\cdot\mathbf{r}'_{nm}$, when the size of the phased array is about 0.1 m by 0.1 m. (a) Amplitude of Φ evaluated via (2.96) and (b) phase of Φ evaluated via (2.96).

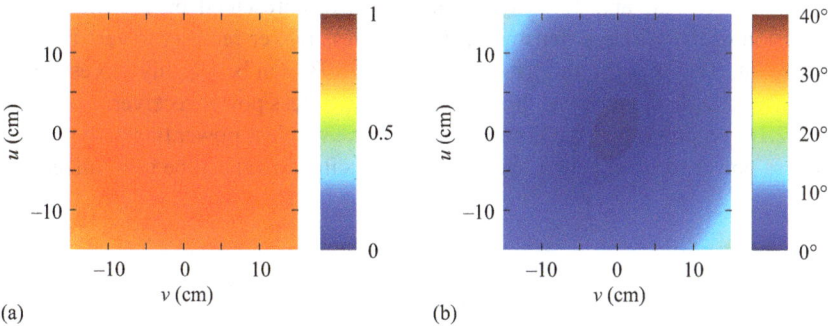

(a)   (b)

Figure 2.16   Numerical results of constructive coefficient Φ with $\psi_{nm} = -k_0\widehat{\mathbf{r}}_o\cdot\mathbf{r}'_{nm}$, when the size of the phased array is about 1 m by 1 m. (a) Amplitude of Φ evaluated via (2.96); (b) phase of Φ evaluated via (2.96).

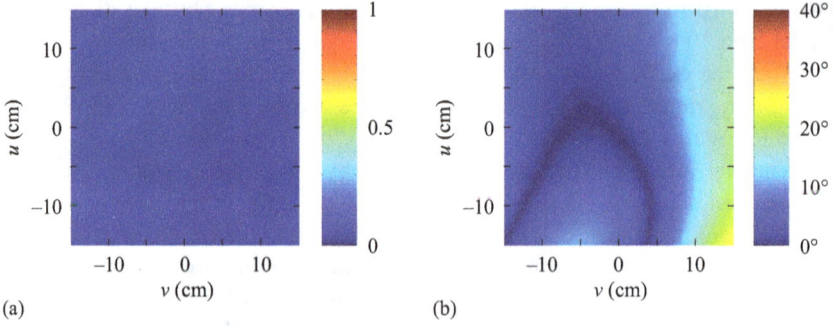

Figure 2.17   Numerical results of constructive coefficient Φ with $\psi_{nm} = -k_0\hat{\mathbf{r}}_o \cdot \mathbf{r}'_{nm}$, when the size of the phased array is about 3 m by 3 m. (a) Amplitude of Φ evaluated via (2.96) and (b) phase of Φ evaluated via (2.96).

The amplitude plots in Figure 2.14 offer tremendous insights toward estimating the power transmission efficiency. It is observed from the amplitude plots of Figure 2.14 that a focal point is centered at ($u = 0$, $v = 0$) and a null appears at ($u = 15$ cm, $v = 15$ cm). Thus, the size of the focal point is roughly (30 cm) × (30 cm). It is reasonable to assume that most of the power radiated by the phased array resides within the (30 cm) × (30 cm) region around $\mathbf{r}_o$. If all the power in this (30 cm) × (30 cm) region is captured by a wireless power receiver, the amount of received power would be almost equal to the amount of transmitted power, in other words, the power transmission efficiency would approach 100%. Based on the numerical calculations around (2.102), the null of the focal point can be determined analytically by letting

$$\frac{N}{2}\xi_x = \frac{N\,k_0}{2\,r_o}[(\mathbf{r} - \mathbf{r}_o) \cdot \hat{\mathbf{x}}]s_x = \pi \text{ or } \frac{M}{2}\xi_z = \frac{M\,k_0}{2\,r_o}[(\mathbf{r} - \mathbf{r}_o) \cdot \hat{\mathbf{z}}]s_z = \pi.$$

(2.103)

In (2.103), $N \times s_x$ and $M \times s_z$ can be roughly represented by D_t, the physical dimension of the wireless transmitter. If a wireless power receiver covers the entire focal point, $[(\mathbf{r} - \mathbf{r}_o) \cdot \hat{\mathbf{x}}]$ and $[(\mathbf{r} - \mathbf{r}_o) \cdot \hat{\mathbf{z}}]$ in (2.103) can be roughly represented by $D_r/2$, where D_r is the physical dimension of the wireless power receiver. In addition, r_o in (2.103) represents d, the distance between the wireless power transmitter and the wireless power receiver. Then, the two equations in (2.103) can be generalized to be

$$\frac{D_t D_r}{d\lambda_0} = 2.$$

(2.104)

Since the wireless power receiver is assumed to cover the entire focal point, the power transmission efficiency corresponding to (2.104) would approach 100%. It is consistent with the findings of [6,7]. In [6,7], it was found that the power transmission efficiency would approach 100% when $\tau = \sqrt{A_t A_r}/(\lambda_0 d)$ is greater than 2, where A_t and A_r are the surface areas of the wireless power transmitter's aperture and the wireless power receiver's aperture, respectively. Because $\sqrt{A_t} \approx D_t$ and $\sqrt{A_r} \approx D_r$, $\tau \approx (D_t D_r)/(\lambda_0 d)$. Therefore, the numerical example in this section provides excellent evidence to support the theoretical analysis of [6,7].

The data displayed in Figure 2.14 agree with Equation (1.10) (derived in Chapter 1) as well. Based upon Equation (1.10), the power transmission efficiency is calculated to be approximately 1% in Section 1.3 when a receiving antenna with a gain value of $G_r = 2$ is placed at $\mathbf{r}_o$. According to (2.92), "$G_r = 2$" corresponds to an effective aperture of $(\lambda_0)^2/(2\pi) \approx (0.4\lambda_0) \times (0.4\lambda_0) = (2 \text{ cm}) \times (2 \text{ cm})$ if assumed to have a square shape. As the power transmission efficiency approaches 100% when a wireless power receiver's aperture covers $(30 \text{ cm}) \times (30 \text{ cm})$, it is reasonable that the power transmission efficiency is 1% when the effective aperture reduces to $(2 \text{ cm}) \times (2 \text{ cm})$ (after the nonuniformity of power density distribution in Figure 2.14 is taken into account).

It is observed from Figure 2.14 that the electromagnetic fields behave as a plane wave within a $(10 \text{ cm}) \times (10 \text{ cm})$ region around $\mathbf{r}_o$, in which the value of $Q_{amplitude}$ is close to NM and the value of Q_{phase} is close to 1. The impacts of $Q_{amplitude}$ and Q_{phase} on the "plane wave behavior" are examined separately next.

Following the reasoning of (2.103), the value of $Q_{amplitude}$ is close to NM if

$$\frac{N}{2}\xi_x = \frac{N k_0}{2 r_o}[(\mathbf{r} - \mathbf{r}_o) \cdot \hat{\mathbf{x}}]s_x \ll \pi \text{ and } \frac{M}{2}\xi_z = \frac{M k_0}{2 r_o}[(\mathbf{r} - \mathbf{r}_o) \cdot \hat{\mathbf{z}}]s_z \ll \pi.$$

$$(2.105)$$

Due to the arguments presented after (2.103), the two inequalities in (2.105) can be rewritten as

$$\frac{D_t D_r}{2 d \lambda_0} \ll 1. \tag{2.106}$$

When D_t, d, and λ_0 are fixed, Equation (2.106) would hold true as long as D_r is small enough. If the wireless power receiver is a mobile electronic device and it only includes one regular antenna, it would be reasonable to assume that $D_r \cong \lambda_0$. With the assumption of $D_r \cong \lambda_0$, Equation (2.106) becomes $D_t \ll 2d$, which is essentially the same as the condition of geometrical far zone.

From the expression of Q_{phase} in (2.80), its phase stays close to 0 if

$$k_0 \frac{|\mathbf{r} - \mathbf{r}_b|^2}{2 r_b} = k_0 \frac{|\mathbf{r} - \mathbf{r}_o|^2}{2 r_o} \ll 2\pi. \tag{2.107}$$

Apparently, the value of $|\mathbf{r} - \mathbf{r}_o|$ does not exceed $D_r/2$. Then, (2.107) can be rewritten as

$$d \gg \frac{(D_r)^2}{8 \lambda_0}. \tag{2.108}$$

It is interesting to note that (2.108) resembles the far-field criterion in (2.24) albeit (2.108) does not involve D_t. When d and λ_0 are fixed, Equation (2.108) would hold true as long as D_r is small enough. If $D_r \cong \lambda_0$, Equation (2.108) becomes $D_r \ll 8d$, which is essentially the same as the condition of geometrical far zone too.

In conclusion, under the condition of geometrical far zone and with the assumption of $D_r \cong \lambda_0$, $Q_{amplitude}$ is close to NM, Q_{phase} is close to 1, and the electromagnetic fields radiated by the phased array behave as a plane wave around $\mathbf{r}_o$. In other words, it is always possible to identify a "plane wave region" to accommodate a wireless power receiver with a physical dimension on the order of one wavelength. As analyzed in the previous section (i.e., Section 2.4), "wireless power receiver's size being smaller than the size of plane wave region" is one of the four conditions to keep the Friis transmission equation valid in the electrical near zone.

The numerical setup in Figure 2.11 is also used to examine the high-order terms neglected in (2.72), which are

$$
|\mathbf{r} - \mathbf{r}'_{nm}| - |\mathbf{r}_o - \mathbf{r}'_{nm}| \quad - \hat{\mathbf{r}}_o \cdot (\mathbf{r} - \mathbf{r}_o) - \frac{|(\mathbf{r} - \mathbf{r}_o) - \mathbf{r}'_{nm}|^2}{2r_o} + \frac{|\mathbf{r}'_{nm}|^2}{2r_o}
$$

$$
= \left\{ -\frac{r_o}{8} \left[\frac{2\mathbf{r}_o \cdot [(\mathbf{r} - \mathbf{r}_o) - \mathbf{r}'_{nm}]}{(r_o)^2} + \frac{|(\mathbf{r} - \mathbf{r}_o) - \mathbf{r}'_{nm}|^2}{(r_o)^2} \right]^2 + \frac{r_o}{8} \left[-\frac{2\mathbf{r}_o \cdot \mathbf{r}'_{nm}}{(r_o)^2} + \frac{|\mathbf{r}'_{nm}|^2}{(r_o)^2} \right]^2 \right\}
$$

$$
+ \left\{ \frac{r_o}{16} \left[\frac{2\mathbf{r}_o \cdot [(\mathbf{r} - \mathbf{r}_o) - \mathbf{r}'_{nm}]}{(r_o)^2} + \frac{|(\mathbf{r} - \mathbf{r}_o) - \mathbf{r}'_{nm}|^2}{(r_o)^2} \right]^3 - \frac{r_o}{16} \left[-\frac{2\mathbf{r}_o \cdot \mathbf{r}'_{nm}}{(r_o)^2} + \frac{|\mathbf{r}'_{nm}|^2}{(r_o)^2} \right]^3 \right\}
$$

$$
+ \cdots
$$

$$
\text{(2.109)}
$$

On the basis of the numerical setup in Figure 2.11, the size of the phased array is fixed to be 3 m by 3 m, r_o is fixed to be 10 m, and the size of the observation region is fixed to be 0.1 m by 0.1 m around $\mathbf{r}_o$ (which is the "plane wave region" according to Figure 2.14). When n varies from 1 to N, m varies from 1 to M, and $\mathbf{r}$ varies within the observation region, the following phase angle is evaluated.

$$
\chi(\mathbf{r}, \mathbf{r}'_{nm}) = k_0 \left| |\mathbf{r} - \mathbf{r}'_{nm}| - |\mathbf{r}_o - \mathbf{r}'_{nm}| - \hat{\mathbf{r}}_o \cdot (\mathbf{r} - \mathbf{r}_o) - \frac{|(\mathbf{r} - \mathbf{r}_o) - \mathbf{r}'_{nm}|^2}{2r_o} + \frac{|\mathbf{r}'_{nm}|^2}{2r_o} \right|.
$$

$$
\text{(2.110)}
$$

Most of the χ values evaluated through (2.110) are very small. The largest phase angle of χ is $\chi_{\max} = 12.6°$ when the observation point $\mathbf{r}$ is located at ($u = 5$ cm, $v = -5$ cm) and $\mathbf{r}'_{nm}$ is located at ($x = -1.5$ m, $z = 1.5$ m).

The Taylor series in (2.109) is with respect to $1/r_o$. When r_o is large (which is true in microwave power transmission applications, generally speaking), the leading terms on the right-hand side of (2.109) are examined below.

$$
\left\{ -\frac{r_o}{8} \left[\frac{2\mathbf{r}_o \cdot [(\mathbf{r} - \mathbf{r}_o) - \mathbf{r}'_{nm}]}{(r_o)^2} + \frac{|(\mathbf{r} - \mathbf{r}_o) - \mathbf{r}'_{nm}|^2}{(r_o)^2} \right]^2 + \frac{r_o}{8} \left[-\frac{2\mathbf{r}_o \cdot \mathbf{r}'_{nm}}{(r_o)^2} + \frac{|\mathbf{r}'_{nm}|^2}{(r_o)^2} \right]^2 \right\}
$$

$$
\xrightarrow{\mathbf{r}_o \cdot (\mathbf{r} - \mathbf{r}_o) = 0} \frac{r_o}{8} \left[-\frac{2\hat{\mathbf{r}}_o \cdot \mathbf{r}'_{nm}}{r_o} + \frac{|(\mathbf{r} - \mathbf{r}_o) - \mathbf{r}'_{nm}|^2}{(r_o)^2} \right]^2 + \frac{r_o}{8} \left[-\frac{2\hat{\mathbf{r}}_o \cdot \mathbf{r}'_{nm}}{r_o} + \frac{|\mathbf{r}'_{nm}|^2}{(r_o)^2} \right]^2
$$

$$
\text{(2.111)}
$$

The right-hand side of (2.111) is further expanded to be

$$-\frac{r_o}{8}\left[-\frac{2\hat{\mathbf{r}}_o\cdot\mathbf{r}'_{nm}}{r_o}+\frac{|(\mathbf{r}-\mathbf{r}_o)-\mathbf{r}'_{nm}|^2}{(r_o)^2}\right]^2+\frac{r_o}{8}\left[-\frac{2\hat{\mathbf{r}}_o\cdot\mathbf{r}'_{nm}}{r_o}+\frac{|\mathbf{r}'_{nm}|^2}{(r_o)^2}\right]^2$$

$$=-\left[\frac{(\hat{\mathbf{r}}_o\cdot\mathbf{r}'_{nm})^2}{2r_o}+\frac{(\hat{\mathbf{r}}_o\cdot\mathbf{r}'_{nm})\,|(\mathbf{r}-\mathbf{r}_o)-\mathbf{r}'_{nm}|^2}{2(r_o)^2}-\frac{|(\mathbf{r}-\mathbf{r}_o)-\mathbf{r}'_{nm}|^4}{8(r_o)^3}\right]$$

$$+\left[\frac{(\hat{\mathbf{r}}_o\cdot\mathbf{r}'_{nm})^2}{2r_o}-\frac{(\hat{\mathbf{r}}_o\cdot\mathbf{r}'_{nm})\,|\mathbf{r}'_{nm}|^2}{2(r_o)^2}+\frac{|\mathbf{r}'_{nm}|^4}{8(r_o)^3}\right]$$

$$=-\left[\frac{(\hat{\mathbf{r}}_o\cdot\mathbf{r}'_{nm})^2}{2r_o}+\frac{(\hat{\mathbf{r}}_o\cdot\mathbf{r}'_{nm})(|\mathbf{r}-\mathbf{r}_o|^2-2(\mathbf{r}-\mathbf{r}_o)\cdot\mathbf{r}'_{nm}+|\mathbf{r}'_{nm}|^2)}{2(r_o)^2}-\frac{|(\mathbf{r}-\mathbf{r}_o)-\mathbf{r}'_{nm}|^4}{8(r_o)^3}\right]$$

$$+\left[\frac{(\hat{\mathbf{r}}_o\cdot\mathbf{r}'_{nm})^2}{2r_o}-\frac{(\hat{\mathbf{r}}_o\cdot\mathbf{r}'_{nm})\,|\mathbf{r}'_{nm}|^2}{2(r_o)^2}+\frac{|\mathbf{r}'_{nm}|^4}{8(r_o)^3}\right]$$

$$(2.112)$$

The two terms involving $(\hat{\mathbf{r}}_o\cdot\mathbf{r}'_{nm})^2/(2r_o)$ cancel each other at the end of (2.112). Consequently, the highest order in (2.112) is $(r_o)^{-2}$ rather than $(r_o)^{-1}$. Actually, multiple terms cancel each other at the end of (2.112). After the cancelations are conducted, the dominant term in (2.112) turns out to be

$$\frac{(\hat{\mathbf{r}}_o\cdot\mathbf{r}'_{nm})[(\mathbf{r}-\mathbf{r}_o)\cdot\mathbf{r}'_{nm}]}{(r_o)^2}. \tag{2.113}$$

When $\mathbf{r}-\mathbf{r}_o=0.05\hat{\mathbf{u}}-0.05\hat{\mathbf{v}}=-0.05\hat{\boldsymbol{\theta}}_o+0.05\hat{\boldsymbol{\phi}}_o$ and $\mathbf{r}'_{nm}=-1.5\hat{\mathbf{x}}+1.5\hat{\mathbf{z}}$,

$$(\mathbf{r}-\mathbf{r}_o)\cdot\mathbf{r}'_{nm}=0.05\times1.5\times\left(\frac{\sqrt{2}}{4}+\frac{\sqrt{3}}{2}\right)\approx0.0915$$

$$\hat{\mathbf{r}}_o\cdot\mathbf{r}'_{nm}=1.5\times\left(\frac{\sqrt{6}}{4}+\frac{1}{2}\right)\approx1.67$$

$$\frac{(\hat{\mathbf{r}}_o\cdot\mathbf{r}'_{nm})(\mathbf{r}-\mathbf{r}_o)\cdot\mathbf{r}'_{nm}}{(r_o)^2}\approx\frac{0.153}{10\times10}$$

Then, χ_{max} is estimated as

$$k_0\frac{(\hat{\mathbf{r}}_o\cdot\mathbf{r}'_{nm})[(\mathbf{r}-\mathbf{r}_o)\cdot\mathbf{r}'_{nm}]}{(r_o)^2}=\frac{360°}{\lambda_0}\times\frac{0.153}{100}\approx11°.$$

It matches the $\chi_{max}=12.6°$ evaluated through (2.110) very well. Therefore, an analytical formulation could be obtained to estimate χ_{max}. To find χ_{max} analytically, $\hat{\mathbf{r}}_o\cdot\mathbf{r}'_{nm}$ can be estimated by $(D_t/2)\times\sqrt{1-(\hat{\mathbf{r}}_o\cdot\hat{\mathbf{n}})^2}$ and $(\mathbf{r}-\mathbf{r}_o)\cdot\mathbf{r}'_{nm}$ can be estimated by $(D_r/2)\times(D_t/2)\times(\hat{\mathbf{r}}_o\cdot\hat{\mathbf{n}})$, where $\hat{\mathbf{n}}$ is the direction normal to the two-dimensional phased array's aperture ($\hat{\mathbf{n}}=\hat{\mathbf{y}}$ in the numerical example of

Figure 2.11) and $\sqrt{1 - (\hat{\mathbf{r}}_o \cdot \hat{\mathbf{n}})^2}$ is the projection of $\hat{\mathbf{r}}_o$ onto the plane in which the phased array resides. As a result, the analytical formulation to estimate χ_{max} is

$$\chi_{max} = k_0 \frac{\sqrt{1 - (\hat{\mathbf{r}}_o \cdot \hat{\mathbf{n}})^2} \frac{D_t}{2} (\hat{\mathbf{r}}_o \cdot \hat{\mathbf{n}}) \frac{D_r}{2} \frac{D_t}{2}}{(r_o)^2} = \frac{2\pi}{\lambda_0} \times \frac{(D_t)^2 D_r}{8d^2} \times (\hat{\mathbf{r}}_o \cdot \hat{\mathbf{n}}) \sqrt{1 - (\hat{\mathbf{r}}_o \cdot \hat{\mathbf{n}})^2},$$

(2.114)

In (2.114), d is used in the place of r_o to represent the distance between the wireless power transmitter and the wireless power receiver.

In Figure 2.18, χ_{max} evaluated via (2.110) and χ_{max} evaluated via (2.114) are compared with each other. In Figure 2.18, the size of the phased array is fixed to be 3 m by 3 m, the size of the observation region is fixed to be 0.1 m by 0.1 m, and the value of d (which is r_o) varies from 3 m to 30 m. The two curves in Figure 2.18 are evaluated via (2.110) and (2.114), respectively. Figure 2.18 demonstrates that χ_{max} could be estimated by the analytical formulation in (2.114) quite precisely.

The phase term in (2.110) constitutes the only difference between (2.96) and (2.97). To be specific, the phase term in (2.110) is included in (2.96) but neglected in (2.97). In Figure 2.14 (with the size of the phased array being 3 m by 3 m and d being 10 m), Φ evaluated via (2.96) is plotted in Subplot (a) and Subplot (b), and Φ evaluated via (2.97) is plotted in Subplot (c) and Subplot (d). The results evaluated by (2.96) and the results evaluated by (2.97) appear almost the same as each other in Figure 2.14. When the Φ value evaluated via (2.96) is denoted as Φ^D (with superscript "D" standing for "definition") and the Φ value evaluated via (2.97) is denoted as Φ^A (with superscript "A" standing for "analytical"), the amplitude of their

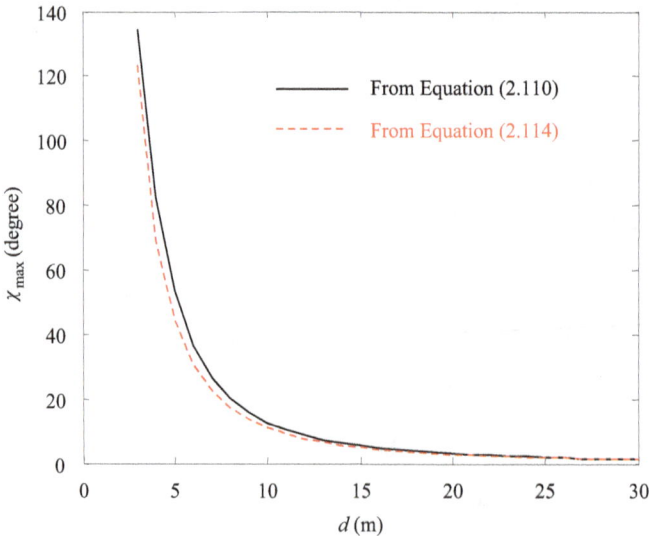

Figure 2.18 χ_{max} evaluated numerically and analytically

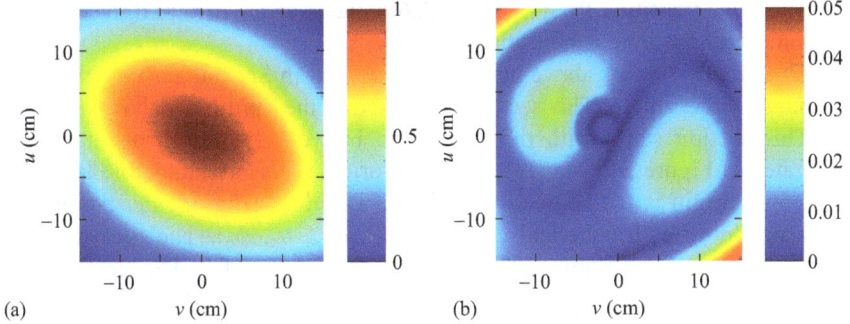

Figure 2.19   Difference between results evaluated by (2.96) and results evaluated by (2.97) when the size of the phased array is 3 m by 3 m and d is 10 m. (a) Amplitude of Φ^D evaluated via (2.96) and (b) Amplitude of $\Phi^D - \Phi^A$.

difference, $|\Phi^D - \Phi^A|$, is plotted in Figure 2.19(b). Figure 2.19(a) is identical to Figure 2.14(a). The largest value of $|\Phi^D - \Phi^A|$ in Figure 2.19(b), which is located at two corners, is not very meaningful as the amplitude of Φ^D is almost zero at the two corners. After the two corners are ignored, the largest value of $|\Phi^D - \Phi^A|$ occurs at $(u = 5 \text{ cm}, v = -5 \text{ cm})$ and $(u = -5 \text{ cm}, v = 5 \text{ cm})$, which is in agreement with the analysis around (2.113). The value of $|\Phi^D - \Phi^A|$ at $(u = 5 \text{ cm}, v = -5 \text{ cm})$ and $(u = -5 \text{ cm}, v = 5 \text{ cm})$ is as small as 0.03. In conclusion, the analytical formulation in (2.97) is reasonably accurate when the size of the phased array is 3 m by 3 m and d is greater than 10 m. However, when d is smaller than 10 m, $\chi_{\max}$ increases drastically (as observed in Figure 2.18) and the accuracy of (2.97) may become unacceptable.

In Figures 2.14 and 2.19, the aperture of the phased array is 3 m by 3 m and the wireless power receiver resides 10 m away, which does not seem to satisfy the condition of geometrical far zone. The next section, Section 2.6, aims to investigate the power transmission efficiency in the "geometrical far zone" and "geometrical near zone."

2.6   Power transmission efficiency in "geometrical far zone" and "geometrical near zone"

The theoretical derivations of Section 2.4 are based on the assumptions of "electrical near zone" and "geometrical far zone." Actually, the condition of geometrical far zone appears three times in Sections 2.4 and 2.5, as articulated below.

First, the size of the "plan wave region" depends on the condition of geometrical far zone, which is a conclusion of the analysis from (2.105) to (2.108).

Second, the high-order terms ignored in (2.72) might become not negligible in the geometrical near zone, as studied at the end of Section 2.5.

Third, the transition from (2.68) to (2.69) may become invalid if the condition of geometrical far zone is not satisfied.

The first two issues above have been inspected in Section 2.5 with the aid of one numerical example. This section is devoted to the third issue above.

This section starts with the interaction between two Hertzian dipoles, as illustrated in Figure 2.20. In free space, a z-orientated Hertzian dipole located at the spatial origin is the source. Its radiation is detected by another z-orientated Hertzian dipole with spherical coordinates of (r_o, θ_o, ϕ_o). The time-harmonic excitation current of the source dipole has a phasor value of I_0. The length of both dipoles is l_0.

The electric field $\mathbf{E}_0$ radiated by the source Hertzian dipole in free space and observed at (r_o, θ_o, ϕ_o) is available in the analytical form [1]:

$$\mathbf{E}_0(r_o, \theta_o, \phi_o) = \widehat{\boldsymbol{\theta}}_o \frac{j\eta_0 k_0 I_0 l_0 \sin(\theta_o) e^{-jk_0 r_o}}{4\pi r_o} \left[1 + \frac{1}{jk_0 r_o} - \frac{1}{(k_0 r_o)^2} \right]$$
$$+ \widehat{\mathbf{r}}_o \frac{\eta_0 I_0 l_0 \cos(\theta_o) e^{-jk_0 r_o}}{2\pi (r_o)^2} \left[1 + \frac{1}{jk_0 r_o} \right] \qquad (2.115)$$

where

$$\widehat{\mathbf{r}}_o = \sin(\theta_o)\cos(\phi_o)\widehat{\mathbf{x}} + \sin(\theta_o)\sin(\phi_o)\widehat{\mathbf{y}} + \cos(\theta_o)\widehat{\mathbf{z}}$$
$$\widehat{\boldsymbol{\theta}}_o = \cos(\theta_o)\cos(\phi_o)\widehat{\mathbf{x}} + \cos(\theta_o)\sin(\phi_o)\widehat{\mathbf{y}} - \sin(\theta_o)\widehat{\mathbf{z}} \qquad (2.116)$$

When $k_0 r_o$ is much greater than 1 (which is generally true in microwave engineering), the term on the order of $(r_o)^{-1}$ in Equation (2.115) is retained and all the other terms are ignored:

$$\mathbf{E}_0(r_o, \theta_o, \phi_o) = \widehat{\boldsymbol{\theta}}_o \frac{j\eta_0 k_0 I_0 l_0}{4\pi} \sin(\theta_o) \frac{e^{-jk_0 r_o}}{r_o}. \qquad (2.117)$$

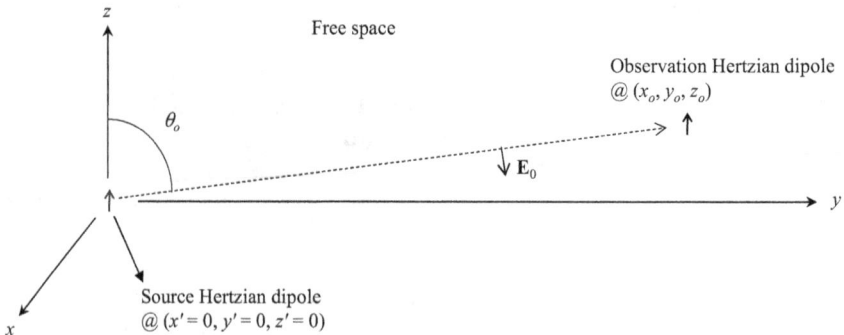

Figure 2.20   Interaction between one source Hertzian dipole and one observation Hertzian dipole

A voltage V_r is developed at the terminal of the observation Hertzian dipole:

$$V_r = (\hat{z} \cdot \mathbf{E}_0)l_0$$
$$= -I_0 \frac{j\eta_0 k_0 (l_0)^2}{4\pi} \sin^2(\theta_o) \frac{e^{-jk_0 r_o}}{r_o} \tag{2.118}$$

If the observation Hertzian dipole is terminated by a matched load, the power received by the matched load is

$$P_r = \frac{|V_r|^2}{8\text{Re}\{Z_0\}} = \frac{|V_r|^2}{8R_0} \tag{2.119}$$

where Z_0 is the input impedance of the observation Hertzian dipole and R_0 is the radiation resistance of the observation Hertzian dipole [1]

$$R_0 = \text{Re}\{Z_0\} = \eta_0 \frac{2\pi}{3} \left(\frac{l_0}{\lambda_0}\right)^2. \tag{2.120}$$

The power transmitted by the source Hertzian dipole is

$$P_t = \frac{1}{2} |I_0|^2 \text{Re}\{Z_0\} = \frac{1}{2} |I_0|^2 R_0 \tag{2.121}$$

The power transmission efficiency between a pair of Hertzian dipoles, $(PTE)_0$, is

$$(PTE)_0 = \frac{P_r}{P_t}$$
$$= \left(\frac{\eta_0 k_0 (l_0)^2}{8\pi R_0}\right)^2 \frac{\sin^4(\theta_o)}{(r_o)^2} \tag{2.122}$$

After the expression of R_0 in (2.120) is substituted, it is easy to re-arrange (2.122) to be

$$(PTE)_0 = \{1.5\sin^2(\theta_o)\} \times \{1.5\sin^2(\theta_o)\} \times \left(\frac{\lambda_0}{4\pi r_o}\right)^2. \tag{2.123}$$

The power transmission efficiency in (2.123) matches the Friis transmission equation, as $1.5\sin^2(\theta_o)$ is the gain value of the source dipole toward the observation point and is also the gain value of the observation dipole toward the spatial origin.

Next, consider a two-dimensional array of Hertzian dipoles as depicted in Figure 2.21. There are $N \times M$ Hertzian dipoles with spacing s_x along the x direction and spacing s_z along the z direction. The Hertzian dipoles are identical to the dipoles in Figure 2.20. The physical dimensions of the array are approximately $L = N \times s_x$ and $W = M \times s_z$. The location of the Hertzian dipoles is

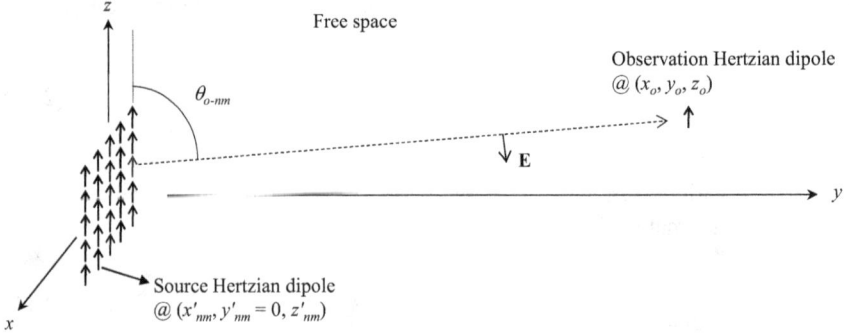

Figure 2.21   *Interaction between an array of Hertzian dipoles and one observation Hertzian dipole*

$\mathbf{r}'_{nm} = (x'_{nm}, y' = 0, z'_{nm})$, $n = 1, 2, \ldots, N$, $m = 1, 2, \ldots, M$. Each dipole is excited by current $I_{nm} = X_0 e^{j\psi_{nm}}$ (that is, with a uniform amplitude of X_0 and with individual phase value of ψ_{nm}). The total electric field observed at the observation Hertzian dipole is

$$\mathbf{E}(r_o, \theta_o, \phi_o) = \hat{\boldsymbol{\theta}}_{o-nm} \frac{j\eta_0 k_0 X_0 l_0}{4\pi} \sum_{n=1}^{N} \sum_{m=1}^{M} \frac{e^{j\psi_{nm}} \sin(\theta_{o-nm}) e^{-jk_0 r_{o-nm}}}{r_{o-nm}}, \quad (2.124)$$

where

$$\hat{\boldsymbol{\theta}}_{o-nm} = \cos(\theta_{o-nm})\cos(\phi_{o-nm})\hat{\mathbf{x}} + \cos(\theta_{o-nm})\sin(\phi_{o-nm})\hat{\mathbf{y}} \\ -\sin(\theta_{o-nm})\hat{\mathbf{z}} \quad (2.125)$$

$$\sin(\theta_{o-nm}) = \frac{\sqrt{(x_o - x'_{nm})^2 + (y_o)^2}}{\sqrt{(x_o - x'_{nm})^2 + (y_o)^2 + (z_o - z'_{nm})^2}} \quad (2.126)$$

$$\cos(\theta_{o-nm}) = \frac{z_o - z'_{nm}}{\sqrt{(x_o - x'_{nm})^2 + (y_o)^2 + (z_o - z'_{nm})^2}} \quad (2.127)$$

$$\sin(\phi_{o-nm}) = \frac{y_o}{\sqrt{(x_o - x'_{nm})^2 + (y_o)^2}} \quad (2.128)$$

$$\cos(\phi_{o-nm}) = \frac{x_o - x'_{nm}}{\sqrt{(x_o - x'_{nm})^2 + (y_o)^2}} \quad (2.129)$$

$$r_{o-nm} = \sqrt{(x_o - x'_{nm})^2 + (y_o)^2 + (z_o - z'_{nm})^2} \tag{2.130}$$

$$x_o = r_o \sin(\theta_o)\cos(\phi_o) \tag{2.131}$$

$$y_o = r_o \sin(\theta_o)\sin(\phi_o) \tag{2.132}$$

$$z_o = r_o \cos(\theta_o) \tag{2.133}$$

The voltage V_r at the terminal of the observation Hertzian dipole is

$$V_r = (\hat{\mathbf{z}} \cdot \mathbf{E})l_0 = -X_0 \frac{j\eta_0 k_0 (l_0)^2}{4\pi} \sum_{n=1}^{N}\sum_{m=1}^{M} \frac{e^{j\psi_{nm}}\sin^2(\theta_{o-nm})e^{-jk_0 r_{o-nm}}}{r_{o-nm}} \tag{2.134}$$

When the excitation phases are selected to be

$$\psi_{nm} = k_0 r_{o-nm}, \tag{2.135}$$

Equation (2.134) becomes

$$V_r = -X_0 \frac{j\eta_0 k_0 (l_0)^2}{4\pi} \sum_{n=1}^{N}\sum_{m=1}^{M} \frac{\sin^2(\theta_{o-nm})}{r_{o-nm}}, \tag{2.136}$$

in other words, the excitation phases are selected to ensure that the contributions from the $N \times M$ source dipoles are in phase at the observation dipole.

The term $\sin^2(\theta_{o-nm})/r_{o-nm}$ in (2.136) includes three items, actually. Their physical meanings are described below one by one.

(i) $1/r_{o-nm}$ stands for the attenuation between the observation dipole and the source dipole with indices nm.

(ii)   One $\sin(\theta_{o-nm})$ is due to the radiation pattern produced by the source dipole with indices nm and observed at the observation dipole.

(iii)   The other $\sin(\theta_{o-nm})$ results from a dot product, which is between $\hat{\mathbf{z}}$ and the polarization direction of the electric field produced by the source dipole with indices nm.

In the geometrical near zone, there are discrepancies among the $N \times M$ antenna elements in terms of their distances, radiation patterns, and polarizations. The impact of these discrepancies on power transmission efficiency is evaluated in this section.

The total transmitted power is

$$P_t = (NM)\frac{1}{2}(X_0)^2 R_0 \tag{2.137}$$

After the received power P_r is evaluated via (2.119), the power transmission efficiency is

$$\begin{aligned}(\text{PTE}) &= \frac{P_r}{P_t} \\ &= \frac{1}{NM}\left(\frac{\eta_0 k_0 (l_0)^2}{8\pi R_0}\right)^2 \left\{\sum_{n=1}^{N}\sum_{m=1}^{M}\frac{\sin^2(\theta_{o-nm})}{r_{o-nm}}\right\}^2\end{aligned} \tag{2.138}$$

The ratio between (PTE) and $(NM)(PTE)_0$ is defined as the *power constructive coefficient*:

$$\frac{(PTE)}{NM(PTE)_0} = \frac{1}{N^2 M^2} \frac{\left\{ \sum\limits_{n=1}^{N} \sum\limits_{m=1}^{M} \frac{\sin^2(\theta_{o-nm})}{r_{o-nm}} \right\}^2}{\frac{\sin^4(\theta_o)}{(r_o)^2}} \tag{2.139}$$

When $r_o \to \infty$, it is easy to show that $r_{o-nm} \to r_o$, $\theta_{o-nm} \to \theta_o$, and (PTE) $\to$ $(NM) \times (PTE)_0$. In other words, the electric fields contributed by the $N \times M$ source dipoles are fully constructive among one another when the observation dipole is in the geometrical far zone and when the excitation phases are selected as in (2.135). When the observation dipole is in the geometrical near zone, the $N \times M$ source dipoles' contributions are no longer fully constructive among one another, and the power transmission efficiency (PTE) would be smaller than $(NM) \times (PTE)_0$. Therefore, the power constructive coefficient in (2.139) indicates the loss of (PTE) with respect to $(NM) \times (PTE)_0$.

After making use of (2.126) and (2.130), the two-fold summation in (2.139) is

$$\sum_{n=1}^{N} \sum_{m=1}^{M} \frac{\sin^2(\theta_{o-nm})}{r_{o-nm}} = \sum_{n=1}^{N} \sum_{m=1}^{M} \frac{(x_o - x'_{nm})^2 + (y_o)^2}{\left[(x_o - x'_{nm})^2 + (y_o)^2 + (z_o - z'_{nm})^2 \right]^{\frac{3}{2}}} \tag{2.140}$$

In engineering, it is a general practice to evaluate a two-fold integral using a two-fold discrete summation as

$$\int_{-L/2}^{+L/2} \int_{-W/2}^{+W/2} f(x', z') dx' dz' \cong s_x s_y \sum_{n=1}^{N} \sum_{m=1}^{M} f(x'_{nm}, z'_{nm}), \tag{2.141}$$

where $f(\cdot, \cdot)$ is a function with two arguments. Because of (2.141), the two-fold sum in (2.140) can be represented by a two-fold integration:

$$\sum_{n=1}^{N} \sum_{m=1}^{M} \frac{(x_o - x'_{nm})^2 + (y_o)^2}{\left[(x_o - x'_{nm})^2 + (y_o)^2 + (z_o - z'_{nm})^2 \right]^{\frac{3}{2}}}$$

$$\cong \frac{1}{s_x s_z} \int_{-L/2}^{+L/2} \int_{-W/2}^{+W/2} \frac{(x_o - x')^2 + (y_o)^2}{\left[(x_o - x')^2 + (y_o)^2 + (z_o - z')^2 \right]^{\frac{3}{2}}} dx' dz' \tag{2.142}$$

$$= \frac{NM}{LW} \int_{-L/2}^{+L/2} \int_{-W/2}^{+W/2} \frac{(x_o - x')^2 + (y_o)^2}{\left[(x_o - x')^2 + (y_o)^2 + (z_o - z')^2 \right]^{\frac{3}{2}}} dx' dz'$$

The two-fold integral in (2.142) can be evaluated analytically. The analytical expression of the two-fold integral is derived in Appendix A of this book. Actually, the two-fold sum and two-fold integral in (2.142) are very close to each other with fairly small values of N and M. As shown in Appendix A, the relative difference between the two-fold sum and two-fold integral is below 10^{-3} when $x_o = 5$ m, $y_o = 5$ m, $z_o = 5$ m, $L = 3$ m, $W = 3$ m, $N = 10$, and $M = 10$.

With the two-fold sum in (2.139) replaced by the two-fold integral in (2.142), the power constructive coefficient in (2.139) is evaluated as

$$\frac{(PTE)}{NM(PTE)_0} = \frac{1}{L^2 W^2} \frac{\left\{ \int_{-L/2}^{+L/2} \int_{-W/2}^{+W/2} \frac{(x_o-x')^2+(y_o)^2}{\left[(x_o-x')^2+(y_o)^2+(z_o-z')^2\right]^{\frac{3}{2}}} dx'dz' \right\}^2}{\frac{\sin^4(\theta_o)}{(r_o)^2}}$$

(2.143)

Some numerical results of the power constructive coefficient in (2.143) are presented in Figure 2.22. The value of y_o is 20 m, 10 m, 5 m, and 3 m in the four

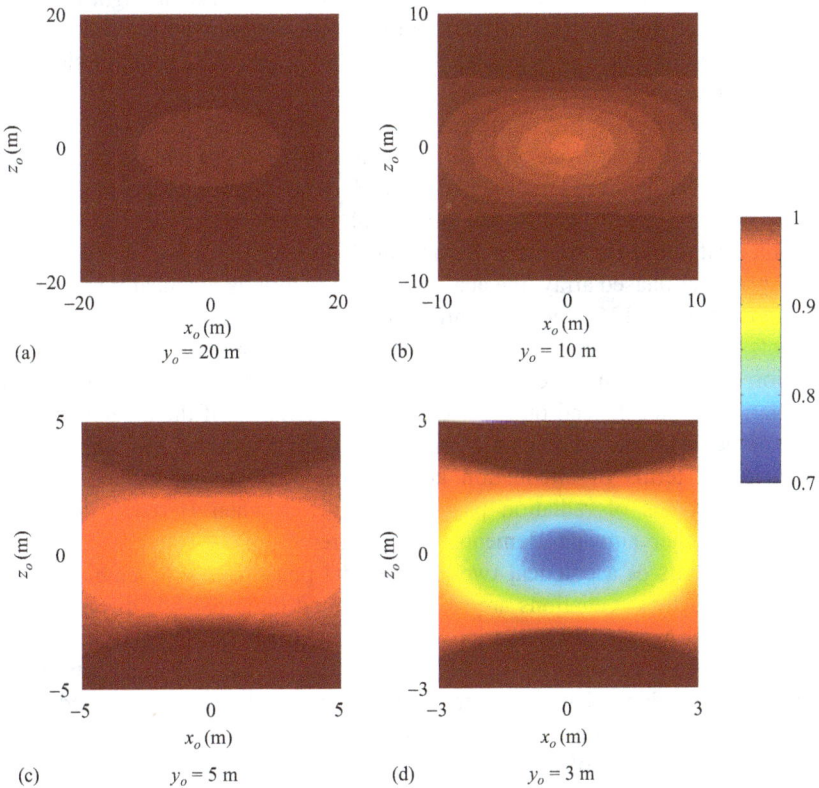

Figure 2.22   Numerical results of power constructive coefficient when (a) $y_o = 20$ m, (b) $y_o = 10$ m, (c) $y_o = 5$ m, and (d) $y_o = 3$ m, with $L = W = 3$ m.

subplots of Figure 2.22, respectively. In Figure 2.22, the power constructive coefficient is plotted with respect to x_o and z_o. The range in the x-z plane is (40 m) × (40 m) when y_o = 20 m, (20 m) × (20 m) when y_o = 10 m, (10 m) × (10 m) when y_o = 5 m, and (6 m) × (6 m) when y_o = 3 m, respectively. The values of L and W are fixed to be 3 m in Figure 2.22.

Apparently, "y_o = 20 m" is far enough geometrically, as the power constructive coefficient is very close to 1 in Figure 2.22(a). Though "y_o = 10 m" does not appear far geometrically, the power constructive coefficient remains close to 1 in Figure 2.22(b): The smallest power constructive coefficient in Figure 2.22(b) is 0.97. When y_o is 5 m, the effect of geometrical nearness becomes visible. The smallest power constructive coefficient in Figure 2.22(c) is 0.894 around x_o = 0 and z_o = 0, meaning that the effect of geometrical nearness could be translated to a power loss of 10% when y_o is 5 m. When y_o is as small as 3 m, the power loss reaches 25%, as the smallest power constructive coefficient is about 0.75. It is noted that the physical dimensions of the phased array are L = 3 m by W = 3 m in Figure 2.22. When a microwave frequency is employed as the carrier of wireless power transmission, the distance of wireless power transmission ought to be considerably larger than the size of the microwave power transmitter. Therefore, the scenario of "y_o = 3 m, L = 3 m, and W = 3 m" in Figure 2.22 is not of significant concern to this book.

In this section, the power transmission efficiency in the "geometrical far zone" and "geometrical near zone" is studied when the antenna elements in a two-dimensional phased array are assumed to be Hertzian dipoles. Hertzian dipoles are chosen in this section because the analytical expressions of their radiation are available. Although Hertzian dipoles are much simpler than the antenna elements of any practical phased array, the analysis of this section is practically valuable. As discussed below (2.136), there are three types of discrepancies among the antenna elements in a phased array, which are in terms of attenuation, polarization, and radiation pattern, respectively. The discrepancies in terms of attenuation and polarization do not depend on the specific characteristics of the antenna element; that is, when Hertzian dipoles are replaced by other antenna elements, the discrepancies in terms of attenuation and polarization stay unchanged. The discrepancy in terms of radiation pattern does differ when Hertzian dipoles are replaced by other antenna elements, but not tremendously. This is because the antenna elements of a practical two-dimensional phased array resemble isotropic radiators when their radiations are observed around the broadside direction of the array. In other words, the radiation pattern of a practical antenna element has smooth variations around the broadside direction as the radiation pattern of a Hertzian dipole does. The formulations and numerical results presented in this section are therefore anticipated to be able to help with estimating the power transmission efficiency of practical phased arrays in the geometrical near zone.

The numerical results of this section indicate that the effect of geometrical nearness in the scenario of Figure 2.14 (i.e., when the physical size of the phased array is 3 m by 3 m and a wireless power receiver is 10 m away) is not significant enough to invalidate the Friis transmission equation in (2.93). As predicted by the

Friis transmission equation, the power transmission efficiency is approximately 1% when the physical size of the phased array is 3 m by 3 m and a wireless power receiver is 10 m away, which may enable numerous practical applications in the context of Internet of Things as discussed in Chapter 4.

Suppose a phased array is centered at the spatial origin, its aperture is in the x-z plane, and its aperture has the physical dimensions of 3 m × 3 m. Also, suppose an observation point is located at $x_o = 0$, $y_o = 10$ m, and $z_o = 0$. The distance measured from ($x_o = 0$, $y_o = 10$ m, $z_o = 0$) to one corner of the phased array is $\sqrt{10^2 + (1.5 \times \sqrt{2})^2} = 10.22$ m, only 2% different from 10 m. Therefore, there is no significant ambiguity in terms of distance d when the Friis transmission equation in (2.93) is applied with the physical size of the phased array being 3 m by 3 m and a wireless power receiver being 10 m away.

2.7 Mutual coupling among antenna elements in a phased array

In the first six sections of this chapter, it has been assumed that there is no mutual coupling among antenna elements in a phased array. However, the mutual coupling is inevitable in practice, especially when the spacing among antenna elements is small. In this section, the impact of mutual coupling on the performance of a phased array is analyzed.

As depicted in Figure 2.23, there is only one antenna element radiating in free space. Electrical power is supplied to the antenna element by a time-harmonic transmitter circuit. Over the circuit interface between the transmitter circuit and the

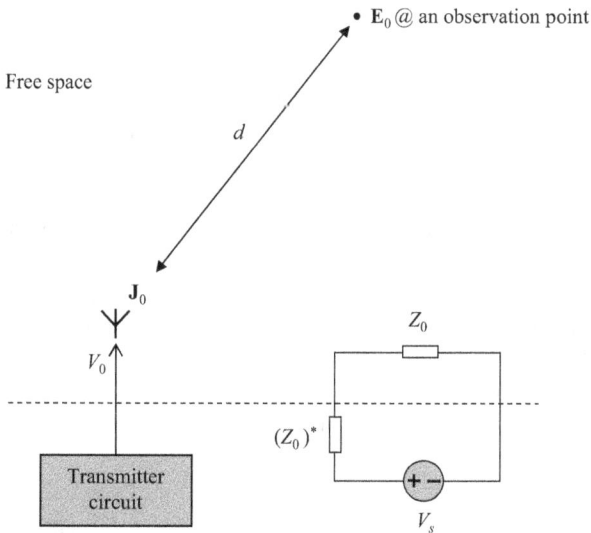

Figure 2.23 Illustration of one antenna element radiating in free space

antenna element, the antenna element can be represented by an equivalent impedance Z_0. Meanwhile, the transmitter circuit can be represented by a voltage source V_s in series with an impedance, based on the Thevenin's theorem. Usually, the condition of conjugate matching is satisfied between the transmitter circuit and the antenna element, to maximize the power transfer from the transmitter circuit to the antenna element. Under the condition of conjugate matching, the impedance value in the equivalent circuit model of the transmitter circuit is $(Z_0)^*$, the complex conjugate of Z_0. Suppose the voltage over the circuit port of the antenna element is V_0. The power delivered from the transmitter circuit to the antenna element is defined as the transmitted power P_t:

$$P_t = P_0 = \frac{|V_0|^2}{2} \mathrm{Re}\left\{\frac{1}{Z_0}\right\} \tag{2.144}$$

If the antenna element is lossless, the entire P_t is radiated by the antenna element. In response to the excitation voltage V_0, electrical current $\mathbf{J}_0$ is induced over the antenna body if it is made of conductors. The electromagnetic fields in space are produced by $\mathbf{J}_0$ primarily. The electric field observed at a certain observation point is denoted as $\mathbf{E}_0$. The antenna gain at the observation point is defined as

$$G_0 = \frac{\frac{|\mathbf{E}_0|^2}{2\eta_0}}{\frac{P_t}{4\pi d^2}} = \frac{\frac{|\mathbf{E}_0|^2}{2\eta_0}}{\frac{P_0}{4\pi d^2}}, \tag{2.145}$$

where d is the distance from the observation point to the antenna element. The gain value has a direct correlation with the power transmission efficiency, as discussed throughout this chapter. One of the main benefits of employing the phased array technique is to maximize the gain value.

Next, consider an array with N antenna elements in free space as shown in Figure 2.24. Each antenna element in the array is assumed to be identical to the antenna element in Figure 2.23. One transmitter circuit is attached to each individual antenna element. The transmitter circuits are identical to the transmitter circuit in Figure 2.23. Moreover, it is assumed that the phase value of each transmitter circuit's voltage source can be adjusted individually; specifically, the source voltages of the N transmitter circuits are $V_s e^{j\psi_1}, V_s e^{j\psi_2}, \ldots, V_s e^{j\psi_N}$ with phase values of $\psi_1, \psi_2, \ldots, \psi_N$ tunable individually. If there is no mutual coupling among the antenna elements, the adjustment in one transmitter circuit only impacts one antenna element. In other words, the N voltages at the antenna elements' circuit port become $V_0 e^{j\psi_1}, V_0 e^{j\psi_2}, \ldots, V_0 e^{j\psi_N}$. Under the condition of "no mutual coupling," the electric currents over the N antenna elements are $\mathbf{J}_0 e^{j\psi_1}, \mathbf{J}_0 e^{j\psi_2}, \ldots,$ $\mathbf{J}_0 e^{j\psi_N}$ correspondingly. The electric field observed at the observation point is the linear superposition of the N electric fields radiated by $\mathbf{J}_0 e^{j\psi_1}, \mathbf{J}_0 e^{j\psi_2}, \ldots, \mathbf{J}_0 e^{j\psi_N}$, respectively. In this section, the observation point is assumed to reside in the geometrical far zone, such that the N electric fields share the same magnitude $|\mathbf{E}_0|$. The phase values of the N electric fields are denoted as $\gamma_1, \gamma_2, \ldots, \gamma_N$. When the N

$$\mathbf{E}_0 e^{j\gamma_1} + \mathbf{E}_0 e^{j\gamma_2} + \mathbf{E}_0 e^{j\gamma_3} + \cdots + \mathbf{E}_0 e^{j\gamma_N}$$

- @ an observation point

Free space

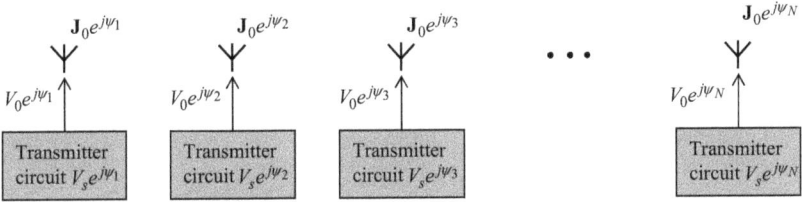

Figure 2.24   Illustration of an antenna array radiating in free space without mutual coupling among antenna elements

transmitter circuits are adjusted to achieve $\gamma_1 = \gamma_2 = \cdots = \gamma_N$, the sum of the N electric fields has the magnitude of $|N\mathbf{E}_0|$. The power transmitted by each antenna element is $P_0 = \frac{|V_0|^2}{2} \mathrm{Re}\left\{\frac{1}{Z_0}\right\}$, and the total amount of transmitted power is $P_t = NP_0$. Consequently, the gain value of the array at the observation point is

$$G_{array} = \frac{\frac{|N\mathbf{E}_0|^2}{2\eta_0}}{\frac{P_t}{4\pi d^2}} = \frac{\frac{N^2|\mathbf{E}_0|^2}{2\eta_0}}{\frac{NP_0}{4\pi d^2}} = NG_0, \tag{2.146}$$

that is, the gain value is enhanced by a factor of N.

When the mutual coupling among antenna elements is not negligible, the scenario in Figure 2.24 turns to be Figure 2.25. Due to the mutual coupling, the input impedance at each antenna element's circuit port deviates from Z_0. When the source voltages of the N transmitter circuits are $V_s e^{j\psi_1}, V_s e^{j\psi_2}, \ldots, V_s e^{j\psi_N}$, the voltages at the N antenna elements' circuit port $\{V_1, V_2, \ldots, V_N\}$ would not be $\{V_0 e^{j\psi_1}, V_0 e^{j\psi_2}, \ldots, V_0 e^{j\psi_N}\}$. The total amount of transmitted power is no longer NP_0. The electric currents over the N antenna elements $\{\mathbf{J}_1, \mathbf{J}_2, \ldots, \mathbf{J}_N\}$ are different from $\{\mathbf{J}_0 e^{j\psi_1}, \mathbf{J}_0 e^{j\psi_2}, \ldots, \mathbf{J}_0 e^{j\psi_N}\}$. When the N electric fields produced by the N antenna elements respectively, which are denoted by $\mathbf{E}_1, \mathbf{E}_2, \ldots, \mathbf{E}_N$ in Figure 2.25, are in phase at the observation point, the magnitude of the total electric field would not be $|N\mathbf{E}_0|$. Finally, the gain value of the antenna array at the observation point deviates from NG_0; in other words, Equation (2.146) does not hold true due to the mutual coupling among antenna elements. Intuitively, all the deviations above would grow when the mutual coupling gets stronger.

In the rest of this section, the impact of mutual coupling is assessed using some numerical examples with Hertzian dipoles as the antenna elements of a phased array.

$$\mathbf{E}_1 + \mathbf{E}_2 + \mathbf{E}_3 + \cdots + \mathbf{E}_N$$
• @ an observation point

Free space

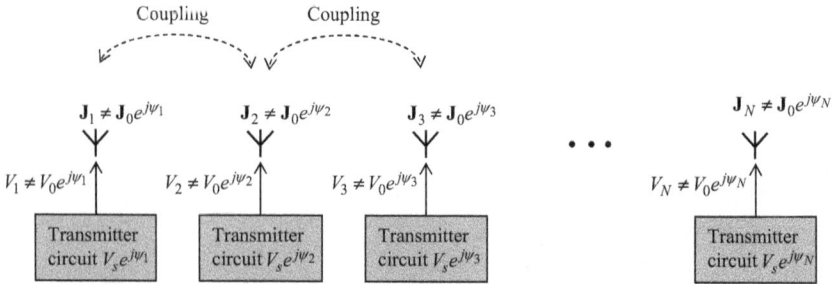

Figure 2.25   Illustration of an antenna array radiating in free space with mutual coupling among antenna elements

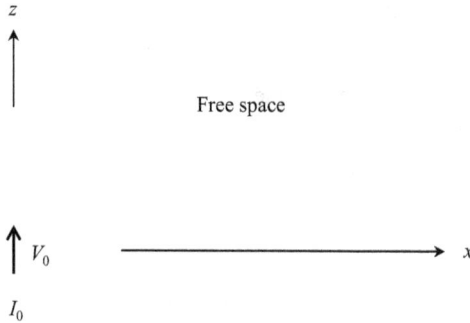

Figure 2.26   Illustration of one Hertzian dipole radiating in free space

If there is only one Hertzian dipole orientated along the z direction in free space (as illustrated by Figure 2.26), the electromagnetic fields radiated by the Hertzian dipole are available analytically [1]. Specifically, the electric field radiated by the Hertzian dipole with time-harmonic current I_0 and length l_0 is formulated in (2.115). The voltage over the Hertzian dipole is $V_0 = I_0 Z_0$. The resistive part of the input impedance Z_0 is

$$R_0 = \mathrm{Re}\{Z_0\} = \eta_0 \frac{2\pi}{3} \left(\frac{l_0}{\lambda_0}\right)^2. \tag{2.147}$$

The power transmitted by the Hertzian dipole is $P_t = |I_0|^2 R_0 / 2$.

Consider an array composed of two Hertzian dipoles in Figure 2.27. Both dipoles are orientated along the z direction and with a length of l_0. One dipole is located at the spatial origin, and the other dipole is located at ($x = -s$, $y = 0$, $z = 0$), with s denoting the spacing between the two dipoles. The dipole at the spatial origin is excited by electric current I_0, and the dipole on the left of Figure 2.27 is excited by electric current $I_0 e^{j\psi}$. The voltage developed over the dipole at the spatial origin is

$$V_0 + V_c = I_0 Z_0 + I_0 e^{j\psi} Z_c. \tag{2.148}$$

In (2.148), V_0 is generated by the dipole at the spatial origin itself, and V_c is generated by the dipole on the left of Figure 2.27. Because of (2.115), the E_z field produced by the dipole on the left and observed at the spatial origin is

$$E_z = -\frac{j\eta_0 k_0 I_0 l_0 e^{j\psi} e^{-jk_0 s}}{4\pi s} \left[1 + \frac{1}{jk_0 s} - \frac{1}{(k_0 s)^2} \right] \tag{2.149}$$

The voltage V_c is

$$V_c = E_z l_0 = -\frac{j\eta_0 k_0 I_0 (l_0)^2 e^{j\psi} e^{-jk_0 s}}{4\pi s} \left[1 + \frac{1}{jk_0 s} - \frac{1}{(k_0 s)^2} \right] \tag{2.150}$$

The mutual impedance Z_c is

$$Z_c = \frac{V_c}{I_0 e^{j\psi}} = -\frac{j\eta_0 k_0 (l_0)^2 e^{-jk_0 s}}{4\pi s} \left[1 + \frac{1}{jk_0 s} - \frac{1}{(k_0 s)^2} \right] \tag{2.151}$$

The power transmitted by the Hertzian dipole at the spatial origin is $|I_0|^2 \mathrm{Re}\{Z_0 + e^{j\psi} Z_c\}/2$ rather than $|I_0|^2 \mathrm{Re}\{Z_0\}/2 = |I_0|^2 R_0/2$. Since ψ may take any value between 0 and 360°, the value of $\mathrm{Re}\{e^{j\psi} Z_c\}$ may be either positive or negative (while R_0 is always positive). Therefore, it is possible that $|I_0|^2 \mathrm{Re}\{Z_0 + e^{j\psi} Z_c\}/2$ is

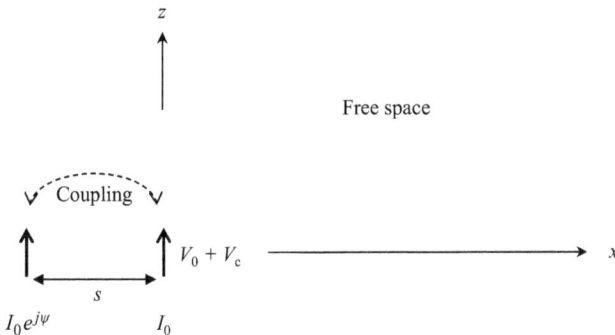

Figure 2.27   Illustration of an array of two Hertzian dipoles radiating in free space

greater than $|I_0|^2 R_0/2$ and it is also possible that $|I_0|^2 \text{Re}\{Z_0 + e^{j\psi}Z_c\}/2$ is smaller than $|I_0|^2 R_0/2$. If $R_{c,\,max}$ is defined as the maximal amplitude of $\text{Re}\{e^{j\psi}Z_c\}$

$$R_{c,\,max} = \frac{\eta_0 k_0 (l_0)^2}{4\pi s} \times \left| 1 + \frac{1}{jk_0 s} - \frac{1}{(k_0 s)^2} \right|, \tag{2.152}$$

the ratio between $R_{c,max}$ and R_0 is a good indicator of the mutual coupling's magnitude

$$\frac{R_{c,\,max}}{R_0} = \frac{3}{4\pi} \times \left| \frac{\lambda_0}{s} + \frac{1}{j2\pi} \left(\frac{\lambda_0}{s}\right)^2 - \frac{1}{4\pi^2} \left(\frac{\lambda_0}{s}\right)^3 \right|. \tag{2.153}$$

It is interesting to note that the ratio in (2.153) solely depends on s/λ_0, which is the "electrical spacing." In Figure 2.28, the ratio between $R_{c,max}$ and R_0 is plotted when s/λ_0 varies between 0.3 and 3. Unsurprisingly, $R_{c,\,max}/R_0$ drops with the increase of spacing. When the spacing is half wavelength, $R_{c,\,max}/R_0$ is as large as 0.45, indicating a quite strong mutual coupling. When the spacing is greater than $2.5\lambda_0$, $R_{c,\,max}/R_0$ drops to be smaller than 0.1, which might be considered negligible.

The far zone gain in the setup of Figure 2.27 is calculated and plotted in Figures 2.29 and 2.30, to reveal the impact of mutual coupling on (2.146). Specifically, the electric field **E** radiated by the two Hertzian dipoles is evaluated through (2.115) at an observation point with spherical coordinates of (r_o, θ_o, ϕ_o). It is assumed that $r_o \to \infty$. The gain of the 2-element array, G, is found using

$$G(\theta_o, \phi_o) = \frac{\dfrac{|\mathbf{E}(r_o,\theta_o,\phi_o)|^2}{2\eta_0}}{\dfrac{P_{rad}}{4\pi(r_o)^2}}, \tag{2.154}$$

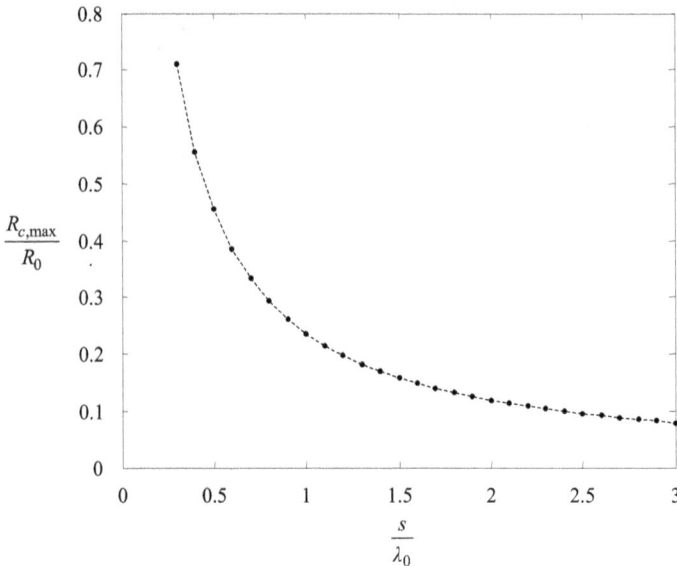

Figure 2.28   Numerical results of mutual coupling's magnitude with respect to the spacing between the two Hertzian dipoles in Figure 2.27

where the total amount of radiated power P_{rad} is obtained by

$$P_{rad} = \int_0^\pi \int_0^{2\pi} \frac{|\mathbf{E}(r_o, \theta_o, \phi_o)|^2}{2\eta_0} (r_o)^2 \sin(\theta_o) d\phi_o d\theta_o. \qquad (2.155)$$

The gain value of G does not rely on the amplitude of I_0, but it is sensitive to ψ.

The two radiation patterns in Figure 2.29 are calculated when $s = \lambda_0/2$. The only difference between the two plots is that $\psi = 0$ in Figure 2.29(a) and $\psi = 180°$ in Figure 2.29(b). The radiation patterns in Figure 2.29 are plotted in the x-y plane. As expected, there are two beams along the y-axis when $\psi = 0$, and there are two beams along the x-axis when $\psi = 180°$. The far zone gain value associated with one Hertzian dipole is 1.5 in the x-y plane [1]. When the mutual coupling is assumed to be negligible, the peak gain value of a 2-element array is $2 \times 1.5 = 3$ according to (2.146). In Figure 2.29, the gain value at the beams' center when $\psi = 0$ is greater than 3, while the gain value at the beams' center when $\psi = 180°$ is smaller than 3. Apparently, the deviations from 3 reveal that the mutual coupling between the two dipoles is not negligible when the spacing is half wavelength.

One phenomenon in Figure 2.29 is worth noting, although it is irrelevant to mutual coupling. The two beams in Figure 2.29(a) are narrower than those in

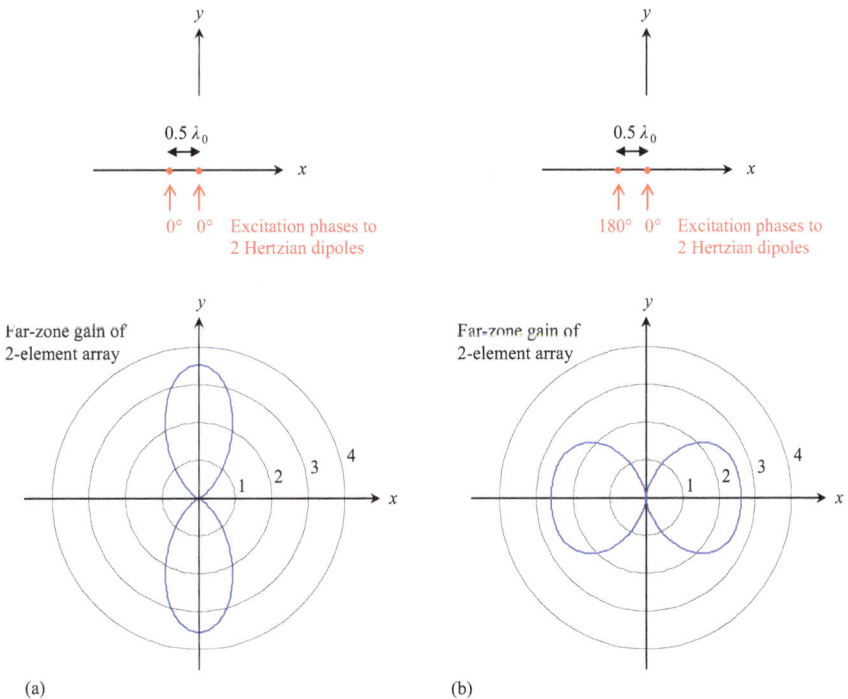

(a)   (b)

Figure 2.29   *Numerical results of far zone gain (in linear scale) of an array including two Hertzian dipoles when s = 0.5λ₀. (a) Two dipoles are excited by the same phase and (b) Difference of excitation phases to dipoles is 180°.*

Figure 2.29(b). This phenomenon can be explained by (2.25), which formulates the electric field in the electrical far zone. Equation (2.25) resembles the mathematical expressions of Fourier series or Fourier transformation. As a result, the relationship between the spatial current distribution over an antenna and the far zone radiation pattern of the antenna is analogous to the relationship between a signal's time domain waveform and frequency domain waveform. The radiation beams in Figure 2.29(a) can be considered as a result of Fourier transforming the current distribution along x, while the radiation beams in Figure 2.29(b) can be considered as a result of Fourier transforming the current distribution along y. It is well known that, if a signal's waveform is narrow in the frequency domain, its waveform in the time domain must be broad. As an analogy, a narrow beam in a radiation pattern must correspond to a large physical antenna aperture generally. In Figure 2.29, the 2-element array covers a certain physical aperture along the x-axis. In contrast, the physical size of the 2-element array is zero if measured along the y-axis. Therefore, the beams in Figure 2.29(a) are narrower than the beams in Figure 2.29(b).

In Figure 2.30, the spacing between the two Hertzian dipoles is as large as $2.5\lambda_0$. It is observed from Figure 2.28 that the mutual coupling is quite weak when $s = 2.5\lambda_0$. This observation is confirmed by the two radiation pattern plots in Figure 2.30. Since the spacing is large, it is not surprising that many beams appear

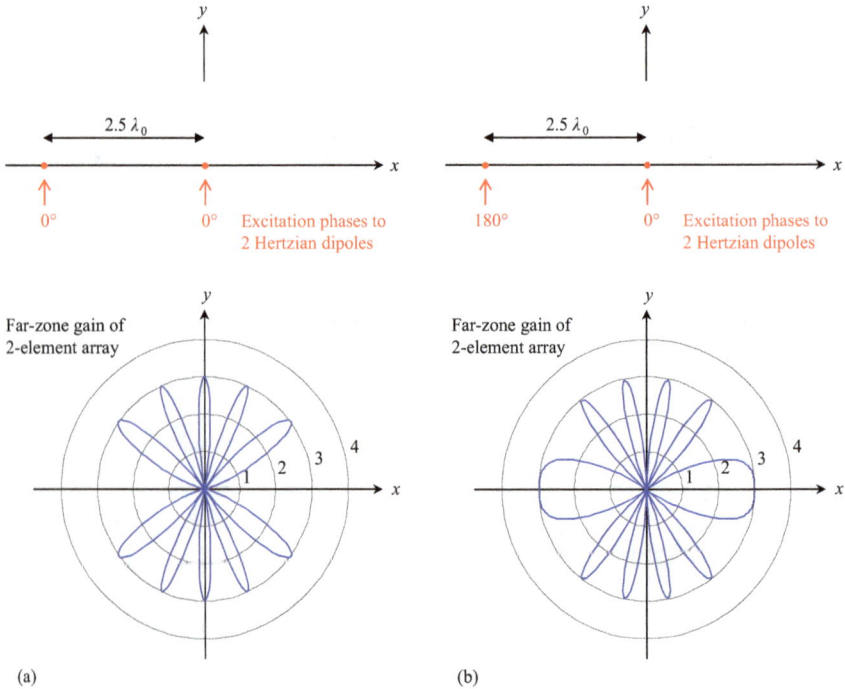

(a)   (b)

Figure 2.30   Numerical results of far zone gain (in linear scale) of an array including two Hertzian dipoles when $s = 2.5\lambda_0$. (a) Two dipoles are excited by the same phase and (b) Difference of excitation phases to dipoles is 180°.

in the plots of Figure 2.30. Meanwhile, because the coupling between the two dipoles is weak, the value of gain is 3 consistently at the center of each beam, which follows (2.146).

In this section, the impact of mutual coupling is visualized using Hertzian dipoles as antenna elements of a phased array. As a matter of fact, the mutual coupling among most practical antenna elements is weaker than that among Hertzian dipoles. Indeed, the magnitude of mutual coupling demonstrated by numerous planar antenna elements, such as the slot antennas proposed in [8], is much weaker than Hertzian dipoles. Thus, although the mutual coupling between two Hertzian dipoles appears pretty strong when the spacing is half wavelength in Figures 2.28 and 2.29, it does not lead to the conclusion that half wavelength is an unacceptably small spacing in practice. In the experimental examples of this book, the value of spacing is chosen between half wavelength and one wavelength, and with regular microstrip patches as the antenna elements, the impact of mutual coupling is not significant. Generally speaking, mutual coupling among antenna elements creates complications and thus should be minimized in a phased array. Nevertheless, it must be noted that mutual coupling is actually an enabling factor (rather than an undesirable factor) in the design of some antennas such as Yagi–Uda antennas and log-periodic antennas [1].

References

[1] Balanis C.A. *Antenna Theory: Analysis and Design*. 3rd ed. New York: Wiley-Interscience; 2005.

[2] Balanis C.A. *Advanced Engineering Electromagnetics*. 2nd ed. Hoboken, NJ: John Wiley & Sons; 2012.

[3] Lui H.-S., Hui H.T. 'Direction-of-arrival estimation: Measurement using compact antenna arrays under the influence of mutual coupling'. *IEEE Antennas and Propagation Magazine*. 2015;57(6):62–8.

[4] Yuan Q., Chen Q., Sawaya K. 'Accurate DOA estimation using array antenna with arbitrary geometry'. *IEEE Transactions on Antennas and Propagation*. 2005;53(4):1352–7.

[5] Garnica J., Chinga R.A., Lin J. 'Wireless power transmission: From far field to near field'. *Proceedings of the IEEE*. 2013;101(6):1321–31.

[6] Goubau G. 'Microwave power transmission from an orbiting solar power station'. *Journal of Microwave Power*. 1970;5(4):224–31.

[7] Goubau G., Schwering F. Free space beam transmission. In: Okress EC (ed.). *Microwave Power Engineering: Generation, Transmission, Rectification*. New York: Academic Press (Electrical Science); 1968.

[8] Lu M., Billo R.E., *inventors; Wireless Power Transmission*. United States Patent 9030161. 2015.

Chapter 3

Theoretical principles and practical implementation of retro-reflective beamforming technique for microwave power transmission

Based on the previous chapter (i.e., Chapter 2), the retro-reflective beamforming technique is portrayed in this chapter with the aim of accomplishing efficient microwave power transmission. Two *far zone conditions* are defined and elucidated in Chapter 2: *Condition of electrical far zone* and *condition of geometrical far zone*. This chapter is organized according to the two far zone conditions to a large extent. In Sections 3.1 and 3.2, the retro-reflective beamforming technique in the electrical far zone and retro-reflective beamforming technique in the electrical near zone are described, respectively. Comparative studies are conducted between retro-reflective beamforming in the electrical far zone and retro-reflective beamforming in the electrical near zone using several numerical and experimental examples in Section 3.3. The retro-reflective beamforming technique is interpreted by a circuit model in Section 3.4 (which is irrelevant to the two far zone conditions). In Section 3.5, the performance of retro-reflective beamforming is investigated in the geometrical far zone and geometrical near zone. The practical implementations of retro-reflective beamforming are discussed in the last section of this chapter, Section 3.6.

3.1 Retro-reflective beamforming in "electrical far zone"

The basic principles of retro-reflective beamforming technique for microwave power transmission are illustrated in Figure 3.1. The wireless power transmitter includes an array of antenna elements. In Figure 3.1, the antenna elements are assumed to be deployed along the x-axis with uniform spacing s between any two adjacent elements. In this section, a target (i.e., a wireless power receiver) is assumed to be located in the electrical far zone of the wireless power transmitter. Other than the wireless power transmitter and wireless power receiver, there is free space in the entire space. The target receives wireless power from the wireless power transmitter via the following two steps.

Step (i) The target broadcasts a time-harmonic pilot signal at frequency f. The pilot signal is received and analyzed by all the antenna elements of the wireless power transmitter.

Free space

Wireless power receiver $@\mathbf{r}_b$
(in electrical far zone)

Pilot signal

Equi-phase
surfaces planar

ϕ_b

y

x

Wireless power transmitter
(antenna array)

-4Δ -3Δ -2Δ $-\Delta$ Phase 0

(a)

Free space

Wireless power receiver $@\mathbf{r}_b$
(in electrical far zone)

$\hat{\mathbf{r}}_b$

Wireless power

ϕ_b

Equi-phase
surfaces planar

y

x

Wireless power transmitter
(antenna array)

4Δ 3Δ 2Δ Δ Phase 0

(b)

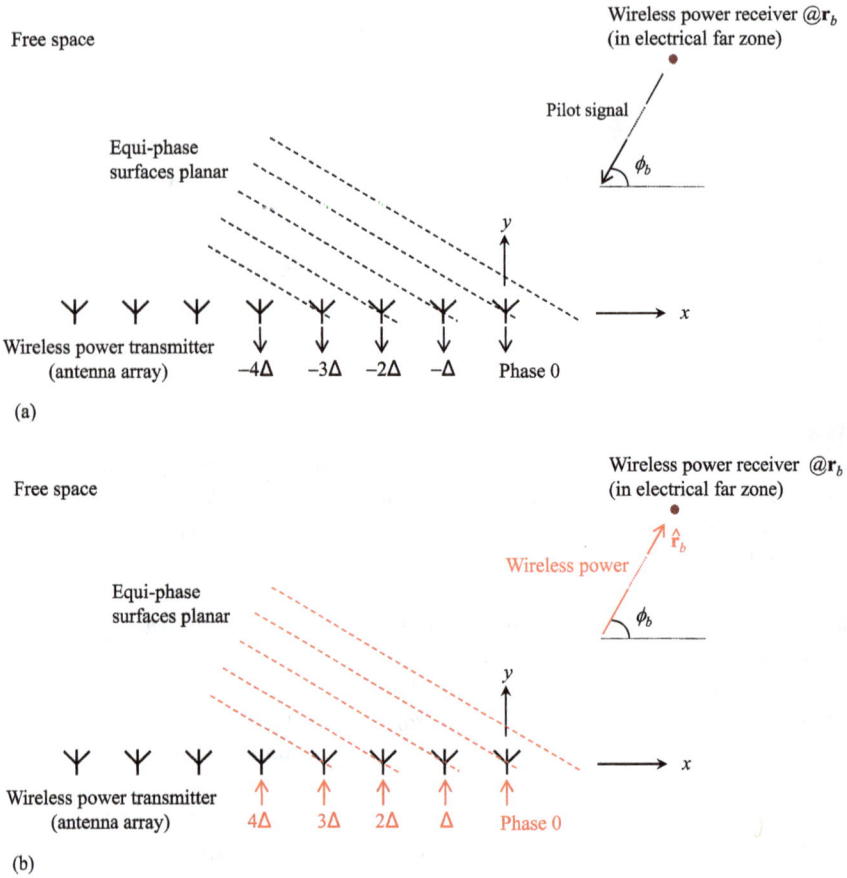

Figure 3.1   Illustration of retro-reflective beamforming for wireless power transmission in "electrical far zone." (a) The first step: propagation of pilot signal and (b) the second step: propagation of wireless power. Reproduced from [1], courtesy of the Electromagnetics Academy.

Step (ii) Based on the outcome of analyzing the pilot signal, the antenna elements of the wireless power transmitter jointly construct a time-harmonic microwave power beam toward the target at frequency f, the frequency of the pilot signal.

The propagation and reception of the pilot signal, depicted in Figure 3.1(a), were studied in Section 2.2. Because the target is assumed to reside in the electrical far zone, the pilot signal behaves as a plane wave when it reaches the wireless power transmitter. Specifically, the wavefronts of the pilot signal propagation are planar: The phase of pilot signal propagation is a constant over each planar wavefront. Suppose there is no mutual coupling among the antenna elements such that every antenna element interacts with the plane wave independently. As argued in Section 2.2, the pilot signals received by the antenna elements exhibit a uniform amplitude and a linear phase pattern. If the pilot signals detected by the antenna elements are denoted by

$$A_n e^{j\alpha_n}, n = 1, 2, 3, \ldots, \tag{3.1}$$

their amplitudes and phases satisfy

$$A_1 = A_2 = A_3 = \cdots = A_0, \tag{3.2}$$

$$\alpha_n - \alpha_{n-1} = \Delta = k_0 s \cos(\phi_b), \tag{3.3}$$

where k_0 is the wavenumber in free space and ϕ_b is the angle of the plane wave's propagation direction with respect to the $+x$ axis. When the pilot signal's phase detected by the rightmost element is defined as the reference phase zero, the pilot signal's phases detected by the other elements are $-\Delta$, -2Δ, -3Δ, $\ldots$, as specified in Figure 3.1(a).

After detecting and analyzing the pilot signal, the antenna array constructs a wireless power beam toward the target in Figure 3.1(b). In Figure 3.1(b), the antenna elements are excited by

$$X_0 e^{j\psi_n}, n = 1, 2, 3, \ldots \tag{3.4}$$

As shown in (3.4), the antenna elements are excited by a uniform amplitude. The constant amplitude X_0 is much greater than A_0 because the power level of wireless power in the second step is much higher than that of the pilot signal in the first step of retro-reflective beamforming. Following (2.26) with $\mathbf{r}_s = \mathbf{0}$, the excitation phases ψ_n are 0, Δ, 2Δ, 3Δ, ..., as specified in Figure 3.1(b). In other words,

$$\psi_n = -\alpha_n, n = 1, 2, 3, \ldots \tag{3.5}$$

The selection of excitation phases in (3.5) leads to retro-reflection: A wireless power beam propagating toward $\hat{\mathbf{r}}_b$ (which is opposite to the incoming direction of the pilot signal) is generated.

The retro-reflective beamforming technique presented above heavily relies on the phased array technique. The antenna array employed by the wireless power transmitter plays two roles in the retro-reflective beamforming technique: It is a receiving antenna for receiving the pilot signal in the first step, and it is a transmitting antenna for transmitting wireless power in the second step of retro-reflective beamforming. The two steps are linked to each other via (3.5). Because of (3.5), the technique portrayed in Figure 3.1 is sometimes termed as *phase conjugation antenna array technique* [2]. In Figure 3.1, the interaction between the wireless power transmitter and wireless power receiver is along one explicit propagation direction. To be specific, the wireless power propagates toward $\hat{\mathbf{r}}_b$ in the second step and the pilot signal propagates toward $-\hat{\mathbf{r}}_b$ in the first step of retro-reflective beamforming. Therefore, the retro-reflective beamforming technique in Figure 3.1 is termed as *retro-directive beamforming technique* by many researchers [3,4].

In Figure 3.1, the retro-reflective beamforming technique is described using a one-dimensional antenna array. The basic principles of Figure 3.1 can be extended to a two-dimensional antenna array straightforwardly. Suppose there is a two-dimensional antenna array in the x-z plane. When a pilot signal broadcasted by a target in the electrical far zone is received by the antenna elements in the first step of retro-reflective beamforming, the phase profile of the received pilot signals exhibits a linear pattern along both the x direction and z direction. In the second step of retro-reflective beamforming, exciting each antenna element with a phase value negative to the phase value of the pilot signal received in the first step would generate a wireless power beam toward the target.

3.2   Retro-reflective beamforming in "electrical near zone"

As discussed in Section 2.3, in microwave power transmission applications it is very probable that a target (i.e., a wireless power receiver) stays in the electrical near zone of the wireless power transmitter to achieve a reasonably high power transmission efficiency. When a target does not reside in the electrical far zone of the wireless power transmitter, the scenario in Figure 3.1 evolves to Figure 3.2. In the first step of retro-reflective beamforming, the pilot signal no longer behaves as a plane wave when it reaches the wireless power transmitter; rather, its wavefronts are curved as illustrated in Figure 3.2(a). Consequently, the phase profile of pilot signals detected by the antenna elements does not exhibit any linear pattern. Denote the pilot signals detected by the antenna elements in the first step of retro-reflective beamforming as

$$A_n e^{j\alpha_n}, n = 1, 2, 3, \ldots \tag{3.6}$$

Because the wavefronts are not planar in Figure 3.2, the linear relationship in (3.3) is no longer valid. When the phase detected by the rightmost element is defined as the reference phase zero, the phase detected by an element in the middle of the array is denoted as $-\Delta$. The phase difference of $-\Delta$ corresponds to the extra path specified by a piece of thick line segment in Figure 3.2(a). If the rightmost element is excited by phase zero and the middle element is excited by phase Δ in the second step of retro-reflective beamforming (Figure 3.2(b)), their radiations end up at the wireless power receiver with the same phase, in other words, their radiations are constructive at the wireless power receiver. The above analysis holds true for any element of the array. Therefore, when the excitations to the antenna elements are selected as

$$X_0 e^{j\psi_n} = X_0 e^{-j\alpha_n}, \tag{3.7}$$

the antenna elements' contributions would be in phase at the target's location in the second step of retro-reflective beamforming.

The excitations of wireless power in (3.7) appear the same as those in Section 3.1. In fact, when the target moves further and further away from the wireless power transmitter in Figure 3.2, the wavefronts lose their curvature gradually and Figure 3.2 eventually reverts to Figure 3.1. Therefore, Figure 3.2

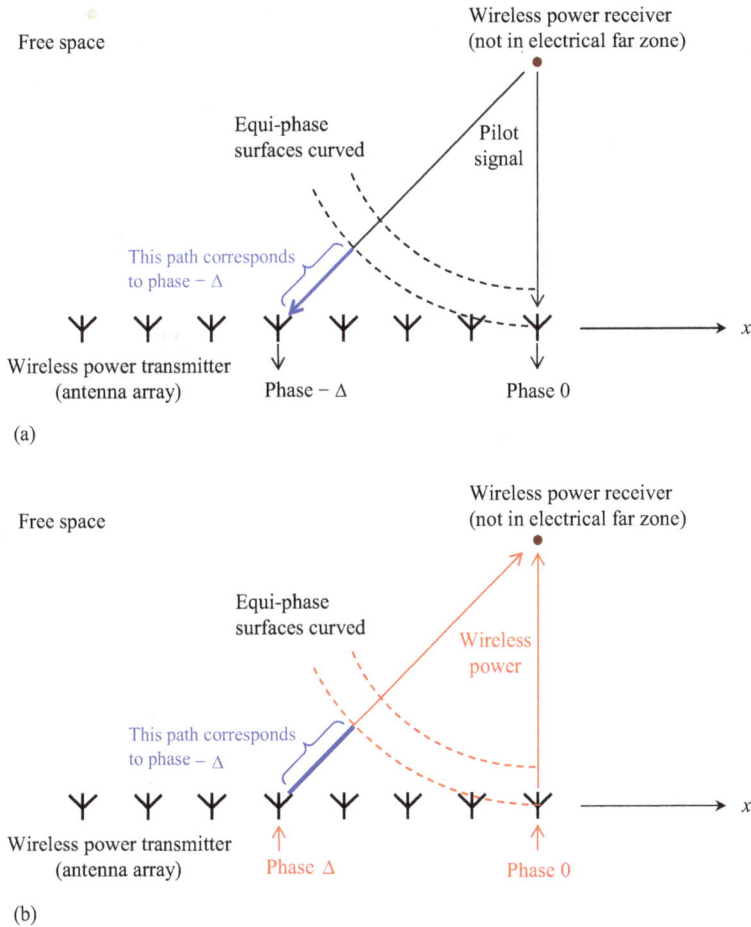

Figure 3.2   *Illustration of retro-reflective beamforming for wireless power transmission in "electrical near zone." (a) The first step: propagation of pilot signal and (b) the second step: propagation of wireless power. Reproduced from [1], courtesy of the Electromagnetics Academy.*

presents a general illustration of the retro-reflective beamforming technique, and the scenario in Figure 3.1 (which is called "retro-directive beamforming" sometimes) is a special case of retro-reflective beamforming. Figures 3.1 and 3.2 can both be characterized by the term "phase conjugation antenna array technique," because the phase profile of wireless power excitation is conjugate to the phase profile of pilot signal reception in both Section 3.1 and this section. It is noted that the amplitudes A_1, A_2, A_3, ... in (3.6) may not be identical to each other since the target is not far away. However, in (3.7), the antenna elements are excited by a uniform amplitude X_0. In other words, the scheme in this section complies with the term "phase conjugation antenna array technique" rigorously: While the phase profile of wireless power excitation is conjugate to the phase profile of pilot signal reception, the amplitude of wireless power excitation does not rely on the pilot signal reception. In this book, the "full conjugation antenna array technique" is used as the name for retro-reflective beamforming schemes in which the amplitude profile of wireless power excitation is determined by the amplitude profile of pilot signal reception. The "full conjugation antenna array technique" is discussed in Sections 3.4 and 3.5.

The one-dimensional retro-reflective antenna array in Figure 3.2 can be extended to the two-dimensional retro-reflective antenna array configuration in Figure 3.3. The antenna elements in the two-dimensional array are deployed into N columns along the x direction and M rows along the z direction, as depicted in Figure 3.3.

In the first step of retro-reflective beamforming, a pilot signal is broadcasted by a target in the electrical near zone, as displayed in Figure 3.3(a). However, the target is in the electrical far zone as far as each antenna element of the

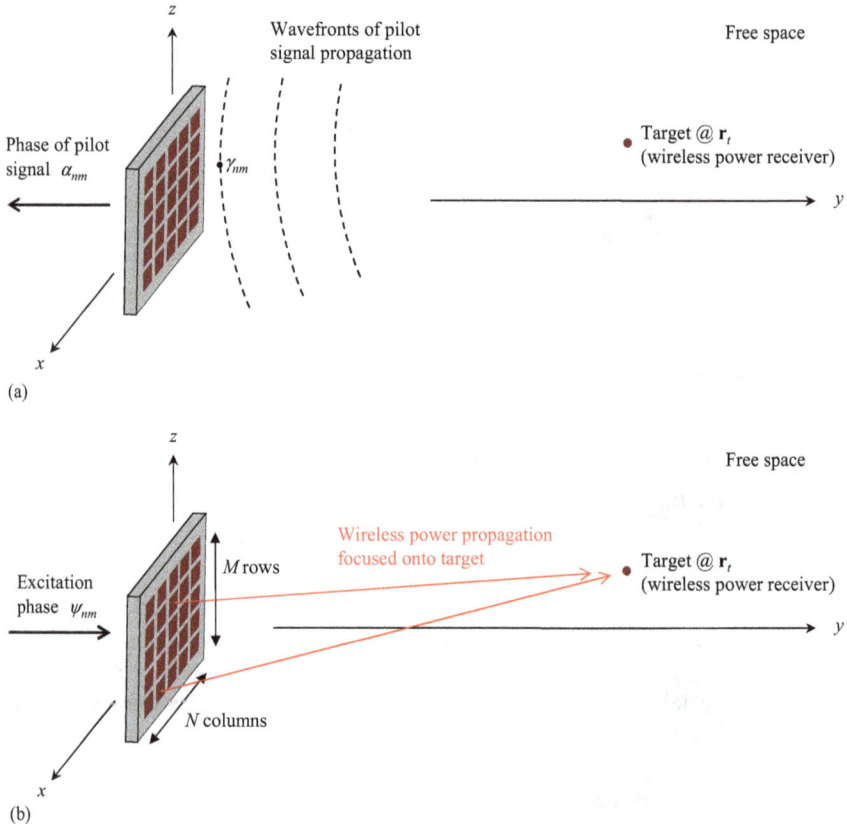

Figure 3.3   Retro-reflective beamforming with a two-dimensional antenna array in the wireless power transmitter and with a target in the electrical near zone. (a) The first step: propagation of pilot signal and (b) the second step: propagation of wireless power.

retro-reflective array is concerned, which is assumed true throughout this book. As a result, the pilot signal propagates as a plane wave when it reaches the antenna element with indices nm (n is the column index and m is the row index, to be specific). Suppose the plane wave propagation of the pilot signal in front of the antenna element with indices nm is γ_{nm}. When the pilot signal originated at the target is defined to have the reference phase zero, $\gamma_{nm} = -k_0 \mid \mathbf{r}_{nm} - \mathbf{r}_t \mid$, where $\mathbf{r}_{nm}$ represents the location of the antenna element with indices nm and $\mathbf{r}_t$ represents the location of the target. In response to the pilot signal, the voltage developed at the circuit port of the antenna element with indices nm is assumed to have a phase α_{nm}. If there is no mutual coupling among the antenna elements of the retro-reflective array, the difference between γ_{nm} and α_{nm}, $\alpha_{mn} - \gamma_{mn} = \kappa_{nm}$, is completely determined by the interaction between the incident pilot signal and antenna element nm. In practice, the antenna elements in a phased array resemble isotropic radiators typically, and thus are insensitive to the curvature of the pilot signal's wavefronts. As a result, the value of κ_{nm} does not depend on n or m approximately. In other words, κ_{nm} can be represented by one constant κ, and $\alpha_{mn} = \gamma_{mn} + \kappa$.

After the phases α_{nm}, $n = 1, 2, 3, ..., N$, $m = 1, 2, 3, ..., M$ are detected, the excitation phases in Figure 3.3(b) are chosen as

$$\psi_{nm} = -\alpha_{nm} = -\gamma_{nm} - \kappa = k_0 \mid \mathbf{r}_{nm} - \mathbf{r}_t \mid -\kappa. \tag{3.8}$$

The excitation phases ψ_{nm} in (3.8) are different from the excitation phases in (2.77) by a constant phase κ. As a matter of fact, the performance of a transmitting phased array does not change when a constant phase is added to the excitation phase of every antenna element. Thus, all the conclusions drawn in Sections 2.4 and 2.5 are applicable to the wireless power propagation shown in Figure 3.3(b). Three of them are articulated below. First, the wireless power radiations of the $N \times M$ antenna elements are in phase at $\mathbf{r}_t$. Second, the propagation of wireless power behaves as a plane wave in a certain "plane wave region" around $\mathbf{r}_t$. Third, the power transmission efficiency can be estimated by the Friis transmission equation in (2.93) if the size of the target is smaller than the size of the "plane wave region."

Tremendous insights can be gained by comparing Figure 3.1 (which illustrates retro-reflective beamforming in the electrical far zone) with Figure 3.2 (which illustrates retro-reflective beamforming in the electrical near zone) [1]. Suppose an observation point is placed in Figure 3.1(b) to measure the wireless power propagation. Also, suppose the observation point is placed along the $\widehat{\mathbf{r}}_b$ direction with a distance of d to the spatial origin. The linear profile of the excitation phases in Figure 3.1(b) ensures that electromagnetic fields contributed by the antenna elements of the retro-reflective array are in phase when $d \to \infty$. When d is not very large, nevertheless, the antenna elements' contributions would not be very constructive among one another. The phenomena in Figure 3.2(b) are somewhat opposite to those in Figure 3.1(b). The excitation phases in Figure 3.2(b) enforce all the antenna elements' radiations to be in phase at the target. However, when d approaches infinity, the antenna elements are no longer constructive among one another because the excitation phases do not have any linear profile. In summary, the wireless power propagation in Figure 3.1(b) demonstrates a beam in the far zone, whereas the wireless power propagation in Figure 3.2(b) demonstrates a focal point in the near zone. The differences between Figures 3.1 and 3.2 are elaborated further in Section 3.3 using some numerical and experimental examples.

3.3   Retro-reflective beamforming in "electrical far zone" and "electrical near zone": comparative studies

In this section, the performance of retro-reflective beamforming in the electrical far zone is compared with that in the electrical near zone, using several numerical and experimental examples.

In the numerical example shown in Figure 3.4, the wireless power transmitter includes six antenna elements. Each element is a Hertzian dipole oriented along the z direction. The 6-element array is deployed along the x-axis. The distance between two adjacent elements is $\lambda_0/2$, with λ_0 denoting the wavelength in free space. The wireless power receiver includes one z-oriented Hertzian dipole, located $40\lambda_0$ away from the spatial origin along the $\phi = 70$ direction. There is only free space other than the wireless power transmitter and wireless power receiver in the entire space. Hertzian dipoles are selected as the antenna elements in this numerical example because the electromagnetic fields radiated by Hertzian dipoles in free space are available analytically [5]. The numerical calculations in this section largely follow the formulations of Section 2.6.

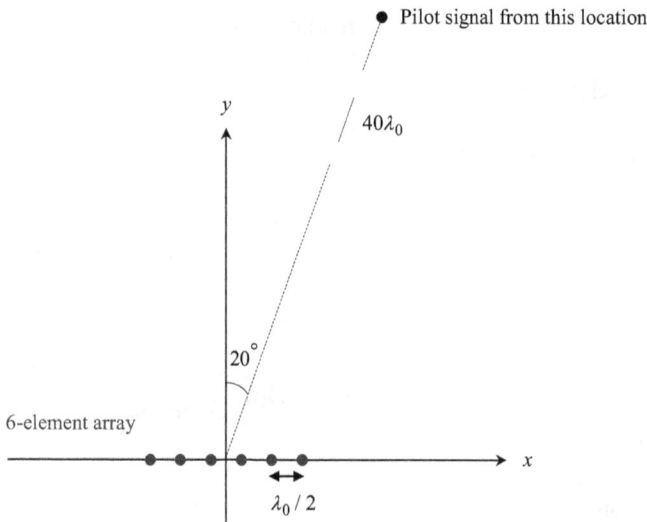

Figure 3.4   Numerical example of an equi-spaced array with a pilot signal from $40\lambda_0$ away. Reproduced from [1], courtesy of the Electromagnetics Academy.

In the first step of retro-reflective beamforming, the wireless power receiver broadcasts a time-harmonic pilot signal at frequency f, where $f = c/\lambda_0$ and $c = 1/\sqrt{\varepsilon_0\mu_0}$ is the speed of light in free space. The six elements of the wireless power transmitter detect the E_z field of the pilot signal, and the detected pilot signals' phase values are denoted as $\alpha_1, \alpha_2, \ldots, \alpha_6$. In the second step of retro-reflective beamforming, the six elements are excited by time-harmonic sources at frequency f with a uniform amplitude and with phases $-\alpha_1, -\alpha_2, \ldots,$ $-\alpha_6$, respectively. The electric field distribution radiated by the 6-element array in a certain region in the x-y plane is plotted in Figure 3.5. Because the electric field's attenuation follows $1/r$ (with r denoting the distance to the spatial origin) in the far zone, $|rE_z|^2$ is plotted in Figure 3.5 in order to clearly visualize the electric field's far-zone behavior. The $|rE_z|^2$ values in Figure 3.5 are normalized by the largest $|rE_z|^2$ in the region of $x \in [-\lambda_0, 10\lambda_0]$ and $y \in [\lambda_0, 10\lambda_0]$. Since the wireless power receiver resides in the electrical far zone of the wireless power transmitter, the 6-element array behaves as a retro-directive beamformer. In Figure 3.5, the $|rE_z|^2$ value increases when the distance toward the spatial origin increases along $\phi = 70°$. This observation agrees with the discussion at the end of Section 3.2: The farther away an observation point is from the retro-directive array, the more constructive the electromagnetic fields contributed by the antenna elements of the retro-directive array are. In other words, a directive beam along $\phi = 70°$ emerges in the far zone. The far zone gain

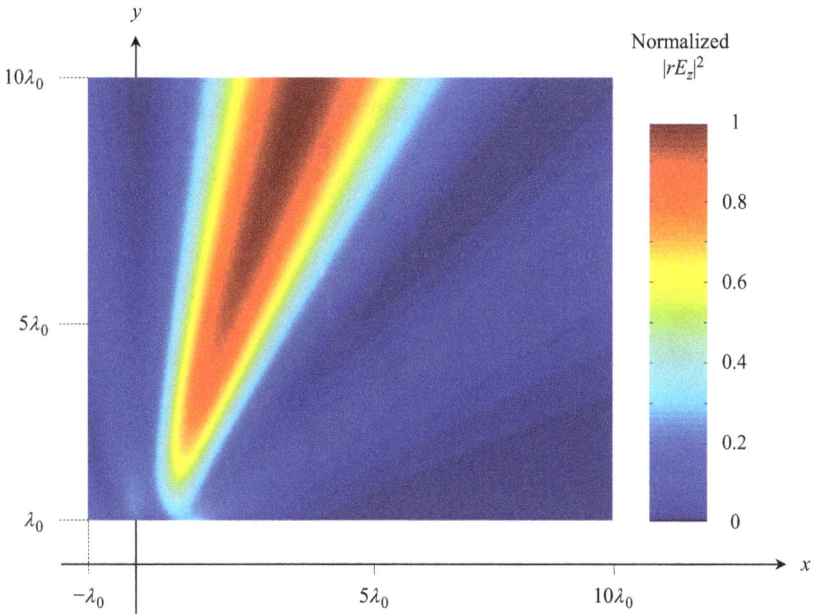

Figure 3.5   Numerical results: Electric field distribution radiated by the antenna array in the setup of Figure 3.4. Reproduced from [1], courtesy of the Electromagnetics Academy.

values of the 6-element retro-directive array are plotted in Figure 3.6. The gain value toward $\phi = 70°$ shown in Figure 3.6 is 10.15 dBi. The gain value of 10.15 dBi is greater than ($6 \times 1.5 = 9$) because of the mutual coupling among the six Hertzian dipoles, as discussed in Section 2.7.

The numerical example in Figure 3.7 has only one difference from the numerical example in Figure 3.4: The wireless power receiver is located $5\lambda_0$ away from the spatial origin along the $\phi - 70°$ direction in Figure 3.7, whereas the wireless power receiver is located $40\lambda_0$ away from the spatial origin along the $\phi = 70°$ direction in Figure 3.4. Figure 3.8 shows that $|rE_z|^2$ reaches the largest value at the location ($x = 1.7\lambda_0$, $y = 4.7\lambda_0$), which corresponds to the cylindrical coordinates of ($\rho = 5\lambda_0$, $\phi = 70°$). In other words, the electric field distribution demonstrates a focal point at the wireless power receiver's location. Meanwhile, at a location far away from the origin along the $\phi = 70°$ direction, the field in Figure 3.8 is not as convergent as that in Figure 3.5. It is also observed from Figure 3.9 that the far zone gain value along $\phi = 70°$ is 9.6 dBi, 0.55 dB lower than the 10.15-dBi gain in Figure 3.6. Thus, the 6-element array in Figure 3.7 behaves as a retro-reflective beamformer; it generates a focal point in the near zone, while its far zone radiation is not as collimated as a retro-directive beamformer's.

In Figure 3.7, the electromagnetic field radiated by the retro-reflective antenna array in the second step of retro-reflective beamforming demonstrates a focal point in the near zone. Focusing electromagnetic fields in the near zone has various applications in addition to wireless power transmission, such as microwave-induced hyperthermia, microwave remote sensing, RFID, and three-dimensional spatial division multiplexing [6]. Numerous techniques have been investigated in order to focus electromagnetic fields in the near zone, including reflector antennas [7], quasi-optical lens antennas [8], planar lenses [9], leaky-wave lenses [10,11], and planar antenna arrays [12–14]. In the retro-reflective beamforming technique, the spatial focusing of the electromagnetic field is

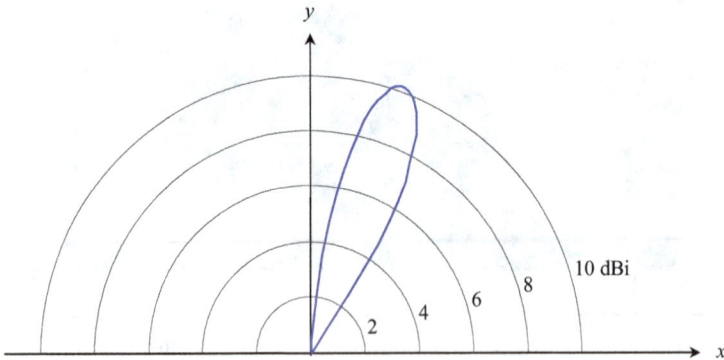

Figure 3.6   Numerical results: Far zone gain of the antenna array in the setup of Figure 3.4. Reproduced from [1], courtesy of the Electromagnetics Academy.

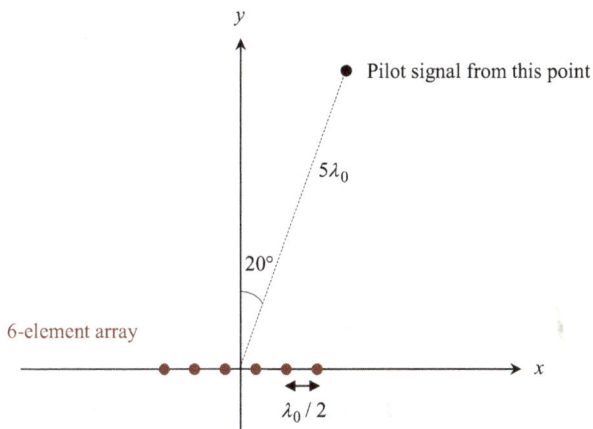

Figure 3.7   *Numerical example of an equi-spaced array with a pilot signal from*
$5\lambda_0$ away. Reproduced from [1], courtesy of the Electromagnetics
Academy.

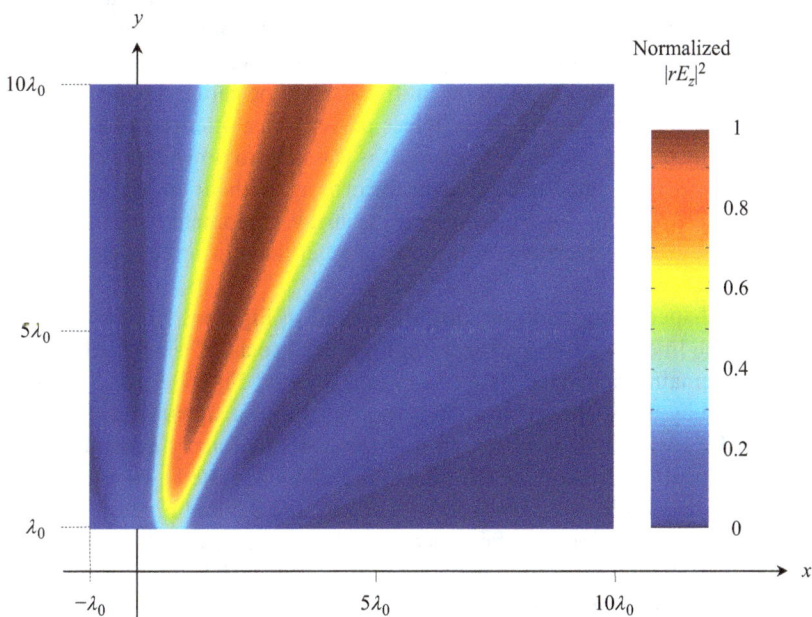

Figure 3.8   *Numerical results: Electric field distribution radiated by the antenna*
array in the setup of Figure 3.7. Reproduced from [1], courtesy of the
Electromagnetics Academy.

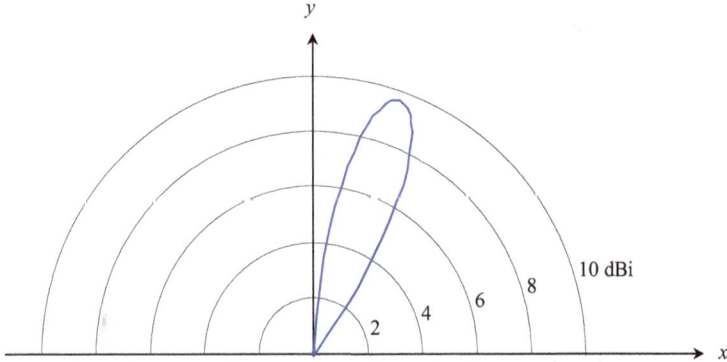

Figure 3.9   Numerical results: Far zone gain of the antenna array in the setup of Figure 3.7. Reproduced from [1], courtesy of the Electromagnetics Academy.

accomplished with the aid of a pilot signal. Specifically in retro-reflective beamforming, the focal point of the electromagnetic field is at the location from which the pilot signal is broadcasted, which is very suitable for wireless power transmission applications. In applications where mobile/portable electronic devices need to be charged wirelessly, the wireless power transmitter should not construct any focal points other than the mobile/portable devices' location. Moreover, the far zone gain of the wireless power transmitter must be minimized to mitigate electromagnetic interference. It is possible to downgrade the far zone gain associated with retro-reflective beamforming without impacting the near zone focusing performance by adjusting the array's geometrical configuration. Numerical results of two array configurations other than linear equi-spaced arrays are studied below.

The V-shaped array configuration illustrated in Figure 3.10 is arrived at by deforming the linear equi-spaced array in Figure 3.7 to a "V" shape. All the other conditions remain unchanged between Figure 3.7 and Figure 3.10. The electric field radiated by the V-shaped array in response to a pilot signal from $5\lambda_0$ away is plotted in Figure 3.11. It is observed that the V-shaped array generates a focal point at the location from which the pilot signal is broadcasted as the equi-spaced array does. Meanwhile, a comparison between Figure 3.8 and Figure 3.11 clearly demonstrates that, at a location far away from the spatial origin along the $\phi = 70°$ direction, the electric field is more divergent in Figure 3.11 than in Figure 3.8. The V-shaped array generates a far zone beam along the $\phi = 70°$ direction, as displayed in Figure 3.12, with a peak gain value of 9.2 dBi, 0.4 dB lower than the maximum gain value of Figure 3.9. In conclusion, deforming an equi-spaced array to a V-shape reduces the maximum far-zone gain without affecting the near-zone focusing performance.

The configuration illustrated in Figure 3.13 is termed a "perturbed array": On the basis of the linear equi-spaced array in Figure 3.7, all the elements' spatial

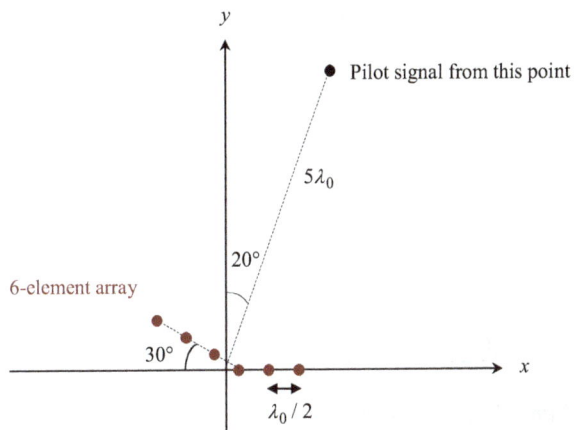

Figure 3.10   *Numerical example of a V-shaped array with a pilot signal from $5\lambda_0$ away. Reproduced from [1], courtesy of the Electromagnetics Academy.*

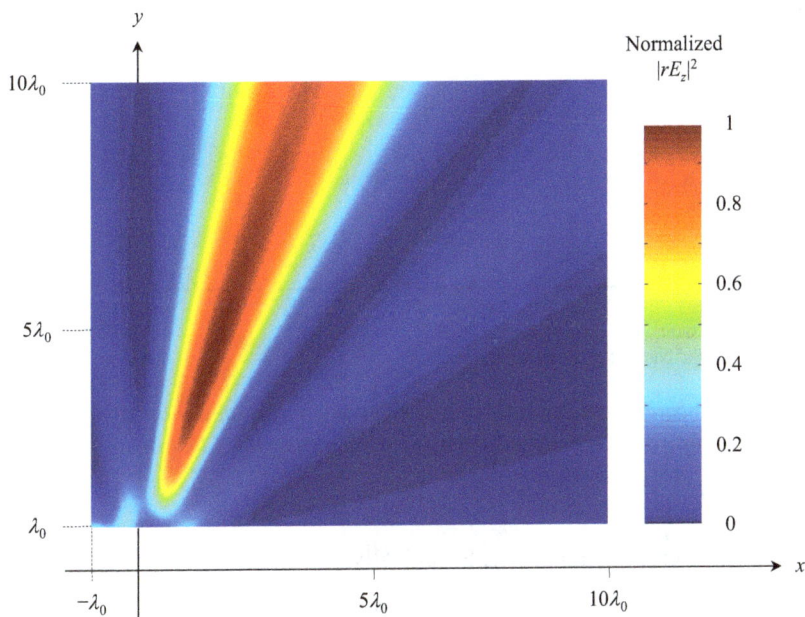

Figure 3.11   *Numerical results: Electric field distribution radiated by the antenna array in the setup of Figure 3.10. Reproduced from [1], courtesy of the Electromagnetics Academy.*

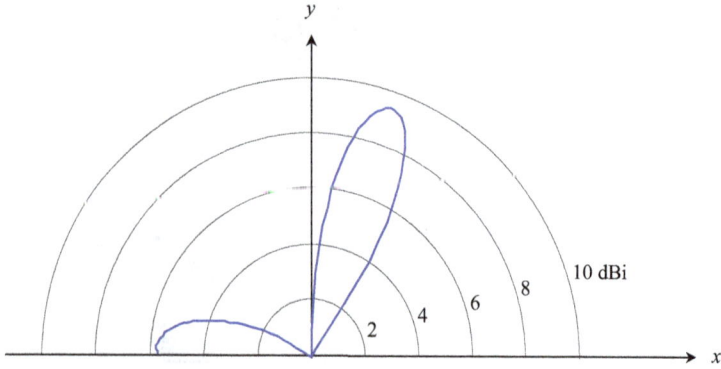

Figure 3.12   Numerical results: Far zone gain of the antenna array in the setup of Figure 3.10. Reproduced from [1], courtesy of the Electromagnetics Academy.

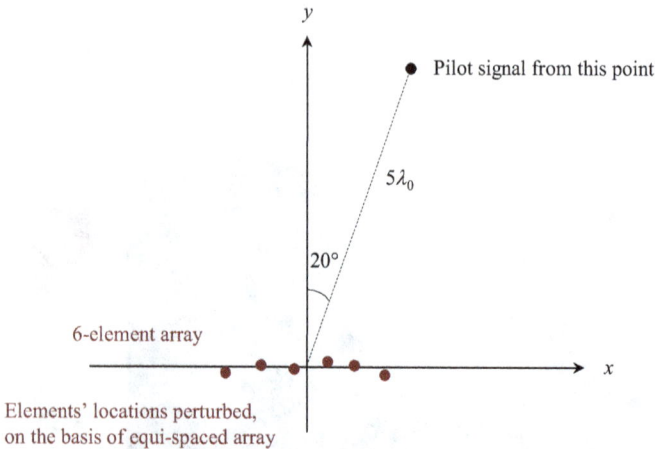

Figure 3.13   Numerical example of a perturbed array with a pilot signal from $5\lambda_0$ away. Reproduced from [1], courtesy of the Electromagnetics Academy.

locations are perturbed randomly. Specifically,

$$
\begin{cases}
x_n = x_n^{(e)} + (\Delta x)_n = x_n^{(e)} + \varsigma_{xn} \times 0.1\lambda_0 \\
y_n = y_n^{(e)} + (\Delta y)_n = y_n^{(e)} + \varsigma_{yn} \times 0.1\lambda_0 \\
z_n = z_n^{(e)} + (\Delta z)_n = z_n^{(e)} + \varsigma_{zn} \times 0.1\lambda_0, \quad n = 1, 2, \ldots, 6,
\end{cases}
\tag{3.9}
$$

where (x_n, y_n, z_n), $n = 1, 2, \ldots, 6$, denote the coordinates of the six elements in the perturbed array, $(x_n^{(e)}, y_n^{(e)}, z_n^{(e)})$, $n = 1, 2, \ldots, 6$, are the coordinates of the six elements in the equi-spaced array in Figure 3.7, and (ς_{xn}, ς_{yn}, ς_{zn}) are random numbers uniformly distributed between -1 and 1. The perturbed array has similar characteristics to the V-shaped array: It reduces the maximum far zone gain without affecting the near zone focusing performance, as shown in Figures 3.14 and 3.15. The gain value along $\phi = 70°$ in Figure 3.15 is 8.7 dBi, 0.9 dB lower than the maximum gain value in Figure 3.9.

The numerical studies in Figures 3.4–3.15 are verified by some experimental results in the rest of this section. In all the experiments of this section, the wireless power receiver includes a monopole antenna made of copper wire with a length of 34.5 mm, and the wireless power transmitter consists of six microstrip antenna elements. The six microstrip antenna elements are identical to each other. They are deployed into three array configurations and studied separately: equi-spaced array, V-shaped array, and perturbed array. One of the experimental setups is shown in Figure 3.16, in which the six antenna elements are configured into a linear equi-spaced array. Each rectangular microstrip patch has dimensions of 31.4 mm along the z direction and 46 mm along the x direction. The microstrip antennas and monopole antenna are all linearly polarized along the z direction. An individual microstrip patch antenna generates a main radiation beam toward $+y$ direction with a peak gain value of 4 dBi at 2.12 GHz (the 4-dBi gain corresponds to a relatively broad beam compared with typical microstrip patch antennas). The monopole

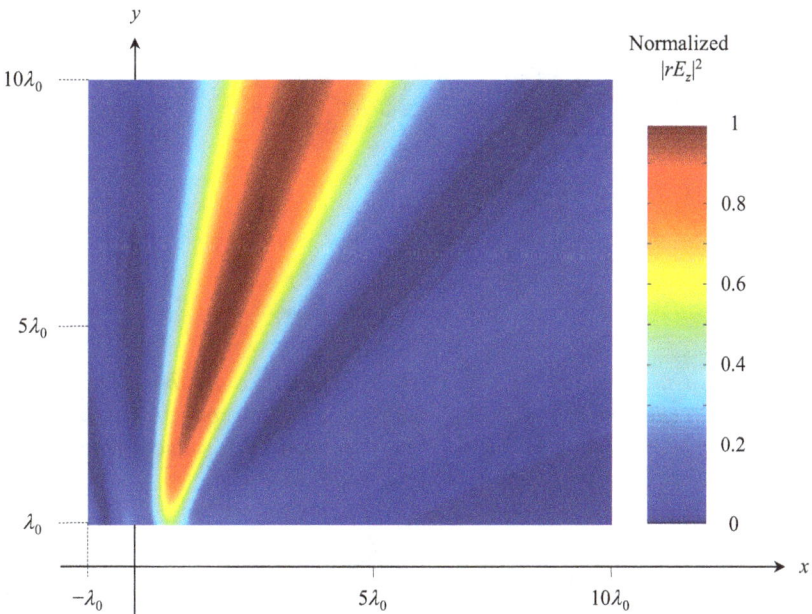

Figure 3.14   Numerical results: Electric field distribution radiated by the antenna array in the setup of Figure 3.13. Reproduced from [1], courtesy of the Electromagnetics Academy.

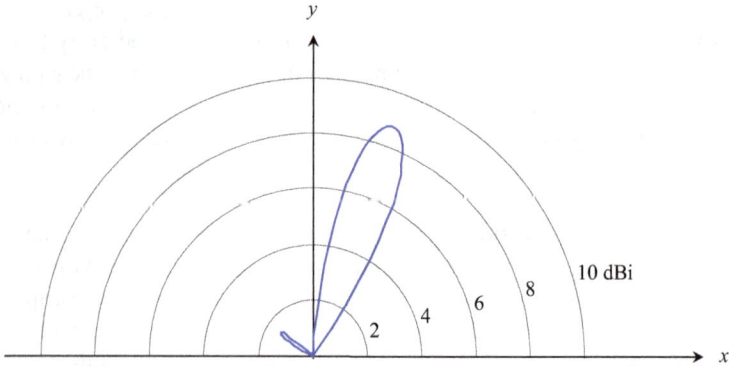

Figure 3.15   *Numerical results: Far-zone gain of the antenna array in the setup of Figure 3.13. Reproduced from [1], courtesy of the Electromagnetics Academy.*

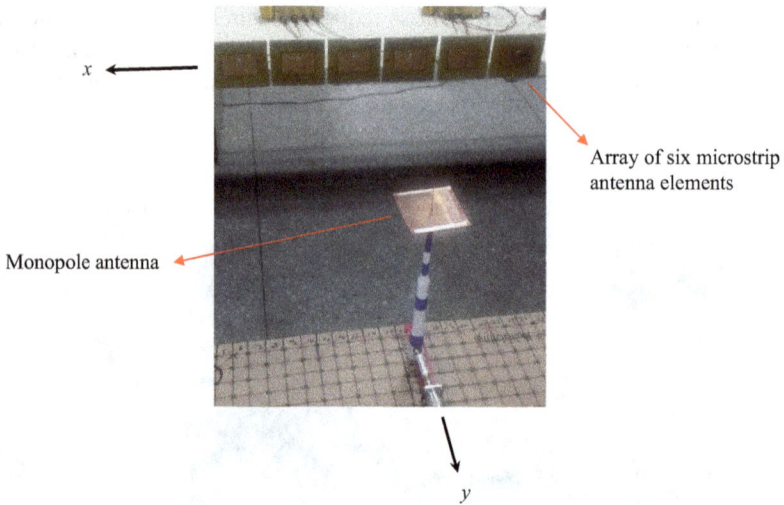

Figure 3.16   *One experimental setup for comparative studies between retro-reflective beamforming in the electrical far zone and retro-reflective beamforming in the electrical near zone. Reproduced from [1], courtesy of the Electromagnetics Academy.*

antenna has an omnidirectional radiation pattern in the *x-y* plane with a gain value of 0.1 dBi at 2.12 GHz (the monopole antenna in the experiments of this section is not optimized; however, its low gain value has no impact on the characteristics of the retro-reflective beamforming arrays).

The primary difference between the experimental setup in Figure 3.16 and the numerical setup in Figures 3.4–3.15 is that microstrip antennas are employed in the experiments whereas Hertzian dipoles are used as antenna elements in the numerical simulations. Hertzian dipoles are adopted in numerical studies largely because the electromagnetic fields radiated by Hertzian dipoles are available analytically. The microstrip patches in Figure 3.16 are designed intentionally to have a broad radiation beam toward the broadside direction (i.e., $+y$ direction), which resembles the radiation pattern toward $+y$ direction of z-orientated Hertzian dipoles.

All of the experiments in this section are conducted at the frequency of $f = 2.12$ GHz. Every experiment consists of two steps. In the first step, a time-harmonic pilot signal at frequency f is broadcasted by the monopole antenna, and the pilot signal is received by the six microstrip patch antennas. The pilot signals' phase values detected by the six microstrip antennas in the first step are denoted as α_1, α_2, ..., α_6. In the second step, time-harmonic wireless power at frequency f is radiated by the six microstrip antennas. The six microstrip antennas are excited with a uniform amplitude in the second step. The power radiated by each microstrip antenna is 21 dBm $= 125$ mW, and the total transmitted power P_t is 125 mW $\times$ 6 $= 750$ mW. In the second step, the six microstrip antennas are excited with phase values of $-\alpha_1$, $-\alpha_2$, ..., $-\alpha_6$, respectively. The monopole antenna moves over a certain region in the x-y plane to detect wireless power in the second step; the value of received power is read by a power meter directly connected to the monopole antenna.

The geometrical configuration of the equi-spaced array is depicted in Figure 3.17. The pilot signal is broadcasted from ($x = 0.8$ m, $y = 4$ m) in the first step. In the second step, the power received by the monopole antenna, P_r, is measured to be -11.51 dBm when the monopole antenna is located at ($x = 0.8$ m, $y = 4$ m). By making use of the Friis transmission equation in (2.93),

$$\frac{P_r}{P_t} = \left(\frac{\lambda_0}{4\pi d}\right)^2 G_{array} G_r,$$

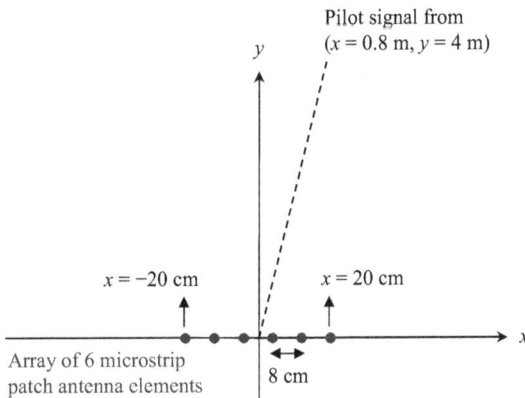

Figure 3.17   *Geometrical configuration of the experimental setup with equi-spaced array and with pilot signal from ($x = 0.8$ m, $y = 4$ m). Reproduced from [1], courtesy of the Electromagnetics Academy.*

G_{array} is found to be

$$G_{array} = \frac{P_r}{P_t} \left(\frac{4\pi d}{\lambda_0}\right)^2 \frac{1}{G_r} = 10.82 \text{ dBi},$$

where $\lambda_0 = c/f$ is the wavelength in free space, $G_r = 0.1$ dBi is the gain value of monopole antenna, and distance $d = \sqrt{4^2 + 0.8^2}$ m. The broadside gain value of each individual microstrip antenna element is 4 dBi. Because the six microstrip antenna elements' radiations are supposed to be completely constructive at $(x = 0.8 \text{ m}, y = 4 \text{ m})$, the gain value should be enhanced by a factor of 6 (that is, by 7.8 dB) when the mutual coupling among the antenna elements is ignored (as discussed in Chapter 2). It is noticed that the measured G_{array} value of 10.82 dBi is slightly lower than (4 dBi + 7.8 dB = 11.8 dBi).

In Figure 3.18, the values of G_{array} obtained based on measurement data are plotted. Obviously, the peak values of G_{array} are located at $(x = 0.2 \text{ m}, y = 1 \text{ m})$, $(x = 0.4 \text{ m}, y = 2 \text{ m})$, $(x = 0.6 \text{ m}, y = 3 \text{ m})$, and $(x = 0.8 \text{ m}, y = 4 \text{ m})$. With the increase of y, the peak G_{array} converges to 10.8 dBi. The experimental results in Figure 3.18 are consistent with the numerical results in Figures 3.5 and 3.6. The 6-element array behaves as a retro-directive beamformer: A far zone beam with a peak gain value of 10.8 dBi opposite to the incoming direction of the pilot signal is generated in reaction to the pilot signal.

Figure 3.19 has only one difference compared to Figure 3.17. The pilot signal is broadcasted from $(x = 0.2 \text{ m}, y = 1 \text{ m})$ in Figure 3.19; in contrast, the pilot signal is broadcasted from $(x = 0.8 \text{ m}, y = 4 \text{ m})$ in Figure 3.17. The measurement results

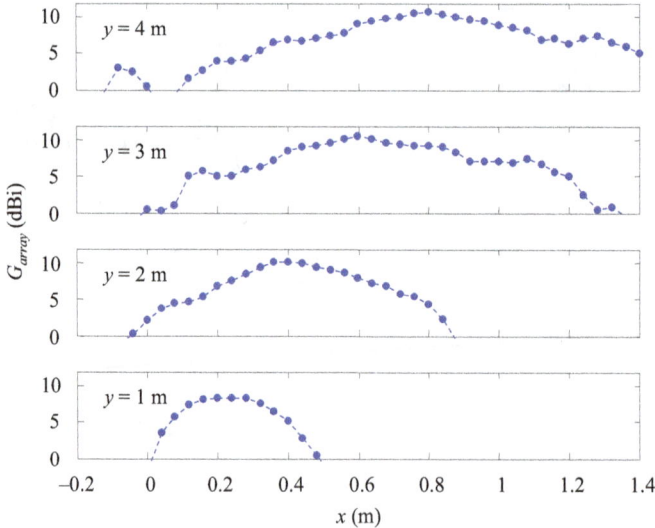

Figure 3.18   *Measurement results for the configuration of Figure 3.17. Reproduced from [1], courtesy of the Electromagnetics Academy.*

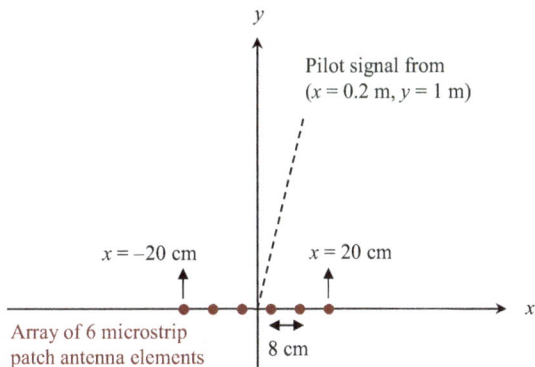

Figure 3.19   Geometrical configuration of the experimental setup with equi-spaced array and with pilot signal from (x = 0.2 m, y = 1 m). Reproduced from [1], courtesy of the Electromagnetics Academy.

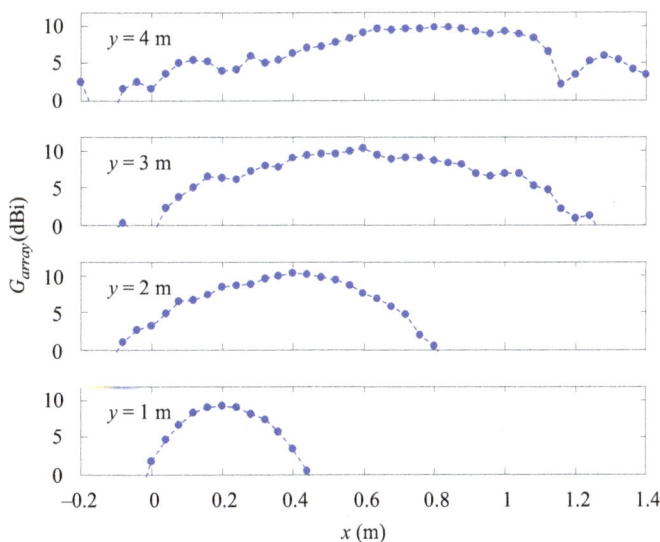

Figure 3.20   Measurement results for the configuration of Figure 3.19. Reproduced from [1], courtesy of the Electromagnetics Academy.

associated with Figure 3.19 are plotted in Figure 3.20. The comparison between Figure 3.18 and Figure 3.20 is analogous to the comparison between Figure 3.5 and Figure 3.8. In terms of the G_{array} value at ($x = 0.2$ m, $y = 1$ m), Figure 3.20 is higher than Figure 3.18 (to be specific, higher by 1.5 dB). In terms of the G_{array} value at ($x = 0.8$ m, $y = 4$ m), Figure 3.18 is higher than Figure 3.20 (to be specific, higher by 1 dB). The 6-element array in Figure 3.19 behaves as a retro-reflective beamformer: It generates a focal point in the near zone, and its peak far zone gain is not as high as that of the retro-directive beamformer in Figure 3.18.

In Figure 3.21, the six microstrip antenna elements are configured into a V-shaped array as in Figure 3.10. All the other conditions remain unchanged between Figure 3.21 and Figure 3.19. The measurement results associated with Figure 3.21 are plotted in Figure 3.22. It is observed from Figure 3.22 that the V-shaped array generates a focal point at ($x - 0.2$ m, $y = 1$ m), the location from which the pilot signal is broadcasted.

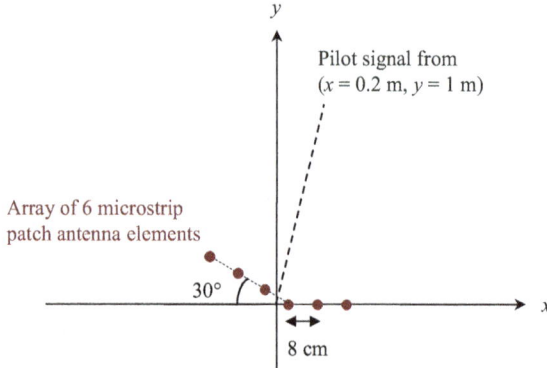

Figure 3.21   *Geometrical configuration of the experimental setup with a V-shaped array and with pilot signal from ($x = 0.2$ m, $y = 1$ m). Reproduced from [1], courtesy of the Electromagnetics Academy.*

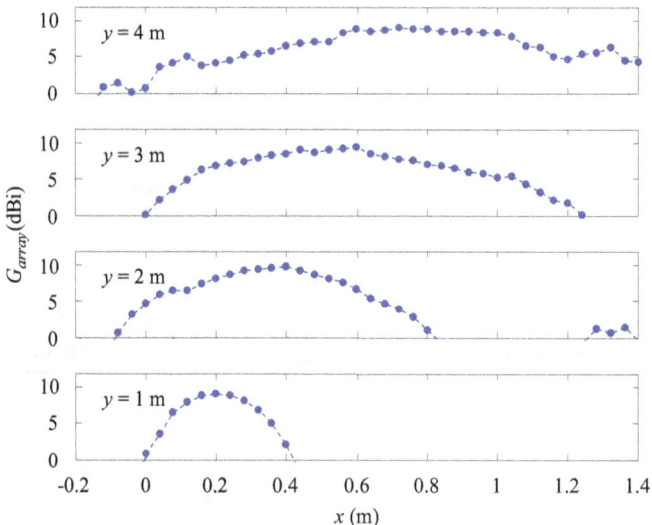

Figure 3.22   *Measurement results for the configuration of Figure 3.21. Reproduced from [1], courtesy of the Electromagnetics Academy.*

Meanwhile, it is observed that the power level at ($x = 0.2$ m, $y = 1$ m) in Figure 3.22 is almost the same as that in Figure 3.20. This means that deforming the equi-spaced array to a V-shape does not impact the array's performance in terms of near-zone focusing. However, Figures 3.22 and 3.20 demonstrate different far-zone radiation performances. Specifically, the far-zone gain at ($x = 0.8$ m, $y = 4$ m) in Figure 3.22 is 1 dB lower than that in Figure 3.20. Therefore, the experimental data in Figure 3.22 verify the conclusions drawn from Figures 3.10–3.12: It is possible to reduce the maximum far-zone gain of retro-reflective beamforming by adjusting the array's geometrical configuration without affecting the near-zone focusing performance.

Figure 3.23 studies the perturbed array, and it corresponds to the numerical example in Figure 3.13. The geometrical configuration in Figure 3.23 is obtained by perturbing the locations of the six microstrip antenna elements in Figure 3.19. In the numerical example of Figure 3.13, the locations of Hertzian dipoles are perturbed along all the three directions (that is, x, y, and z). Nevertheless in the experiments, perturbation is applied along the y and z directions, but not along the x direction. As shown in Figure 3.16, the six microstrip antenna elements almost touch each other along the x direction in the equi-spaced array, and as a result, perturbation along the x direction would create strong undesired mutual coupling among the six antenna elements. Therefore on the basis of the equi-spaced array, the six antenna elements' locations are perturbed along y and z directions only in the experiments. The perturbation follows (3.9) with specific data listed below.

Element 1:	$(\Delta y)_1 = -13$ mm	$(\Delta z)_1 = 0$
Element 2:	$(\Delta y)_2 = -20$ mm	$(\Delta z)_2 = -5$ mm
Element 3:	$(\Delta y)_3 = -6$ mm	$(\Delta z)_3 = 3$ mm
Element 4:	$(\Delta y)_4 = 0$	$(\Delta z)_4 = 0$
Element 5:	$(\Delta y)_5 = 14$ mm	$(\Delta z)_5 = -3$ mm
Element 6:	$(\Delta y)_6 = 7$ mm	$(\Delta z)_6 = 2$ mm

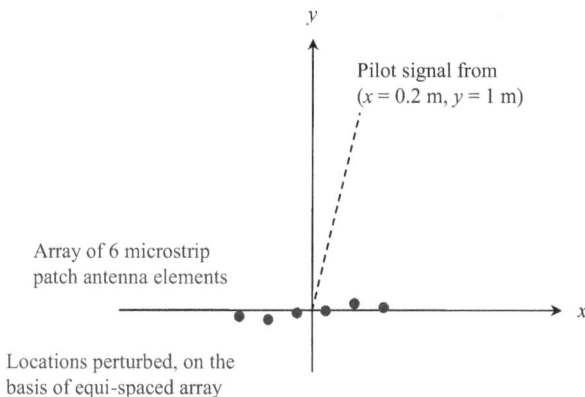

Figure 3.23 *Geometrical configuration of the experimental setup with perturbed array and with pilot signal from ($x = 0.2$ m, $y = 1$ m). Reproduced from [1], courtesy of the Electromagnetics Academy.*

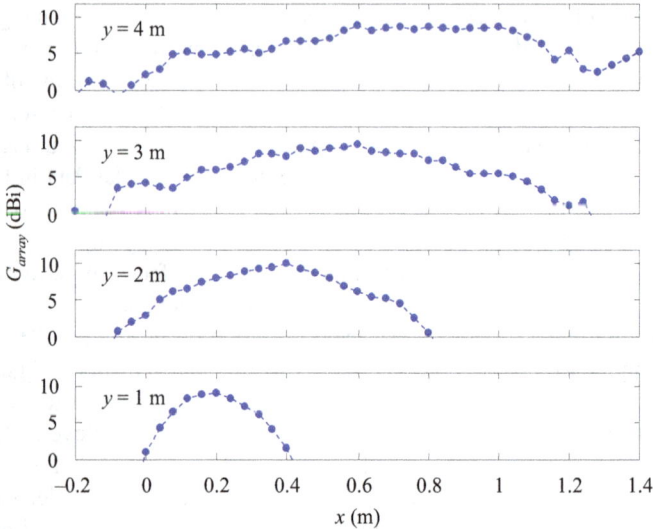

*Figure 3.24   Measurement results for the configuration of Figure 3.23.
Reproduced from [1], courtesy of the Electromagnetics Academy.*

*Table 3.1   Measured peak far-zone gain values of the four experimental examples
studied in Section 3.3. Reproduced from [1], courtesy of the
Electromagnetics Academy*

Case	Configuration	Peak far-zone gain
(i)	Equi-spaced array, with pilot signal from ($x = 0.8$ m, $y = 4$ m)	10.82 dBi
(ii)	Equi-spaced array, with pilot signal from ($x = 0.2$ m, $y = 1$ m)	9.88 dBi
(iii)	V-shaped array, with pilot signal from ($x = 0.2$ m, $y = 1$ m)	9 dBi
(iv)	Perturbed array, with pilot signal from ($x = 0.2$ m, $y = 1$ m)	8.96 dBi

Though perturbation is not conducted along all the three directions, the results in Figure 3.24 verify the conclusions drawn from Figures 3.13–3.15. The power level at the focal point, ($x = 0.2$ m, $y = 1$ m), in Figure 3.24 is the same as those in Figures 3.22 and 3.20. Meanwhile, the far-zone gain at ($x = 0.8$ m, $y = 4$ m) in Figure 3.24 is about 1 dB lower than that in Figure 3.20. Therefore, perturbation of array elements' location reduces the peak far-zone gain of retro-reflective beam-forming without affecting the near-zone focusing performance.

In Table 3.1, the four experimental examples investigated in this section are compared with one another in terms of measured peak far-zone gain value. In Case (i), the 6-element equi-spaced array behaves as a retro-directive beamformer; it generates a gain value as large as 10.82 dBi. In Case (ii), the equi-spaced array is a retro-reflective beamformer; its maximum far-zone gain drops to 9.88 dBi. Deforming the equi-spaced array to a V-shape or perturbed configuration results in further lower far-zone gain values, roughly 9 dBi to be specific.

3.4 Theoretical analysis of retro-reflective beamforming technique using a circuit model

In Sections 3.1 and 3.2, the basic scheme of the retro-reflective beamforming technique was attained based on the theory of phased array. In this section, the retro-reflective beamforming technique portrayed in Sections 3.1 and 3.2 is interpreted by a circuit model. The analysis in this section depends on circuit parameters rather than the electromagnetic laws. It aims to provide more insights of the retro-reflective beamforming technique.

Consider the interaction between a wireless power transmitter and a wireless power receiver as depicted in Figure 3.25. It is assumed that the wireless power transmitter includes an array of N antenna elements while the wireless power receiver has one antenna element. The interaction between the wireless power transmitter and the wireless power receiver can be modeled as a circuit network with $(N + 1)$ circuit ports. The $(N + 1)$ ports correspond to the circuit terminals of $(N + 1)$ antenna elements, respectively. The N circuit ports of the wireless power transmitter's antenna elements are numbered "Port 1," "Port 2," "Port 3" ...., and the circuit port of the wireless power receiver's antenna is defined as "Port t" (with "t" standing for "target").

Suppose all the electrical signal or electrical power transmitted between the wireless power transmitter and wireless power receiver is time-harmonic at frequency f. The circuit network in Figure 3.25 can be described by scattering

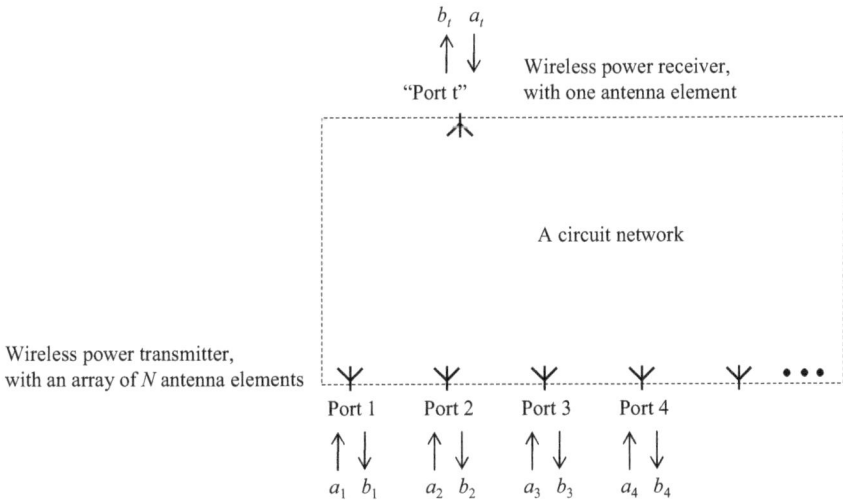

Figure 3.25 *A circuit model of wireless power transmission based on retro-reflective beamforming technique*

parameters at frequency f as

$$
\begin{bmatrix} b_t \\ b_1 \\ b_2 \\ b_3 \\ \vdots \\ b_N \end{bmatrix} = \begin{bmatrix} S_{tt} & S_{t1} & S_{t2} & S_{t3} & \cdots & S_{tN} \\ S_{1t} & S_{11} & S_{12} & S_{13} & \cdots & S_{1N} \\ S_{2t} & S_{21} & S_{22} & S_{23} & \cdots & S_{2N} \\ S_{3t} & S_{31} & S_{32} & S_{33} & \cdots & S_{3N} \\ \vdots & \vdots & \vdots & \vdots & \ddots & \vdots \\ S_{Nt} & S_{N1} & S_{N2} & S_{N3} & \cdots & S_{NN} \end{bmatrix} \times \begin{bmatrix} a_t \\ a_1 \\ a_2 \\ a_3 \\ \vdots \\ a_N \end{bmatrix}. \tag{3.10}
$$

As illustrated in Figure 3.25, a_t, a_1, a_2, ..., a_N are the incoming waves and b_t, b_1, b_2, ..., b_N are the outgoing waves at the corresponding ports. The terms starting with "S" in (3.10) are the scattering parameters.

In the first step of retro-reflective beamforming, a time-harmonic pilot signal at frequency f is transmitted from the wireless power receiver to the wireless power transmitter, as displayed in Figure 3.26. The pilot signal is supplied to the wireless power receiver's antenna element from an oscillator. Each antenna element of the wireless power transmitter is connected to a pilot signal analyzer (the pilot signal analyzers are not shown in Figure 3.26). Suppose the incoming wave at "Port t" is a_t. Also, suppose the N pilot signal analyzers are matched to the corresponding antenna elements respectively, such that $a_1 = a_2 = \cdots = a_N = 0$. The pilot signals detected by the N pilot signal analyzers are denoted as P_n, $n = 1$, 2, 3, ..., N. The detected pilot signal P_n, which is the outgoing wave at Port n, can be found from (3.10):

$$
P_n = S_{nt}a_t, n = 1, 2, 3, \cdots, N. \tag{3.11}
$$

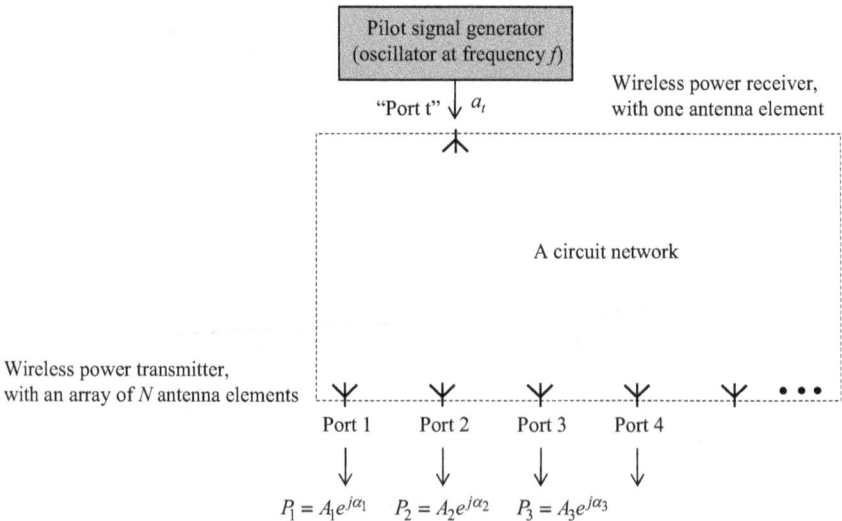

Figure 3.26   Analysis of pilot signal propagation based on circuit parameters

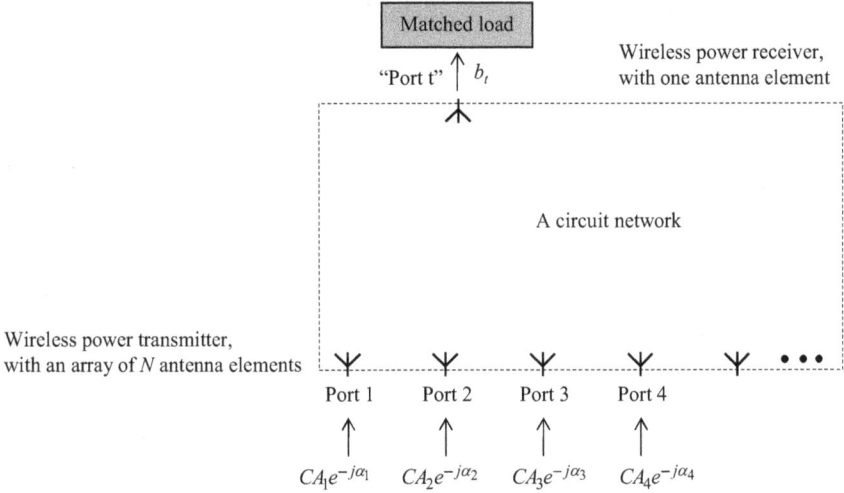

Figure 3.27   Analysis of wireless power propagation based on circuit parameters

Furthermore, denote $P_n = A_n e^{j\alpha_n}$. In other words, A_1, A_2, A_3, ..., A_N represent the amplitude of the N pilot signals detected by the wireless power transmitter, and α_1, α_2, α_3, ..., α_N represent the phase values of the N pilot signals.

In the second step of retro-reflective beamforming, wireless power is transmitted from the wireless power transmitter to the wireless power receiver. As illustrated in Figure 3.27, the incoming wave supplied to the n-th antenna element of the wireless power transmitter is denoted as $X_n e^{j\psi_n}$, $n = 1, 2, 3, ..., N$, where X_n is the amplitude and ψ_n is the phase of the n-th incoming wave. The wireless power receiver's antenna is terminated by a matched load, such that $a_t = 0$. Due to (3.10), the outgoing wave b_t at "Port t" is

$$b_t = \sum_{n=1}^{N} S_{tn}(X_n e^{j\psi_n}). \tag{3.12}$$

Apparently, $|b_t|^2$ embodies the wireless power delivered to the wireless power receiver and ought to be maximized in the second step of retro-reflective beamforming. Mathematically, the right-hand side of (3.12) constitutes an inner product between two vectors: One of them is $[S_{t1}\ S_{t2}\ S_{t3}\ \cdots\ S_{tN}]$ and the other is $[(X_1 e^{j\psi_1})\ (X_2 e^{j\psi_2})\ (X_3 e^{j\psi_3})\ \cdots\ (X_N e^{j\psi_N})]$. It is known that the inner product would reach the maximal magnitude when the two vectors are "parallel to each other" and the inner product's value is zero when the two vectors are "perpendicular to each other." Therefore, in order to maximize $|b_t|^2$, the wireless power excitations should be tailored to be "parallel to" S_{tn}, $n = 1, 2, 3, ..., N$. If the circuit network in Figure 3.25 is a reciprocal network, $S_{tn} = S_{nt}$, $n = 1, 2, 3, ..., N$. Because S_{nt}, $n = 1, 2, 3, ..., N$, are readily available from (3.11), the wireless power excitations should be chosen to be "parallel to" the pilot signals detected in the first step. Specifically, $|b_t|^2$ would be maximized when

$$X_n e^{j\psi_n} = C(P_n)^* = C(A_n e^{j\alpha_n})^* = CA_n e^{-j\alpha_n}, \tag{3.13}$$

where C is a real-valued constant and the superscript "*" stands for the complex conjugation operation. After (3.11) is substituted into (3.13),

$$X_n e^{j\psi_n} = C(P_n)^* = C(a_t S_{nt})^* = C(a_t)^* (S_{nt})^*. \tag{3.14}$$

With the excitations of (3.14), $|b_t|^2$, which is the power received by the wireless power receiver in the second step of retro-reflective beamforming, reaches the maximal value

$$|b_t|^2 = \left| \sum_{n=1}^{N} S_{tn}(X_n e^{j\psi_n}) \right|^2 = \left| \sum_{n=1}^{N} S_{tn} C(a_t)^* (S_{nt})^* \right|^2 = C^2 |a_t|^2 \left(\sum_{n=1}^{N} |S_{tn}|^2 \right)^2. \tag{3.15}$$

In addition, with the excitations of (3.14), the total power transmitted by the wireless power transmitter in the second step of retro-reflective beamforming is

$$\sum_{n=1}^{N} |X_n|^2 = C^2 |a_t|^2 \sum_{n=1}^{N} |S_{nt}|^2. \tag{3.16}$$

The ratio between (3.15) and (3.16) is the optimal power transmission efficiency

$$\sum_{n=1}^{N} |S_{nt}|^2 = |S_{1t}|^2 + |S_{2t}|^2 + \cdots + |S_{Nt}|^2. \tag{3.17}$$

Obviously, Equation (3.13) is equivalent to

$$\begin{cases} X_n = CA_n \\ \psi_n = -\alpha_n \end{cases}. \tag{3.18}$$

If $X_1 = X_2 = \cdots = X_N$, i.e., if the excitations to the antenna elements have a uniform amplitude in the second step of retro-reflective beamforming, Equation (3.18) is identical to (3.5) and (3.7). Thus, the retro-reflective beamforming schemes described in Sections 3.1 and 3.2 are well justified by the circuit model in Figure 3.25. Equations (3.5) and (3.7) were obtained by assuming free space as the wireless propagation medium. The deviation of (3.18) in this section, however, does not rely on the free space assumption. The validity of (3.18) only depends on the condition of reciprocity. Therefore, when the wireless propagation channels involve multi-path (for instance, if certain obstacles exist between the wireless power transmitter and wireless power receiver), Equation (3.18) always leads to the optimal power transmission efficiency as long as the multi-path channels are reciprocal.

Equation (3.18) indicates that, if the pilot signals' amplitude is nonuniform among the antenna elements of the wireless power transmitter, the wireless power excitations' amplitude should be nonuniform among the antenna elements accordingly. The retro-reflective beamforming scheme with the excitations selected as specified in (3.18) is termed the "full conjugation antenna array technique" in this book. In contrast, the schemes with a uniform amplitude among wireless power excitations are termed the "phase conjugation antenna array technique." The "full conjugation antenna array technique" and the "phase conjugation antenna array technique" are compared with each other in Section 3.5. According to the numerical results of Section 3.5, the "full conjugation antenna array technique" does improve the power transmission efficiency compared with the "phase conjugation antenna array technique," but the improvement is not significant for practical purposes. Meanwhile, the "phase conjugation antenna array technique" is much simpler, because detecting pilot signals' amplitude or adjusting wireless power excitations' amplitude is unnecessary. Therefore, it does not seem highly appealing to pursue the "full conjugation antenna array technique" in practice.

The derivation of this section appears fairly similar to that of the matched filter [15]. Whereas a matched filter intends to maximize the signal-to-noise ratio, the retro-reflective beamforming technique aims to maximize the power transmission efficiency. In the applications of wireless power transmission, C in (3.18) is much greater than 1 and its specific value is determined by the power budget of the wireless power transmitter.

When the derivation above is extended from the frequency domain to the time domain, the time-reversal technique is arrived at. Suppose a waveform generator is attached to the wireless power receiver's antenna element in the first step of retro-reflective beamforming, and it generates a pilot signal with a time domain waveform that is not time-harmonic. As depicted in Figure 3.28, the pilot signals in the time domain detected by the N antenna elements of wireless power transmitter are denoted as $p_1(t)$, $p_2(t)$, ..., $p_N(t)$. According to the theory of Fourier transformation,

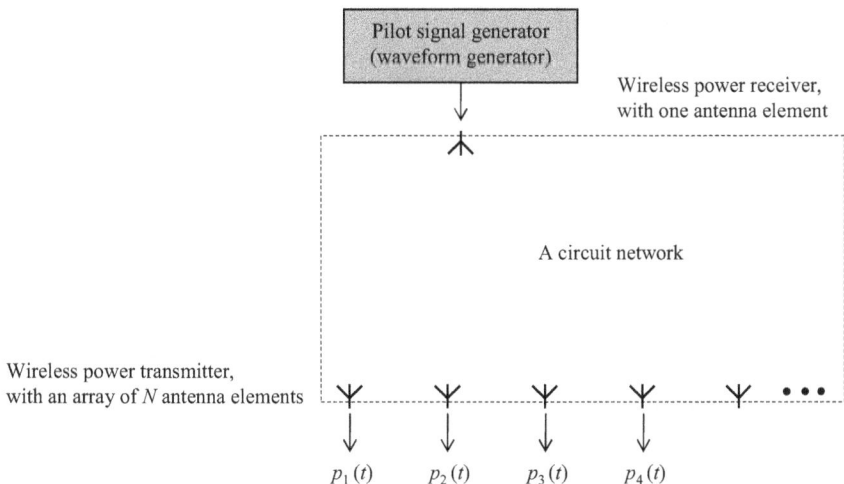

Figure 3.28   Pilot signal propagation in the time domain

each time domain pilot signal $p_n(t)$ can be characterized by its spectrum $P_n(\omega)$ as

$$p_n(t) = \frac{1}{2\pi} \int_{-\infty}^{+\infty} P_n(\omega) e^{j\omega t} d\omega, \tag{3.19}$$

where $\omega = 2\pi f$ is the angular frequency. Denote $P_n(\omega) = A_n(\omega) e^{j\alpha_n(\omega)}$. In the second step of retro-reflective beamforming, suppose the wireless power excitation is prepared "one frequency by one frequency." Specifically, suppose the wireless power excitation to the n-th antenna element of wireless power transmitter has amplitude $X_n(\omega)$ and phase $\psi_n(\omega)$ at angular frequency ω. Following (3.18), the wireless power excitation at ω is prepared as $X_n(\omega) = CA_n(\omega)$ and $\psi_n(\omega) = -\alpha_n(\omega)$. Then, as shown in Figure 3.29, the time domain wireless power excitation to the n-th antenna element is

$$
\begin{aligned}
\frac{1}{2\pi} \int_{-\infty}^{+\infty} X_n(\omega) e^{j\psi_n(\omega)} e^{j\omega t} d\omega &= \frac{1}{2\pi} \int_{-\infty}^{+\infty} CA_n(\omega) e^{j\alpha_n(\omega)} e^{j\omega t} d\omega \\
&= C \frac{1}{2\pi} \int_{-\infty}^{+\infty} [P_n(\omega)]^* e^{j\omega t} d\omega \\
&= C p_n(-t)
\end{aligned}
\tag{3.20}
$$

Comparison between Figure 3.28 and Figure 3.29 indicates that the wireless power excitation to the n-th antenna element in Figure 3.29 is the time-reversed version of the pilot signal detected by the n-th antenna element in Figure 3.28. Thus, the retro-reflective beamforming technique is essentially equivalent to the time-reversal technique. As explained in Section 1.4, the time-reversal technique is like the "rewind" button of a movie player. If the propagation of the pilot signal in the first step of retro-reflective beamforming is considered a movie, the propagation of wireless power in the second step of retro-reflective beamforming appears like the rewound movie. Specifically, the pilot signal propagation diverges in the space whereas the wireless power propagation converges onto the wireless power receiver, from which the pilot signal is broadcasted.

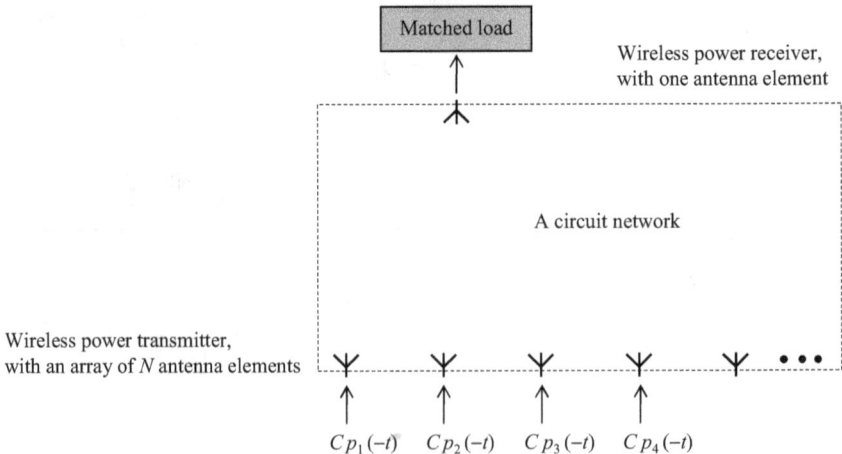

Figure 3.29   Wireless power propagation in the time domain

Equation (3.20) reveals the possibility for the retro-reflective beamforming technique to accommodate diverse waveforms, whereas the retro-reflective beamforming technique is described with the assumption of time-harmonic waveform in Chapter 2 and this chapter. Albeit various waveforms have been pursued by numerous researchers for wireless power transmission applications in recent years [16–19], time-harmonic waveform remains the most investigated waveform in the research (particularly experimental research) of wireless power transmission. Moreover, employing a frequency band rather than a single frequency point for wireless power transmission is likely to lead to more regulatory issues (which are discussed in Section 1.1). Therefore, the narrative of the retro-reflective beamforming technique in this book focuses on the time-harmonic waveform.

3.5　Retro-reflective beamforming in "geometrical near zone" and "geometrical far zone"

Two terms, namely the "full conjugation antenna array technique" and the "phase conjugation antenna array technique," are defined in the first few sections of this chapter. The wireless power excitations of the "full conjugation antenna array technique" follow (3.18) in the second step of retro-reflective beamforming. In the "phase conjugation antenna array technique," the selection of wireless power excitations only follows the second line of (3.18) and the wireless power excitations have a uniform amplitude among the antenna elements of the wireless power transmitter. These two terms are closely related to "geometrical near zone" and "geometrical far zone." When a pilot signal is broadcasted by a wireless power receiver in a wireless power transmitter's geometrical far zone, the pilot signals detected by the antenna elements of the wireless power transmitter share the same amplitude. Then in the second step of retro-reflective beamforming, the wireless power excitations to the antenna elements of the wireless power transmitter would in turn have a uniform amplitude. When the wireless power receiver resides in the wireless power transmitter's geometrical near zone, the pilot signals detected by the antenna elements of the wireless power transmitter do not exhibit the same amplitude. According to the proof in Section 3.4, the amplitude profile of wireless power excitations should follow the amplitude profile of detected pilot signals. The "phase conjugation antenna array technique," in which the wireless power excitations always have a uniform amplitude regardless of the detected pilot signals' amplitude profile, is well known to be sub-optimal in terms of beamforming [20]. Nevertheless, the numerical studies of this section demonstrate that the difference in power transmission efficiency between the "full conjugation antenna array technique" and the "phase conjugation antenna array technique" is not substantial as long as the distance between the wireless power transmitter and the wireless power receiver is not very small with respect to the physical size of the wireless power transmitter. In other words, the "phase conjugation antenna array technique" can be used in place of the "full conjugation antenna array technique" without much decline in power transmission efficiency in practice.

The setup of numerical studies in this section is based on the numerical example of Section 2.6. As depicted in Figure 3.30, a wireless power transmitter

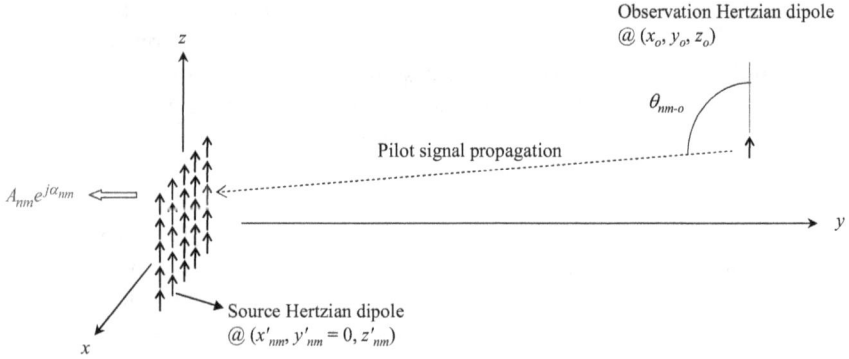

Figure 3.30   The first step of retro-reflective beamforming, pilot signal propagation, in the numerical example of Section 3.5

includes a two-dimensional array of z-orientated Hertzian dipoles. There are $N \times M$ source Hertzian dipoles with spacing s_x along the x direction and spacing s_z along the z direction. The physical dimensions of the array are approximately $L = N \times s_x$ and $W = M \times s_z$. The Hertzian dipoles in the array are located at $\mathbf{r}'_{nm} = (x'_{nm}, y' = 0, z'_{nm})$, $n = 1, 2, \ldots, N$, $m = 1, 2, \ldots, M$. A wireless power receiver has one observation Hertzian dipole orientated along the z direction with spherical coordinates of (r_o, θ_o, ϕ_o). The physical length of every Hertzian dipole in Figure 3.30 is l_0. Many formulations of Section 2.6 are made use of in this section.

   In the first step of retro-reflective beamforming, a pilot signal is broadcasted by the observation Hertzian dipole and detected by all the source Hertzian dipoles. Suppose the excitation current over the observation Hertzian dipole has an amplitude of I_0 and a phase of zero. The electric field detected by the source Hertzian dipole located at $\mathbf{r}'_{nm}$ is

$$\mathbf{E}(\mathbf{r}'_{nm}) = \widehat{\boldsymbol{\theta}}_{nm\text{-}o} \frac{j\eta_0 k_0 I_0 l_0}{4\pi} \frac{\sin(\theta_{nm\text{-}o}) e^{-jk_0 r_{nm\text{-}o}}}{r_{nm\text{-}o}}. \tag{3.21}$$

The z component of the electric field in (3.21) is considered as the pilot signal detected by the source Hertzian dipole with indices nm. When the pilot signal detected by the dipole with indices nm is denoted as $A_{nm} e^{j\alpha_{nm}}$, it is straightforward to obtain the following three equations.

$$A_{nm} e^{j\alpha_{nm}} = \widehat{\mathbf{z}} \cdot \mathbf{E}(\mathbf{r}'_{nm}) = \frac{-j\eta_0 k_0 I_0 l_0}{4\pi} \frac{\sin^2(\theta_{nm\text{-}o}) e^{-jk_0 r_{nm\text{-}o}}}{r_{nm\text{-}o}} \tag{3.22}$$

$$A_{nm} = \frac{\eta_0 k_0 I_0 l_0}{4\pi} \frac{\sin^2(\theta_{nm\text{-}o})}{r_{nm\text{-}o}} \tag{3.23}$$

$$\alpha_{nm} = -k_0 r_{nm\text{-}o} - \frac{\pi}{2} \tag{3.24}$$

In the second step of retro-reflective beamforming, wireless power is transmitted from the source array to the observation dipole, as shown in Figure 3.31. The excitation current over a source Hertzian dipole, $X_{nm}e^{j\psi_{nm}}$, is selected by (3.18) as

$$X_{nm} = CA_{nm} = C\frac{\eta_0 k_0 I_0 l_0}{4\pi}\frac{\sin^2(\theta_{nm-o})}{r_{nm-o}}, \tag{3.25}$$

$$\psi_{nm} = -\alpha_{nm} = k_0 r_{nm-o} + \frac{\pi}{2}. \tag{3.26}$$

The phase values of wireless power excitations in (3.26) are identical to those in Section 2.6 except for a constant phase of 90° (a constant phase in ψ_{nm} has no impact on the performance of retro-reflective beamforming). The amplitude values of wireless power excitations in (3.25) are determined by the amplitude values of detected pilot signals, whereas in Section 2.6 the wireless power excitations have a uniform amplitude. Intuitively, the nonuniform amplitude profile in (3.25) offers better power transmission efficiency than the uniform amplitude profile. When the observation dipole is located along the y-axis, the A_{nm} value detected by a source Hertzian dipole at the center of the array is relatively large, and the A_{nm} value of a source Hertzian dipole at a corner of the array is relatively small. The relatively larger A_{nm} value indicates that the wireless channel between the center source dipole and the observation dipole is "more efficient," and the relatively smaller A_{nm} value indicates that the wireless channel between the corner source dipole and the observation dipole is "less efficient." In the second step of retro-reflective beamforming, it would be more optimal if a relatively larger power is assigned to the center source dipole and a relatively smaller power is assigned to the corner source dipole. In contrast, the uniform amplitude profile is not expected to achieve the optimal efficiency as it does not take advantage of the amplitude information carried by the pilot signals.

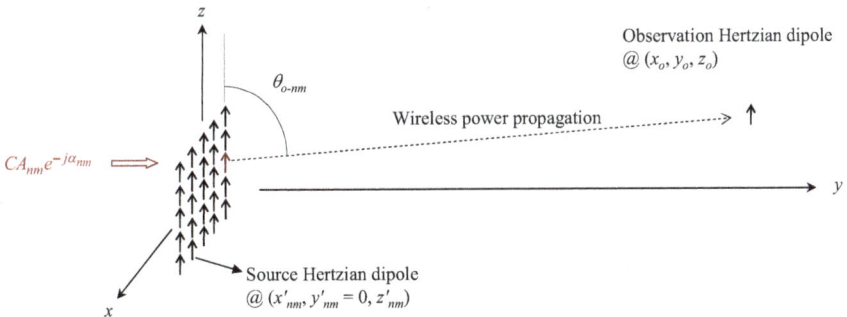

Figure 3.31 *The second step of retro-reflective beamforming, wireless power propagation, in the numerical example of Section 3.5*

In the second step of retro-reflective beamforming, the electric field at the observation Hertzian dipole is

$$
\mathbf{E}(r_o, \theta_o, \phi_o) = \frac{j\eta_0 k_0 l_0}{4\pi} \sum_{n=1}^{N} \sum_{m=1}^{M} \hat{\theta}_{o-nm} \frac{X_{nm} e^{j\psi_{nm}} \sin(\theta_{o-nm}) e^{-jk_0 r_{o-nm}}}{r_{o-nm}}
$$

$$
= -CI_0 \left(\frac{\eta_0 k_0 l_0}{4\pi}\right)^2 \sum_{n=1}^{N} \sum_{m=1}^{M} \hat{\theta}_{o-nm} \frac{\sin^2(\theta_{nm-o}) \sin(\theta_{o-nm})}{r_{nm-o}} \frac{e^{jk_0 r_{nm-o}} e^{-jk_0 r_{o-nm}}}{r_{o-nm}}
$$

$$(3.27)$$

The *z* component of the electric field in (3.27) is

$$
E_z(r_o, \theta_o, \phi_o) = -CI_0 \left(\frac{\eta_0 k_0 l_0}{4\pi}\right)^2 \sum_{n=1}^{N} \sum_{m=1}^{M} \hat{z} \cdot \hat{\theta}_{o-nm} \frac{\sin^2(\theta_{nm-o}) \sin(\theta_{o-nm})}{r_{nm-o}} \frac{e^{jk_0 r_{nm-o}} e^{-jk_0 r_{o-nm}}}{r_{o-nm}}
$$

$$
= CI_0 \left(\frac{\eta_0 k_0 l_0}{4\pi}\right)^2 \sum_{n=1}^{N} \sum_{m=1}^{M} \frac{\sin^2(\theta_{nm-o})}{r_{nm-o}} \frac{\sin^2(\theta_{o-nm})}{r_{o-nm}} e^{jk_0 r_{nm-o}} e^{-jk_0 r_{o-nm}}
$$

$$(3.28)$$

Because $r_{nm-o} = r_{o-nm}$, $\theta_{nm-o} = \pi - \theta_{o-nm}$, and $\sin(\theta_{nm-o}) = \sin(\theta_{o-nm})$, Equation (3.28) becomes

$$
E_z(r_o, \theta_o, \phi_o) = CI_0 \left(\frac{\eta_0 k_0 l_0}{4\pi}\right)^2 \sum_{n=1}^{N} \sum_{m=1}^{M} \frac{\sin^4(\theta_{o-nm})}{(r_{o-nm})^2}.
$$

$$(3.29)$$

In the second step of retro-reflective beamforming, the power received by the observation Hertzian dipole is proportional to $|E_z(r_o, \theta_o, \phi_o)|^2$. Also in the second step of retro-reflective beamforming, the power transmitted by the source array is proportional to $\sum_{n=1}^{N} \sum_{m=1}^{M} (X_{nm})^2$. Following the formulations of Section 2.6, it is not difficult to find the power constructive coefficient

$$
\frac{(\text{PTE})}{NM(\text{PTE})_0} = \frac{\displaystyle\sum_{n=1}^{N} \sum_{m=1}^{M} \frac{\sin^4(\theta_{o-nm})}{(r_{o-nm})^2}}{NM \dfrac{\sin^4(\theta_o)}{(r_o)^2}}
$$

$$(3.30)$$

$$
= \frac{\displaystyle\sum_{n=1}^{N} \sum_{m=1}^{M} \frac{\left[(x_o - x'_{nm})^2 + (y_o)^2\right]^2}{\left[(x_o - x'_{nm})^2 + (y_o)^2 + (z_o - z'_{nm})^2\right]^3}}{NM \dfrac{\sin^4(\theta_o)}{(r_o)^2}}
$$

As in Section 2.6, the two-fold summation in (3.30) can be approximated by a two-fold integration:

$$\sum_{n=1}^{N}\sum_{m=1}^{M}\frac{\left[(x_o - x'_{nm})^2 + (y_o)^2\right]^2}{\left[(x_o - x'_{nm})^2 + (y_o)^2 + (z_o - z'_{nm})^2\right]^3}$$

$$\approx \frac{1}{S_x S_z}\int_{-L/2}^{+L/2}\int_{-W/2}^{+W/2} dx'dz' \frac{\left[(x_o - x')^2 + (y_o)^2\right]^2}{\left[(x_o - x')^2 + (y_o)^2 + (z_o - z')^2\right]^3}$$

(3.31)

With the two-fold summation replaced by a two-fold integration, the power constructive coefficient in (3.30) can be evaluated as

$$\frac{(PTE)}{NM(PTE)_0} = \frac{\frac{1}{LW}\int_{-L/2}^{+L/2}\int_{-W/2}^{+W/2}\frac{\left[(x_o-x')^2+(y_o)^2\right]^2}{\left[(x_o-x')^2+(y_o)^2+(z_o-z')^2\right]^3}dx'dz'}{\frac{\sin^4(\theta_o)}{(r_o)^2}}$$

(3.32)

The two-fold integral in (3.32) can be evaluated with high precision, as detailed in Appendix B.

Some numerical results similar to those in Section 2.6 are presented in Figure 3.32. The value of y_o is 20 m, 10 m, 5 m, and 3 m in Figure 3.32, respectively. In Figure 3.32, the power constructive coefficient is plotted with respect to x_o and z_o. The values of L and W are fixed to be 3 m in Figure 3.32.

When $y_o = 20$ m, the observation dipole is in the geometrical far zone of the source array. As a result, the "full conjugation antenna array technique" and the "phase conjugation antenna array technique" are equivalent to each other, as discussed at the beginning of this section. Therefore unsurprisingly, Figure 3.32(a) is almost the same as Figure 2.22(a). The other three plots in Figure 3.32 appear to have little difference from the counterpart plots in Section 2.6. This means that the "full conjugation antenna array technique" is not superior to the "phase conjugation antenna array technique" when y_o is 10 m or 5 m or 3 m.

Some additional numerical results of the power constructive coefficient are presented in Table 3.2. The numerical results in Table 3.2 are obtained with $x_o = z_o = 0$ and $L = W = 3$ m. When y_o takes various values, the power constructive coefficients in the rightmost column of Table 3.2 are evaluated using (3.32) (i.e., the full conjugation scheme), whereas the power constructive coefficients in the second column of Table 3.2 are evaluated via (2.143) (that is, the phase conjugation scheme). The two schemes could barely be differentiated from each other in Table 3.2 when y_o is greater than 3 m. As argued in Section 2.6, "$y_o < 3$ m" embodies a short distance with respect to the physical dimension of the wireless power transmitter, and hence is not of significant concern to this book. As a matter of fact, when y_o is as small as 2 m, the full conjugation scheme does not yield a substantially higher power transmission efficiency than the phase conjugation scheme in Table 3.2. Therefore, it does not seem highly beneficial to pursue the full

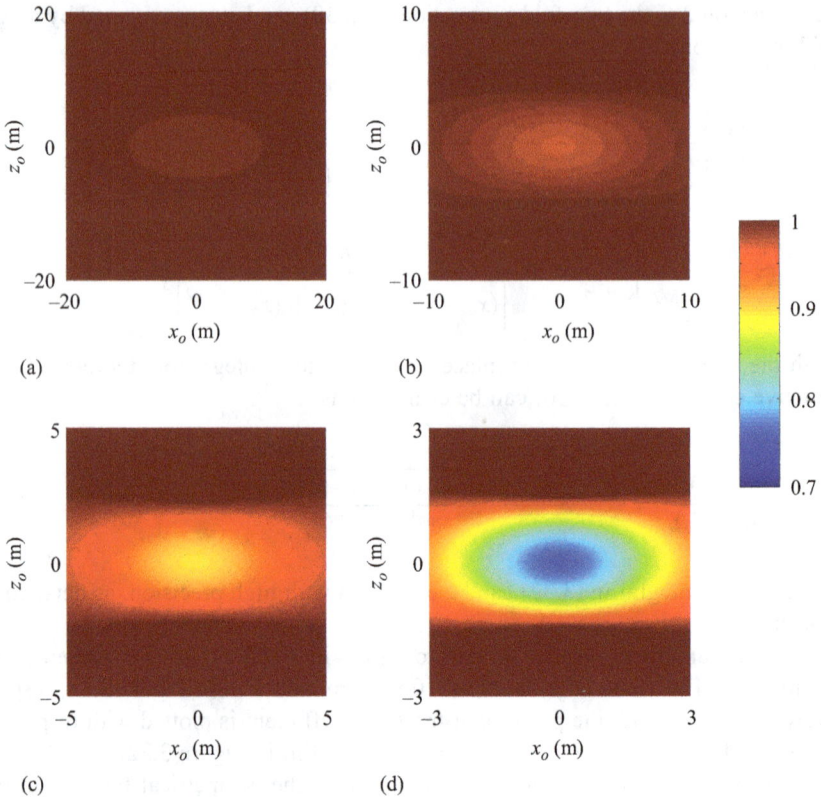

Figure 3.32   Numerical results of power constructive coefficient when
(a) $y_o = 20$ m, (b) $y_o = 10$ m, (c) $y_o = 5$ m, and (d) $y_o = 3$ m, with
$L = W = 3$ m

Table 3.2   Numerical results of power constructive coefficient with various values
of y_o, with $x_o = z_o = 0$ and $L = W = 3$ m

y_o (in meters)	Power constructive coefficient of phase conjugation scheme in Section 2.6	Power constructive coefficient of full conjugation scheme in this section
10	0.971	0.971
9	0.964	0.964
8	0.955	0.956
7	0.942	0.943
6	0.923	0.924
5	0.894	0.895
4	0.843	0.846
3	0.752	0.759
2	0.575	0.594

conjugation scheme in practice, although it offers the optimal power transmission efficiency theoretically. The phase conjugation scheme, in which the wireless power excitations to antenna elements have a uniform amplitude, is sub-optimal in terms of beamforming [20]. However, its practical implementation is relatively simple and low-cost as it does not call for any hardware or software for detecting the pilot signals' amplitude or adjusting wireless power excitations' amplitude. Thus, all the experimental examples in this book follow the phase conjugation scheme rather than the full conjugation scheme.

3.6   Practical implementation of retro-reflective beamforming technique for microwave power transmission

Extensive research efforts on the practical implementation of the retro-reflective beamforming technique have been reported by numerous researchers. The hardware architectures of retro-reflective beamforming for microwave power transmission applications that have been investigated in the literature are reviewed in this section. The practical implementation of retro-reflective beamforming is discussed in this section using hardware block diagrams without detailed schematics. Some schematics of system design, antenna design, and circuit design are presented in Chapters 4, 5, and 6.

According to the fundamental two-step procedure of retro-reflective beamforming technique, the practical implementation of retro-reflective beamforming for microwave power transmission must follow the generic hardware block diagram in Figure 3.33. In Figure 3.33, a wireless power receiver is composed of four components: Circuits for pilot signal transmission, a pilot signal transmitting antenna, a wireless power receiving antenna or an array of wireless power receiving antennas, and circuits for wireless power reception. As depicted in Figure 3.33, a wireless power transmitter also includes four components: An array of pilot signal receiving antenna elements, circuits for pilot signal reception, circuits for wireless power transmission, and an array of wireless power transmitting antenna elements. In the first step of retro-reflective beamforming, a pilot signal is broadcasted by the wireless power receiver and detected by the wireless power transmitter. Based on the outcomes of analyzing the pilot signal, wireless power is delivered from the wireless power transmitter to the wireless power receiver in the second step of retro-reflective beamforming through a microwave power beam. Several specific schemes of Figure 3.33 are discussed in the rest of this section.

The retro-reflective scheme in Figure 3.34 is based on detecting the DOA (direction of arrival) of pilot signal propagation. When the wireless power receiver is in the electrical far zone, the pilot signal broadcasted by the wireless power receiver behaves as a plane wave with an explicit incoming direction when it reaches the wireless power transmitter. The DOA of the pilot signal is detected by the pilot signal receiving antenna elements along with the circuits attached to them. Based on the DOA information, the wireless power transmitting antenna

Wireless power receiver

| Circuits for pilot signal transmission | Circuits for wireless power reception |

Pilot signal

Microwave power beam

Pilot signal receiving antenna elements

Wireless power transmitting antenna elements

Electronic control

| Circuits for pilot signal reception | Circuits for wireless power transmission |

Wireless power transmitter

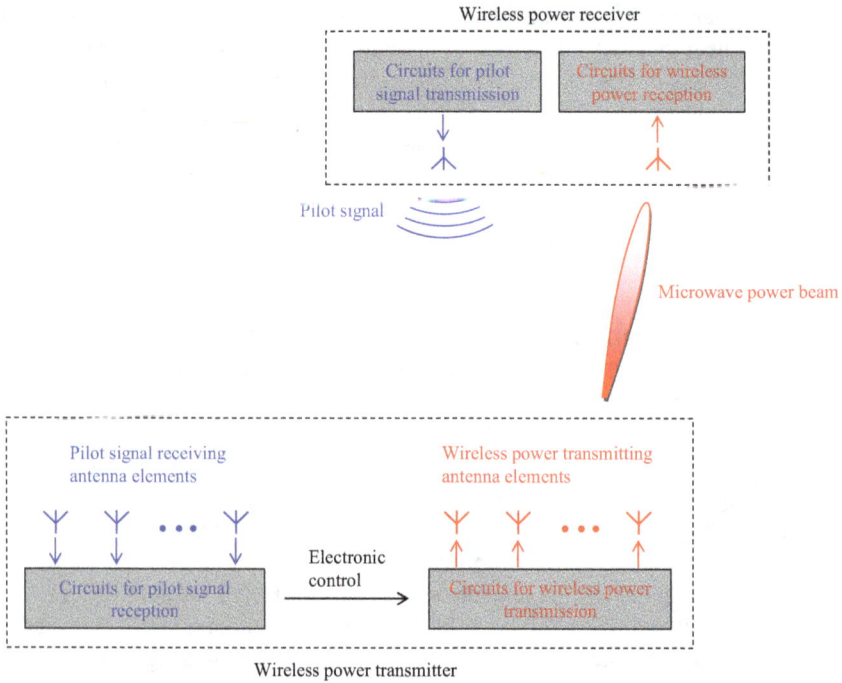

Figure 3.33   Generic hardware block diagram of the retro-reflective beamforming technique for microwave power transmission

elements along with the circuits attached to them generate a microwave power beam toward the direction opposite to the incoming direction of the pilot signal. Many techniques are available for detecting the DOA of the pilot signal and generating the microwave power beam, including but not limited to those described in Sections 2.1 and 2.2. For instance, the monopulse method is utilized for DOA detection in a microwave power transmission experiment reported by a group of Japanese researchers in 2016 [21]. In fact, pilot signal detection and wireless power transmission can be carried out independently in Figure 3.34. As a result, the two arrays (one for pilot signal reception and the other for wireless power transmission) in the wireless power transmitter do not have to have the same number of antenna elements or the same type of antenna elements. As argued in Chapter 2, it is very probable that a wireless power receiver has to reside in the electrical near zone in order to achieve a reasonably high power transmission efficiency in practice. Since the scheme in Figure 3.34 relies on the condition of electrical far zone, its practical application in microwave power transmission is limited. For example, in [21] the DOA detection is not conducted for the entire aperture of the wireless power transmitter; rather, the wireless power transmitter's aperture is divided into four portions and the DOA detection is applied to each portion individually.

Wireless power receiver

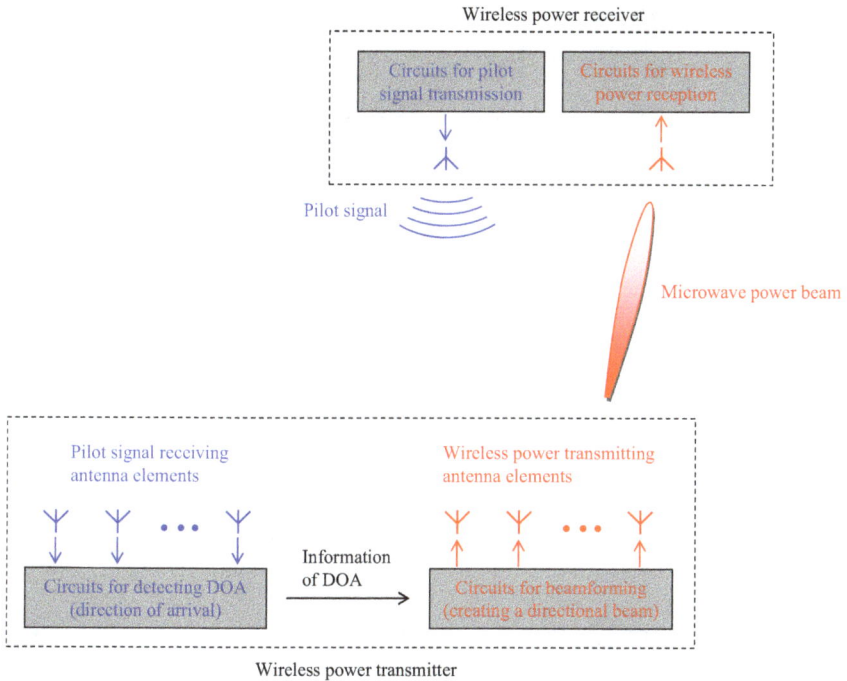

Figure 3.34   A practical implementation scheme of retro-reflective beamforming based on the detection of DOA (direction of arrival)

The retro-reflective beamforming scheme in Figure 3.35 was invented by Van Atta in the 1950s [22]. In Figure 3.35, a one-dimensional 4-element pilot signal receiving array and a one-dimensional 4-element wireless power transmitting array are used for the purpose of illustrating the Van Atta scheme. Both arrays are deployed along the *x*-axis. Suppose a time-harmonic pilot signal is broadcasted by the wireless power receiver. The pilot signals received by the four pilot signal receiving antenna elements are supplied to the four wireless power transmitting antenna elements, respectively. Power amplifiers must be placed over the four interconnection paths between the pilot signal receiving antenna array and the wireless power transmitting antenna array to boost the power level. It is noted that the pilot signal received by the leftmost pilot signal receiving antenna element is fed to the rightmost wireless power transmitting antenna element, and meanwhile, the pilot signal received by the rightmost pilot signal receiving antenna element is fed to the leftmost wireless power transmitting antenna element. When the wireless power receiver is in the electrical far zone, the pilot signals detected by the four pilot signal receiving antenna elements exhibit a linear pattern along the *x*-axis, according to the theory in Section 2.2. In Figure 3.35, the phase values of the detected pilot signals are assumed to be -3Δ, -2Δ, $-\Delta$, and 0 from left to right, respectively.

Wireless power receiver

Circuits for pilot signal transmission

Circuits for wireless power reception

Pilot signal @ *f*

Microwave power beam @ *f*

Pilot signal receiving antenna elements

Wireless power transmitting antenna elements

x

Phase 0   Phase *κ*

−Δ   *κ* − Δ

−2Δ   *κ* − 2Δ

−3Δ   *κ* − 3Δ

Power amplifiers

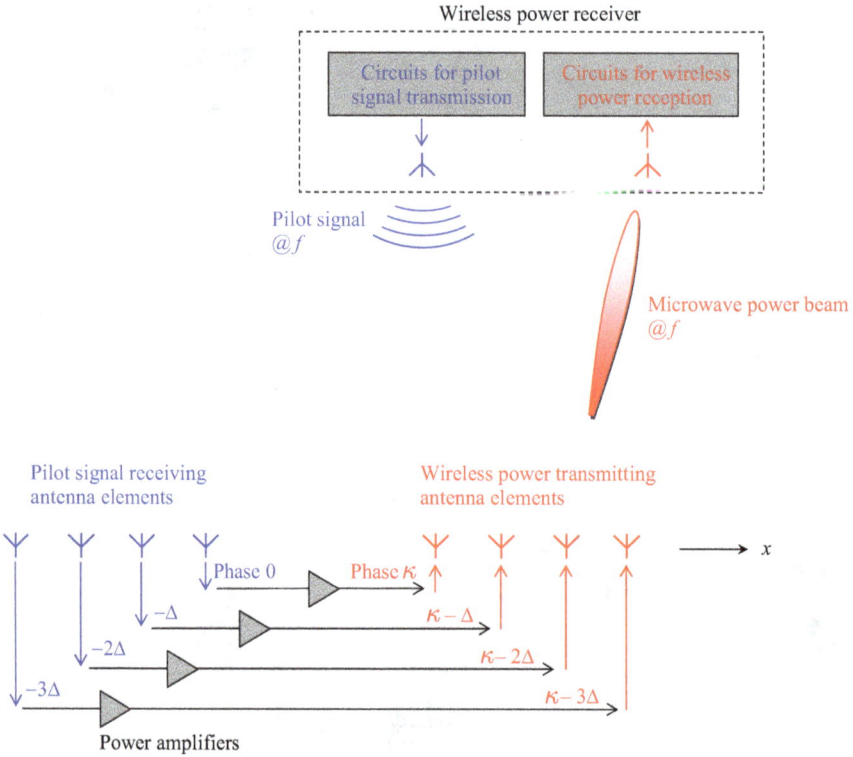

Figure 3.35   A practical implementation scheme of retro-reflective beamforming based on Van Atta array

The interconnections between the pilot signal receiving antenna array and the wireless power transmitting antenna array ensure that a reversed linear phase pattern along the *x*-axis is exhibited over the wireless power transmitting antenna array. Suppose the four interconnection paths share the same phase delay of *κ*. The phase values of the wireless power excitations are *κ*, *κ* − Δ, *κ* − 2Δ, and *κ* − 3Δ from left to right, respectively, which result in a microwave power beam toward the wireless power receiver as analyzed in Section 2.1.1. The Van Atta scheme can be extended to configurations other than the 4-element arrays in Figure 3.35 straightforwardly as long as the condition of electrical far zone is satisfied. The Van Atta method has widespread applications in wireless communication and radar [23]. For instance, when a radar target is equipped with a passive Van Atta array (that is, without any power amplifiers or other active components), the target's visibility would be enhanced significantly over the screen of a mono-static radar, which is the analogy of traffic signs made of retro-reflective surface discussed in Section 1.4. However, due to the requirement of electrical far zone condition, the practical application of Van Atta method in microwave power transmission is limited.

The retro-reflective beamforming scheme in Figure 3.36 was proposed by Pon in the 1960s [24]. In Figure 3.36, a time-harmonic pilot signal at frequency f is broadcasted by the wireless power receiver. Suppose the pilot signal detected by one pilot signal receiving antenna element of the wireless power transmitter (denoted as "PS1" in Figure 3.36) is $c_1 \cos(\omega t + \alpha)$, where $\omega = 2\pi f$ is the angular frequency. The wireless power transmitter includes a local oscillator that generates a time-harmonic signal at the frequency of $2f$. The detected pilot signal and the local oscillator's signal are supplied to a mixer. Suppose the local oscillator's signal supplied to the mixer is $c_2 \cos(2\omega t)$. Since the mixer behaves as a multiplier, the output of the mixer is

$$c_0 c_1 c_2 \cos(\omega t + \alpha)\cos(2\omega t) = \frac{c_0 c_1 c_2}{2} \cos(3\omega t + \alpha) + \frac{c_0 c_1 c_2}{2} \cos(\omega t - \alpha),$$

(3.33)

where c_0 is a constant inherent to the mixer. The time-harmonic component with angular frequency of 3ω in the mixer's output can be easily removed by filters. The component remaining on the right-hand side of (3.33) with the angular frequency of

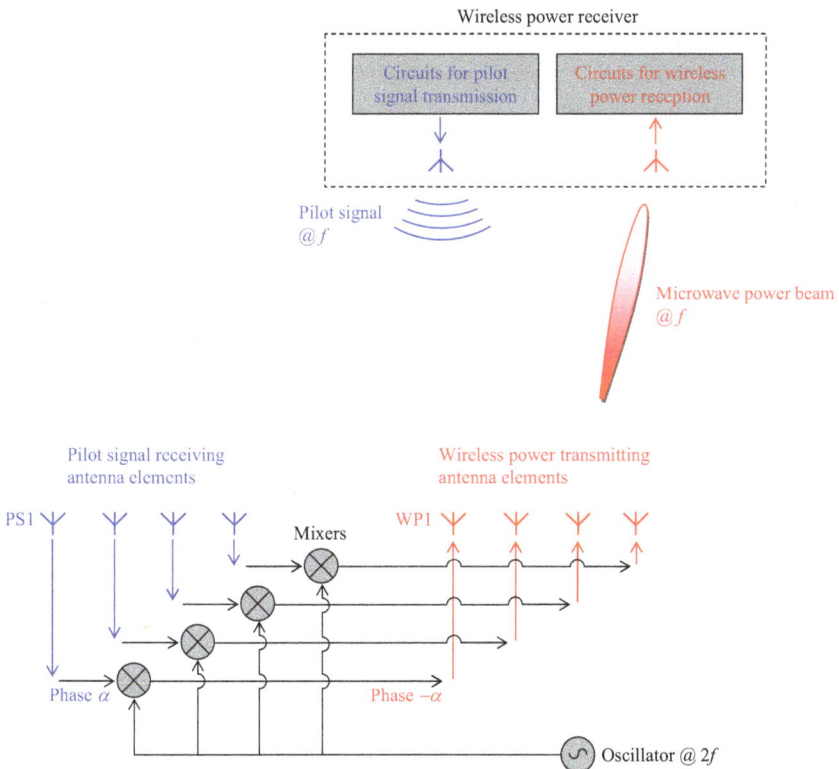

Figure 3.36   A practical implementation scheme of retro-reflective beamforming based on the heterodyne method

ω is then amplified and fed to a wireless power transmitting antenna element denoted as "WP1" in Figure 3.36. For the sake of succinctness, filters and amplifiers are not shown in Figure 3.36. The phase of the pilot signal detected by antenna element "PS1" is conjugate to the phase of wireless power excitation to antenna element "WP1." Similarly, the phase conjugation relationship is enforced for the other pairs in Figure 3.36, with each pair including one pilot signal receiving antenna element and one wireless power transmitting antenna element. The retro-reflective beamforming scheme in Figure 3.36 does not require the wireless power receiver to reside in the electrical far zone. As pointed out in Section 3.4, the phase conjugation relationship leads to retro-reflection under the condition of channel reciprocity. With "PS1" and "WP1" as examples, the reciprocity between the following two wireless channels is a necessary condition of retro-reflection.

(i)   Channel from wireless power receiver's pilot signal transmitting antenna element to "PS1"
(ii) Channel from "WP1" to wireless power receiver's wireless power receiving antenna element

Duplexing is the best means to guarantee the reciprocity between the two channels above. Specifically, the channel reciprocity will be satisfied rigorously when one physical antenna element plays dual roles of "PS1" and "WP1" at the wireless power transmitter and when one physical antenna element serves as both the pilot signal transmitting antenna and wireless power receiving antenna at the wireless power receiver. Of course, duplexing calls for additional hardware and software in order for one antenna element to be shared by pilot signal circuits and wireless power circuits. Various duplexing schemes are discussed at the end of this section. If retro-reflective beamforming is implemented perfectly, the wireless power transmission exhibits a focal point in space with its center located at the pilot signal transmitting antenna, from which the pilot signal originates. Thus, the duplexing at the wireless power receiver ensures that the wireless power receiving antenna coincides with the focal point's center. If the wireless power receiving antenna is off the focal point's center, the power transmission efficiency does not drop much as long as the wireless power receiving antenna is still covered by the focal point (in other words, as long as the focal point's size is able to accommodate the spatial offset between the pilot signal transmitting antenna and the wireless power receiving antenna). Thus, the implementation of duplexing at the wireless power receiver is not absolutely necessary in practice. The duplexing at wireless power transmitters is much more critical. If "PS1" and "WP1" have to be implemented by two respective antenna elements, their phase centers should coincide with each other for the success of retro-reflective beamforming. Obviously, it is because the wireless power transmitter is responsible for detecting the pilot signals' phase and adjusting the wireless power excitations' phase accordingly (in contrast, the wireless power receiver does not need to conduct any jobs pertinent to the phase). Under certain conditions (for instance, when the wireless power receiver's location is roughly known), it is possible to design two antenna elements (one for receiving pilot signal and one for transmitting wireless power) with the same phase center [25].

Neither the Van Atta scheme in Figure 3.35 nor the heterodyne scheme in Figure 3.36 involves any digital circuits. This is because they were proposed more than 50 years ago when the development of digital circuits was not sophisticated. As a result, the circuits for pilot signal reception are connected to the circuits for wireless power transmission directly in Figures 3.35 and 3.36. Indeed, the powerful and low-cost digital circuit modules readily available today could improve the system's functionality and flexibility tremendously when they serve as the interface between the circuits for pilot signal reception and the circuits for wireless power transmission, although avoiding digital circuits could reduce the system's size, weight, cost, and power consumption [26]. As an example, the scheme illustrated in Figure 3.37, with digital signal processing modules included, is highly versatile [27]. At the wireless power transmitter, the phase detectors attached to the pilot signal receiving antenna array obtain the phase values of the pilot signals. The pilot signals' phase values serve as the input of the digital signal processing module in Figure 3.37, and the output of the digital circuits constitutes control signals for the phase shifters. If the algorithm in the digital signal processing module identifies the

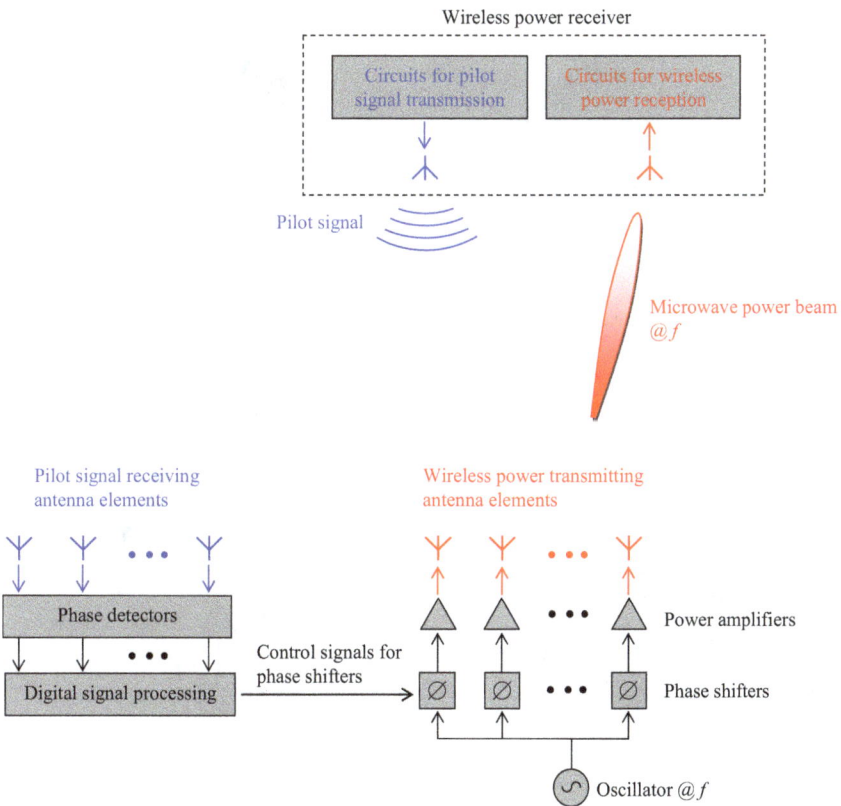

Figure 3.37   A practical implementation scheme of retro-reflective beamforming based on phase detection and phase shifting

DOA (direction of arrival) and adjusts the phase shifters according to the DOA information, the scheme in Figure 3.37 becomes equivalent to the scheme in Figure 3.34. The functionalities of Figures 3.35 and 3.36 can both be implemented by Figure 3.37 straightforwardly through loading appropriate algorithms in the digital signal processing module. In Figure 3.37, the wireless power transmission is time-harmonic but the pilot signal does not have to be time-harmonic, which enables a range of research possibilities [27,28].

The microwave power transmission scheme in Figure 3.38 was invented by Ossia Inc. in recent years [29]. Although Figures 3.38 and 3.37 appear similar to each other, there are some fundamental differences between them. In Figure 3.38, P_r, the value of power received by the wireless power receiver is communicated to the wireless power transmitter, and the wireless power transmitter adjusts the phase shifters with the goal of maximizing P_r. Fundamentally, the scheme in Figure 3.38 is a closed-loop feedback control system. In contrast, the scheme in Figure 3.37 can be considered as an open-loop control system. In Figure 3.37, the wireless power transmission is controlled by the pilot signal reception at the wireless power transmitter, and the pilot signal transmission does

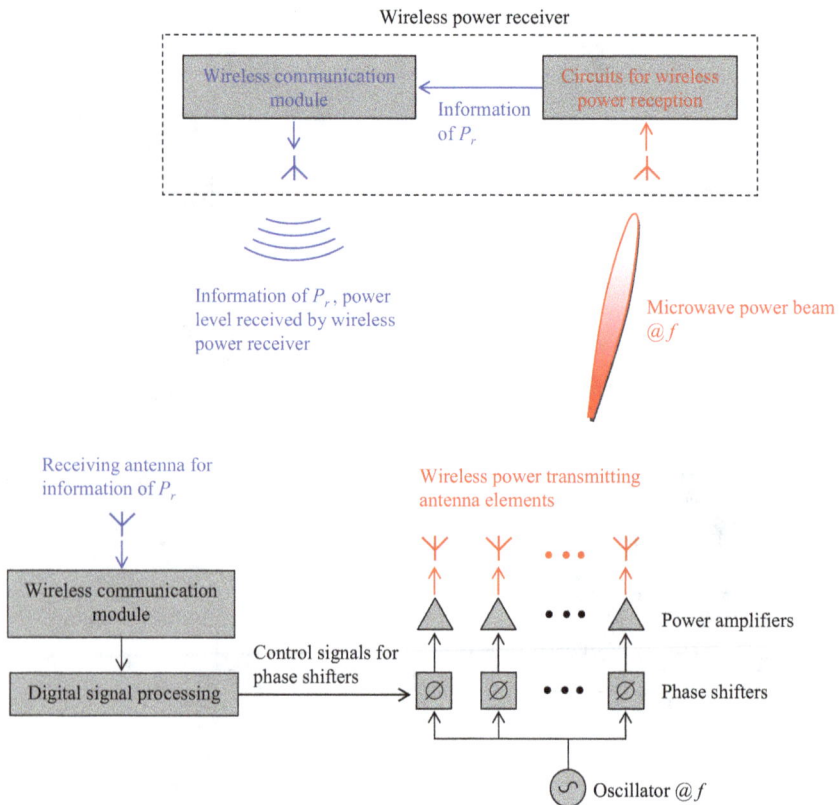

Figure 3.38   A scheme of microwave power transmission based on closed-loop feedback control

not depend on the wireless power reception at the wireless power receiver. In other words, Figure 3.37 does not involve any closed control loop. Compared with the open-loop control system in Figure 3.37, the closed-loop feedback control system in Figure 3.38 may reach better power transmission efficiency but requires sophisticated control algorithms. As Figure 3.38 involves multiple control variables (which correspond to multiple phase shifters), it might be time-consuming for a control algorithm (such as the rotating-element electric-field vector method [21]) to achieve the global maximal power transmission efficiency in practice.

The scheme in Figure 3.37 evolves to Figure 3.39 when digital circuits are utilized to generate and reconfigure the waveforms of wireless power transmission. The waveform of wireless power transmission in Figure 3.39 is not limited to time-harmonic. As a result, the time-reversal scheme discussed in Section 3.4 could be accomplished by the hardware block diagram of Figure 3.39. The digital retro-reflective beamforming scheme in Figure 3.39 resembles the digital beamforming architectures for wireless communication [30], in that they both take full advantage of digital circuits to achieve the utmost flexibility.

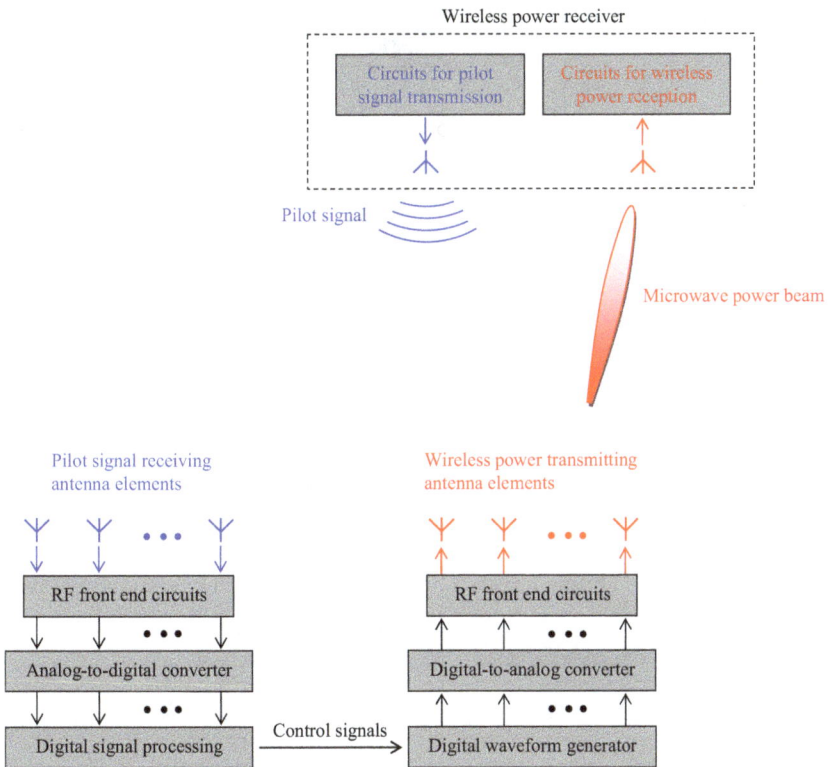

Figure 3.39   A practical implementation scheme of retro-reflective beamforming based on digital signal processing and digital waveform generation

In the generic block diagram of Figure 3.33, the wireless power transmitter includes two antenna arrays, one for pilot signal reception and one for wireless power transmission. When the wireless power receiver is in the electrical near zone, the phase conjugation relationship ought to be enforced between pilot signal reception and wireless power transmission. Since the validity of the phase conjugation scheme depends on channel reciprocity, it is highly desirable to use one physical antenna array to play dual roles of pilot signal receiving antenna array and wireless power transmitting antenna array, as discussed in Figure 3.36. Time-division duplexing is the most straightforward resolution for one antenna array to be shared by pilot signal circuits and wireless power circuits. As depicted in Figure 3.40, all the antenna elements are attached to a single-pole-double-throw switch. The switches are synchronized with one another in time. At each moment, the switches stay at either the *pilot signal state* or the *wireless power state*. When the switches are in the pilot signal state, the pilot signal is transmitted from the wireless power receiver to the wireless power transmitter while all the circuits

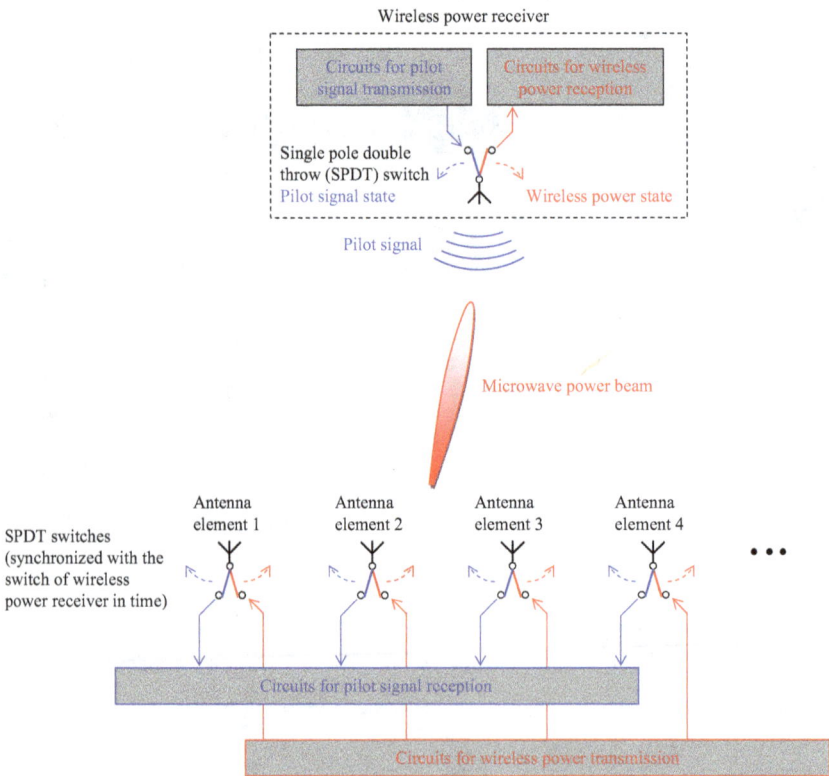

Figure 3.40   A practical implementation scheme of retro-reflective beamforming based on time-division duplexing between pilot signal and wireless power

relevant to wireless power transmission are inactive. Similarly, when the switches are in the wireless power state, wireless power is transmitted from the wireless power transmitter to the wireless power receiver while the circuits relevant to pilot signal transmission are inactive. In practice, the system starts with the pilot signal state. Once the wireless power transmitter finishes analyzing the pilot signals, all the switches are switched to the wireless power state. If the wireless channels remain unchanged, it is unnecessary to switch the system back to the pilot signal state. However, if wireless channels vary in time (when the wireless power receiver is not stationary, for instance), the system must be toggled between the pilot signal state and wireless power state periodically. In order to synchronize the switches in time rigorously, wireless communication must be maintained between the wireless power transmitter and the wireless power receiver. Thus, both the wireless power transmitter and wireless power receiver must be equipped with wireless communication modules. Except for facilitating the synchronization among switches, the wireless communication modules do not influence any parts of the retro-reflective beamforming scheme. Therefore, the wireless communication modules can be implemented independently and are not shown in Figure 3.40. In Figure 3.40, neither the pilot signal nor wireless power has to be time-harmonic in time. In fact, time-division duplexing offers great flexibility in the waveform design of both pilot signal and wireless power [28]. However, two distinctive drawbacks of time-division duplexing make it unsuitable for certain wireless power transmission applications. First, the wireless power supply to the wireless power receiver is not continuous in time because wireless power transmission is paused in the time slots allocated for pilot signal transmission. Second, wireless communication between the wireless power transmitter and wireless power receiver may consume considerable overhead resources especially when multiple wireless power receivers request wireless power simultaneously (which is a topic discussed in Chapter 4).

When time-division is found unsuitable for duplexing, circulators can be employed to accomplish duplexing between the pilot signal and wireless power as portrayed in Figure 3.41 [31]. In practical microwave power transmission applications, high-power wireless power propagation is guided by low-power pilot signal propagation. At the antenna of the wireless power receiver, the power level of pilot signal transmission and the power level of wireless power reception are not drastically different from each other. At the antennas of wireless power transmitter, nevertheless, the power level of wireless power transmission is higher than the power level of pilot signal reception by many orders. Power leakage from the wireless power transmission port to the pilot signal reception port in a circulator might jam the circuits of pilot signal reception if the pilot signal and wireless power share the same frequency f. Thus, customized design must be carried out for circulators to demonstrate adequate isolation between their wireless power transmission port and the pilot signal reception port [32].

As discussed at the end of the previous paragraph, it is possible for the wireless power transmission to jam the pilot signal reception circuits if the pilot signal and wireless power share the same frequency. When a frequency offset is applied between the pilot signal and wireless power, the wireless power transmission could

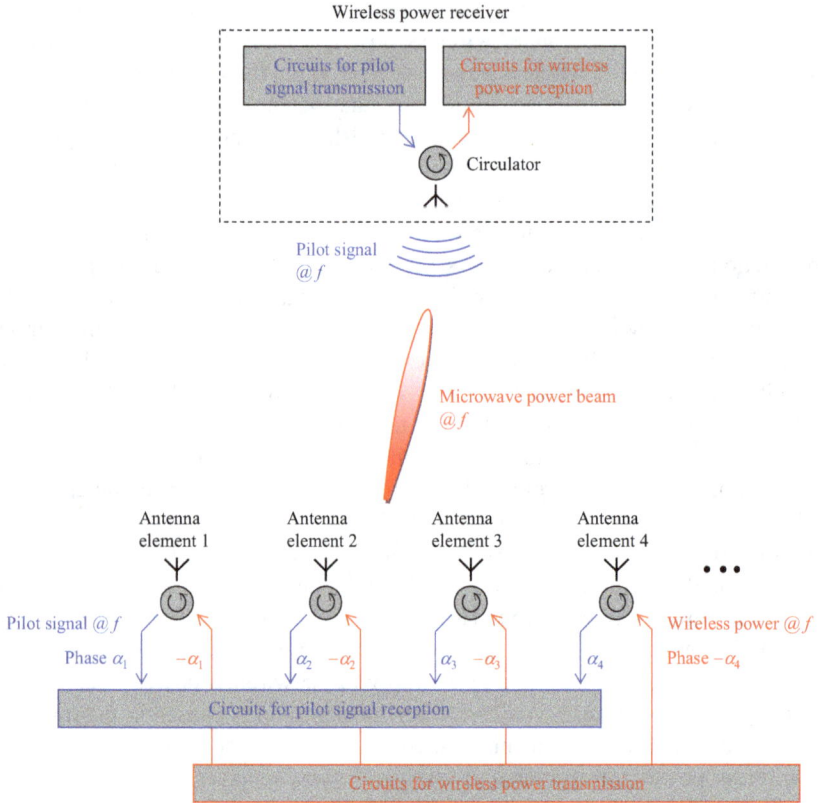

Figure 3.41   A practical implementation scheme of retro-reflective beamforming with circulators for duplexing between pilot signal and wireless power

be well isolated from the pilot signal reception by taking advantage of filters. In Figure 3.42, it is assumed that the pilot signal and wireless power are time-harmonic at frequencies f^{ps} and f^{wp}, respectively. Each antenna element in Figure 3.42 is attached to a diplexer. In each diplexer, there are two filters that pass the pilot signal and the wireless power, respectively. Of course, the filter that passes the pilot signal must sufficiently stop the wireless power, and the filter that passes the wireless power must sufficiently stop the pilot signal. With the frequency offset, the scheme in Figure 3.42 achieves frequency-division duplexing between the pilot signal and wireless power. The frequency offset creates complications in terms of channel reciprocity. At the n-th antenna element of wireless power transmitter, suppose the phase of pilot signal reception is α_n at frequency f^{ps} and the phase of wireless power excitation is ψ_n at frequency f^{wp}. Because the pilot signal propagation and wireless power propagation are at different frequencies, the phase conjugation relationship of "$\psi_n = -\alpha_n$" is no longer valid for the sake of

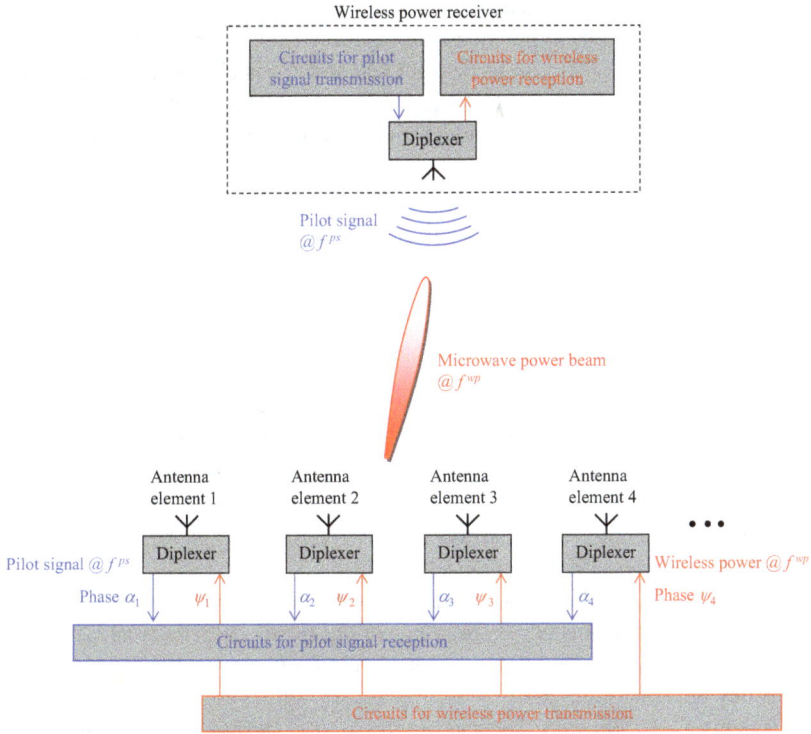

Figure 3.42   *A practical implementation scheme of retro-reflective beamforming with frequency-division duplexing between pilot signal and wireless power*

retro-reflection, although it is possible that the inaccuracy of $\psi_n = -\alpha_n$ is negligibly small when f^{ps} and f^{wp} are close to each other [33,34]. If the wireless channels are assumed to be not dispersive, the phase conjugation relationship between ψ_n and α_n with frequency offset can be adjusted to be

$$\psi_n = -\frac{f^{wp}}{f^{ps}}\alpha_n = -\chi\alpha_n. \tag{3.34}$$

Specifically, under the condition of "wireless channels being not dispersive," the phase of channel transfer functions is linearly proportional to the frequency. Then, if the phase associated with the pilot signal propagation channel is α_n at the frequency of f^{ps}, it would be $\chi\alpha_n = (f^{wp}/f^{ps}) \times \alpha_n$ at the frequency of f^{wp}. Though flawless in theory, the relationship in (3.34) may become erroneous in practice, because ψ_n and α_n are phase angles. In practice, a phase angle is usually obtained after a modulo operation with respect to $360°$ is conducted. For instance, a phase angle of $361°$ is always represented by $1°$ in practice. While $361°$ and $1°$ are equivalent to each other at f^{ps}, $(\chi \times 361°)$ and $(\chi \times 1°)$ would be equivalent to each other at f^{wp} only when χ is a positive integer. In other words, if $\chi = f^{wp}/f^{ps}$ is not an integer, applying (3.34) in practice may lead to errors. As an example of χ being a positive integer, f^{wp} is chosen to be 5.8 GHz and f^{ps} is chosen to be 2.9 GHz in [25]. Nevertheless, as $f^{ps} = 2.9$ GHz and $f^{wp} = 5.8$ GHz are far apart from each other, the assumption that "the wireless channels are not dispersive" is questionable. Moreover, the antenna elements in Figure 3.42 must be dual-band antennas or wide-band antennas in order to cover $f^{ps} = 2.9$ GHz and $f^{wp} = 5.8$ GHz. Actually, two sets of antenna elements are utilized in [25], one for the pilot signal at 2.9 GHz and the other for the wireless power at 5.8 GHz.

References

[1]   Wang X., Ruan B., Lu M. 'Retro-directive beamforming versus retro-reflective beamforming with applications in wireless power transmission'. *Progress In Electromagnetics Research*. 2016;157:79–91.

[2]   Chang Y., Fetterman H.R., Newberg I.L., Panaretos S.K. 'Microwave phase conjugation using antenna arrays'. *IEEE Transactions on Microwave Theory and Techniques*. 1998;46(11):1910–19.

[3]   Strassner B., Chang K. 'Microwave power transmission: historical milestones and system components'. *Proceedings of the IEEE*. 2013;101(6):1379–96.

[4]   Shinohara N. *Wireless Power Transfer via Radiowaves*. ISTE Ltd and John Wiley & Sons, Inc.; 2014.

[5]   Balanis C.A. *Antenna Theory: Analysis and Design*. 3rd ed. New York: Wiley-Interscience; 2005.

[6]   Mazurenko O., Yakornov Y. Focused arrays beamforming. In: Akdagli A (ed.). *Behaviour of Electromagnetic Waves in Different Media and Structures*. Rijeka: InTech; 2011. p. 419–40.

[7]   Kildal P.-S., Davis M.M. 'Characterization of near-field focusing with application to low altitude beam focusing of the Arecibo tri-reflector system'. *IEE Proceedings - Microwaves Antennas and Propagation*. 1996;143(4):284–92.

[8]   Reid D.R., Smith G.S. 'A comparison of the focusing properties of a Fresnel zone plate with a doubly-hyperbolic lens for application in a free-space, focused beam measurement system'. *IEEE Transactions on Antennas and Propagation*. 2009;57(2):499–507.

[9]   Karimkashi S., Kishk A.A. 'Focusing properties of Fresnel zone plate lens antennas in the near-field region'. *IEEE Transactions on Antennas and Propagation*. 2011;59(5):1481–7.

[10]   Gomez-Tornero J.L., Blanco D., Rajo-Iglesias E., Llombart N. 'Holographic surface leaky-wave lenses with circularly-polarized focused near-fields - Part I: Concept, design and analysis theory'. *IEEE Transactions on Antennas and Propagation*. 2013;61(7):3475–85.

[11]   Okuyama T., Monnai Y., Shinoda H. '20-GHz focusing antennas based on corrugated waveguide scattering'. *IEEE Antennas and Wireless Propagation Letters*. 2013;12:1284–6.

[12]   Chou H.-T., Hung T.-M., Wang N.-N., Chou H.-H., Tung C., Nepa P. 'Design of a near-field focused reflectarray antenna for 2.4 GHz RFID reader applications'. *IEEE Transactions on Antennas and Propagation.* 2011;59(3):1013–18.

[13]   Buffi A., Nepa P., Manara G. Design criteria for near-field-focused planar arrays. *IEEE Antennas and Propagation Magazine.* 2012:40–50.

[14]   Alvarez J., Ayestaran R.G., Leon G., *et al.* 'Near field multifocusing on antenna arrays via non-convex optimisation'. *IET Microwaves, Antennas & Propagation.* 2014;8(10):754–64.

[15]   Turin G. 'An introduction to matched filters'. *IRE Transactions on Information Theory.* 1960;6(3):311–29.

[16]   Ibrahim R., Voyer D., Breard A., *et al.* 'Experiments of time-reversed pulse waves for wireless power transmission in an indoor environment'. *IEEE Transactions on Microwave Theory and Techniques.* 2016;64(7):2159–70.

[17]   Ku M.-L., Han Y., Lai H.-Q., Chen Y., Liu K.J.R. 'Power waveforming: wireless power transfer beyond time reversal'. *IEEE Transactions on Signal Processing.* 2016;64(22):5819–34.

[18]   Li B., Liu S., Zhang H.-L., Hu B.-J., Zhao D., Huang Y. 'Wireless power transfer based on microwaves and time reversal for indoor environments'. *IEEE Access.* 2019;7:114897–908.

[19]   Park H.S., Hong S.K. 'A performance predictor of beamforming versus time-reversal based far-field wireless power transfer from linear array'. *Scientific Reports.* 2021;11:22743.

[20]   Massa A., Oliveri G., Viani F., Rocca P. 'Array designs for long-distance wireless power transmission: State-of-the-art and innovative solutions'. *Proceedings of the IEEE.* 2013;101(6):1464–81.

[21]   Takahashi T., Sasaki T., Homma Y., *et al.* 'Phased array system for high efficiency and high accuracy microwave power transmission'. *Presented at IEEE International Symposium on Phased Array Systems and Technology*; Waltham, MA, 2016.

[22]   Van Atta L.C., inventor Electromagnetic reflector. United States patent 2908002. 1955.

[23]   Tseng W.-J., Chung S.-J., Chang K. 'A planar Van Atta array reflector with retrodirectivity in both E-plane and H-plane'. *IEEE Transactions on Antennas and Propagation.* 2000;48(2):173–5.

[24]   Pon C. 'Retrodirective array using the heterodyne technique'. *IEEE Transactions on Antennas and Propagation.* 1964;12:176–80.

[25]   Hsieh L.H., Strassner B.H., Kokel S.J., *et al.* 'Development of a retrodirective wireless microwave power transmission system'. *Presented at IEEE International Symposium on Antennas and Propagation*; Columbus, OH. 2003.

[26]   Miyamoto R.Y., Itoh T. 'Retrodirective arrays for wireless communications'. *Microwave Magazine.* 2002;3:71–9.

[27] Wang X., Sha S., He J., Guo L., Lu M. 'Wireless power delivery to low-power mobile devices based on retro-reflective beamforming'. *IEEE Antennas and Wireless Propagation Letters*. 2014;13:919–22.

[28] Lu M., Billo R.E., inventors; Wireless Power Transmission. United States patent 9030161. 2015.

[29] Zeine H., inventor Wireless power transmission system. United States patent US8446248B2. 2013.

[30] Hong W., Jiang Z.H., Yu C., *et al.* 'Multibeam antenna technologies for 5G wireless communications'. *IEEE Transactions on Antennas and Propagation*. 2017;65(12):6231–49.

[31] Ettorre M., Alomar W.A., Grbic A. '2-D Van Atta array of wideband, wideangle slots for radiative wireless power transfer systems'. *IEEE Transactions on Antennas and Propagation*. 2018;66(9):4577–85.

[32] Ngo T.-B., Do Q.-H., Yoon S.-W. 'A wideband circulator leakage canceler for retro-directive RF system'. *IEEE Microwave and Wireless Components Letters*. 2022;32(10):1211–14.

[33] Chiu L., Yum T.Y., Chang W.S., Xue Q., Chan C.H. 'Retrodirective array for RFID and microwave tracking beacon applications'. *Microwave and Optical Technology Letters*. 2006;48(2):409–11.

[34] Buchanan N.B., Fusco V.F. 'Triple mode PLL antenna array'. *Presented at International Microwave Symposium*; Fort Worth, TX. 2004.

Chapter 4

Retro-reflective beamforming technique for microwave power transmission in Internet of Things applications

As elucidated in Chapter 1, no commercial products based on microwave power transmission have been developed due to various practical restrictions. In the context of the *Internet of Things*, nevertheless, these restrictions do not appear prohibitive. Meanwhile, there are strong demands for supplying wireless power to low-power mobile electronic devices in Internet of Things applications. Thus, commercial products of microwave power transmission may emerge for the Internet of Things applications in the near future. This chapter discusses the potential of the retro-reflective beamforming technique to enable efficient microwave power transmission in the Internet of Things. In Section 4.1, the practical significance and feasibility of microwave power transmission in the Internet of Things are narrated. Section 4.2 presents the basic implementation scheme of applying retro-reflective beamforming to charge low-power mobile devices wirelessly in the Internet of Things. Two specific experimental examples are described in Section 4.3. Section 4.4 is devoted to the theoretical analysis of retro-reflective beamforming when multiple mobile devices require wireless power simultaneously. The theoretical analysis of Section 4.4 is verified by some experimental results in Section 4.5.

4.1 Microwave power transmission in Internet of Things

Internet of Things is one of the fastest-growing markets in the world [1,2]. The development of the Internet of Things intends to incorporate physical objects into the Internet. As foreseen by the US National Intelligence Council, by 2025 all kinds of everyday things will become "nodes" on the Internet, such as food packages, furniture, paper documents, and certain personnel [3]. The impact of the Internet of Things on industry, business, and individual people is expected to be tremendous and profound.

"Tag" is a crucial element of the Internet of Things. A tag attached to an object keeps track of the object's status (model number and manufacturing date, for instance). Once the tag is read by "tag readers" ubiquitously installed over the society's infrastructures (such as factories, warehouses, and supermarkets), the object's entity is converted to the digital format; furthermore, since all the tag readers can be connected to the Internet, the object can be traced by other members of the Internet. Although conventional tags comprised of optical barcodes have been employed for decades, the Internet of Things is based upon another type of tag: Radio-frequency identification (RFID) tags. RFID tags can be considered "electronic barcodes." Compared with optical barcodes, RFID tags can be read from a much longer distance, and in addition, reading RFID tags does not require line-of-sight. These distinctive advantages make RFID tags a better candidate for the Internet of Things. Today, RFID tags are always integrated with various sensors (such as temperature sensors and humidity sensors), resulting in RFID sensor tags.

Ideally, RFID tags should be small, light, low-cost, and readable from long distances (10 m or longer, for instance). In addition, RFID tags are often required to accommodate various sensors and be able to conduct certain complex operations like encryption. It is highly challenging for an RFID tag to satisfy all the above-mentioned requirements. Consequently, there are two categories of RFID tags.

- Passive RFID tags do not have onboard batteries. They are small, light, and low-cost. However, their reading range is limited to a few meters typically and they are unable to carry out complex jobs. Passive tags are suitable for applications with strict demands regarding the minimization of size, weight, and cost.
- Active RFID tags have onboard batteries. Their reading range can reach hundreds of meters and they can accommodate sophisticated functionalities. Nevertheless, the onboard battery increases the tags' size, weight, and cost. Also, the tags' lifetime is limited by the battery's capacity. Active tags are applicable when size, weight, cost, and lifetime are not the end users' top concerns.

Table 4.1 presents metrics to compare optical barcodes, passive RFID tags, and active RFID tags among one another.

With a "wireless charging module" incorporated, a new type of RFID tag would exhibit the advantages of passive tags and active tags simultaneously, and thus would satisfy the requirements of the Internet of Things better. The new type of RFID tag has a small rechargeable battery onboard. While the RFID tag communicates information to an RFID reader, the RFID reader transmits wireless power to charge the tag's battery. Since the proposed tags can acquire power from the reader on demand, a small onboard battery suffices (many commercially-available thin-film and solid-state batteries are good candidates [4,5]). As a result, the new type of tag is almost as small, light, and low-cost as passive RFID tags. Because of the onboard battery, the tags with "wireless charging module" also have the advantages of active RFID tags such as long reading range and sophisticated functionalities.

Wireless charging based on microwave power transmission is especially suitable for RFID tags due to the following three reasons. First, the number of RFID tags is huge in practice. Typically, a person has little difficulty charging his/her cell phone using a wired adapter when the cell phone's battery is near depletion.

Table 4.1   Comparison among optical barcode, passive RFID tag, and active RFID tag

	Optical barcode	Passive RFID tag	Active RFID tag
Reading distance	Less than 1 meter	A few meters	Up to hundreds of meters
Line-of-sight?	Yes	No	No
Functionalities	Simple	Simple	Sophisticated
Cost	Low	Low	High
Size and weight	Low	Low	High
Onboard battery?	No	No	Yes

However, if one RFID tag is attached to every book on a bookshelf, it would be almost impossible for him/her to keep track of the tags and charge them in a timely manner. Wireless charging is undoubtedly an excellent resolution to manage the rechargeable battery of a large number of devices. Second, mobility is one of the most outstanding merits of RFID tags. The microwave power transmission technology is capable of providing wireless power to RFID tags remotely without sacrificing their mobility and therefore is desirable in practice. Third, RFID tags are not power-hungry and do not require powerful wireless power transmitters. As calculated by one numerical example in Section 1.3 of this book, it is feasible for microwave power transmission to achieve 1% of power transmission efficiency in an indoor environment. If an RFID tag is in need of 10 mW of power, the power transmitted by a wireless power transmitter is 10 mW ÷ 1% = 1 Watt. Obviously, all the practical concerns pertinent to microwave power transmission, such as potential biological hazards, are negligible when the transmitted power is as low as 1 Watt. In summary, although every mobile electronic device may benefit from microwave power transmission in principle, the feasibility of applying microwave power transmission technology to the mobile electronic devices of the Internet of Things appears especially high.

4.2   Retro-reflective beamforming scheme for microwave power transmission in the Internet of Things

The retro-reflective beamforming technique has the potential to enable efficient microwave power transmission to RFID tags, as illustrated in Figure 4.1. Suppose there are thousands of containers in a warehouse and an RFID tag is attached to

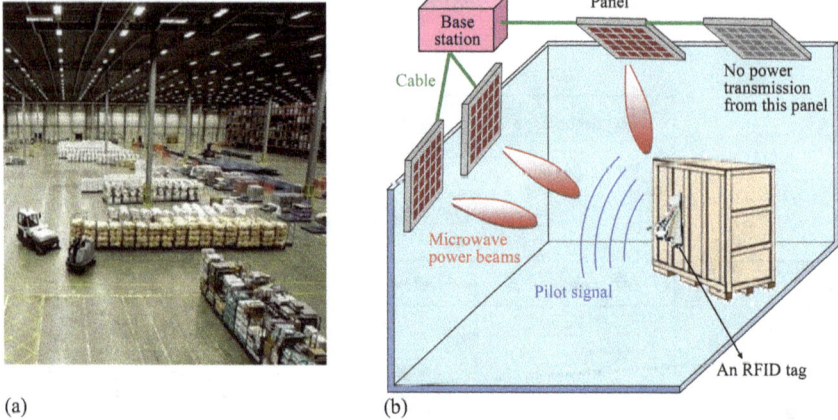

(a)   (b)

Figure 4.1   Illustration of the retro-reflective beamforming technique for microwave power transmission to RFID tags. (a) A warehouse as an application environment and (b) technical schematic.

each container. An RFID reader system is installed in the warehouse, consisting of a base station and multiple panels. The panels are mounted over the ceiling or walls of the warehouse. Each panel includes an array of planar antennas. The multiple panels work collaboratively to communicate with the tags, localize the tags, and supply wireless power to the tags. As illustrated in Figure 4.1, the two-step retro-reflective beamforming scheme can be applied to accomplish the third goal above (i.e., charging the tags wirelessly):

Step (i) One or more than one tag(s) broadcasts pilot signals;
Step (ii) In response to the pilot signals, the panels jointly construct microwave power beam(s) onto the target tag(s).

A panel transmits power only if it has clear line-of-sight interaction with the target tag; if the line-of-sight path is blocked by any obstacle, the panel is deactivated such that the obstacle, which might be a human being, is not illuminated by microwave power beams directly [6].

The timing sequence of retro-reflective beamforming for microwave power transmission to an RFID tag is depicted by a flow chart in Figure 4.2. Interactions between a wireless power transmitter and a tag are toggled among three modes: Communication mode, radar mode, and charging mode. The process in Figure 4.2 starts when a tag communicates a "charging request" signal to the wireless power transmitter. Once the wireless power transmitter acknowledges the "charging request," the system enters radar mode in which the tag broadcasts a pilot signal and the wireless power transmitter prepares for beamforming by analyzing the pilot signal. When the wireless power transmitter is ready, both the wireless power transmitter and tag march into the charging mode and wireless power is delivered to the tag through microwave power beam(s). In practice, the environment may change during the charging process and the beamforming plan must be adjusted

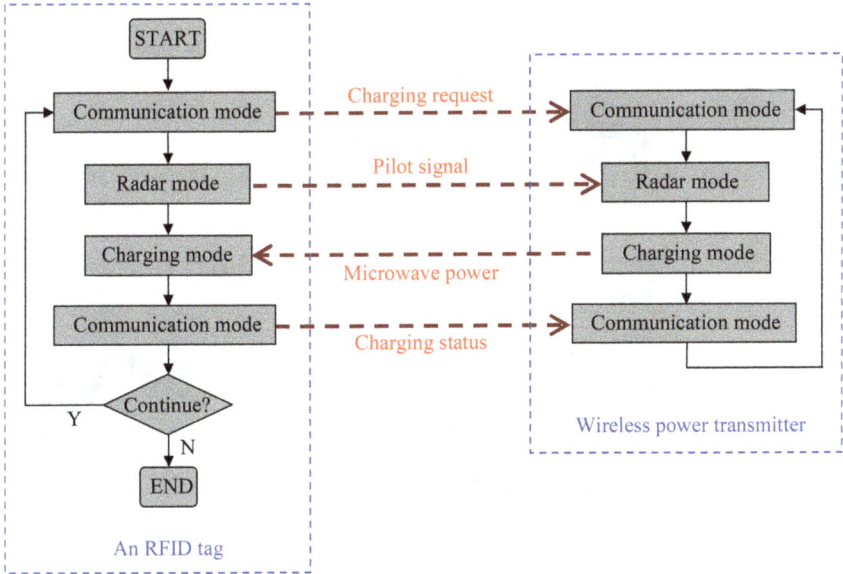

Figure 4.2   Timing sequence of retro-reflective beamforming for microwave power transmission to an RFID tag

accordingly; for example, the tag may move and/or another tag may request charging. To accommodate these situations, the system is switched from charging mode to communication mode and radar mode periodically such that the system would be reconfigured in reaction to the environmental changes.

A hardware block diagram of the retro-reflective beamforming scheme for microwave power transmission to an RFID tag is plotted in Figure 4.3. The tag includes one antenna and two simple circuit blocks. The "pilot signal generator" block of the tag generates the pilot signal. The other circuit block of the tag, "microwave-to-DC converter," converts microwave power received from the wireless power transmitter to DC power. The wireless power transmitter is comprised of multiple antenna elements coordinated by a base station. Behind each antenna element, there are two circuit blocks: "Pilot signal analyzer" and "microwave power generator." The "pilot signal analyzer" analyzes the pilot signal received from the wireless power receiver, and the "microwave power generator" generates microwave power based on the outcome of analyzing the pilot signal. Pilot signal propagation and wireless power propagation are isolated from each other via time-division duplexing. There is a wireless communication module over the wireless power transmitter as well as the tag. Through wireless communication, the wireless power transmitter and tag are synchronized with each other in time. Specifically, the switches in Figure 4.3 are controlled by the wireless communication between the wireless power transmitter and the tag to ensure that the system is toggled among the communication mode, radar mode, and charging mode

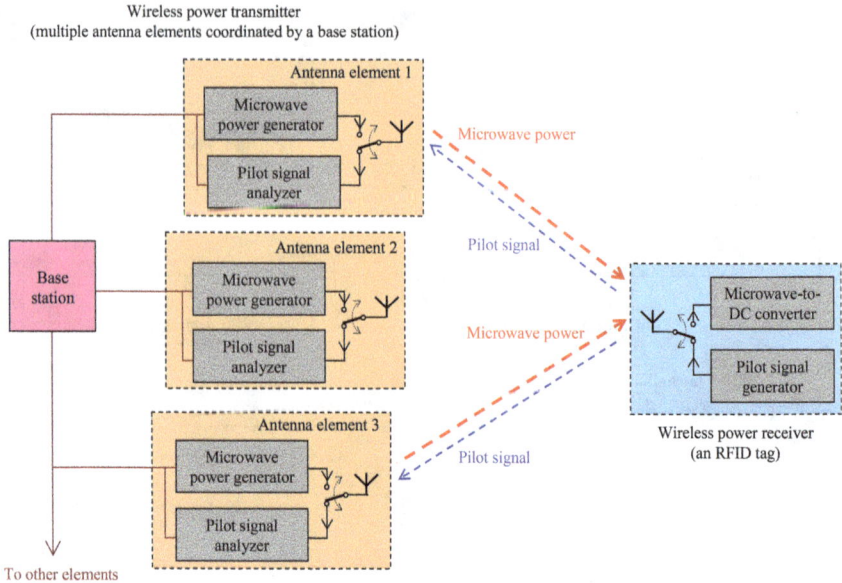

Figure 4.3   *Hardware block diagram of retro-reflective beamforming for microwave power transmission to an RFID tag*

appropriately. Other than controlling the switches, the wireless communication modules have no relevance to pilot signal propagation or wireless power propagation. Thus, the wireless communication modules are not shown in Figure 4.3. In fact, the existing wireless communication hardware and protocols between RFID tags and RFID readers can be taken advantage of for the communication mode in Figure 4.2.

In Section 3.6, several system architectures are discussed for the practical implementation of retro-reflective beamforming, and one of them is the time-division duplexing scheme. The block diagram based on time-division duplexing in Figure 4.3 is believed the best candidate in the context of the Internet of Things applications, for the following three reasons. First, the RFID tags would probably reside in the electrical near zone of a wireless power transmitter in an Internet of Things application environment with one example shown in Figure 4.1. As a result, several architectures discussed in Section 3.6 that rely on the condition of electrical far zone are not applicable. Second, channel reciprocity must be satisfied for the success of retro-reflective beamforming under the condition of electrical near zone, as argued in Section 3.6. The time-division duplexing scheme depicted in Figures 4.2 and 4.3 guarantees channel reciprocity even when the wireless channels between the wireless power transmitter and tags are dispersive. The frequency-division duplexing scheme described in Section 3.6, in contrast, requires the wireless channels to be non-dispersive, which is not likely in practical Internet of Things applications. Third, multiple RFID tags could be accommodated by the time-division duplexing scheme straightforwardly when they simultaneously call for wireless charging. The topic of one wireless power transmitter supplying wireless power to multiple wireless power receivers simultaneously is addressed in Section 4.4.

4.3   Two experimental examples of the retro-reflective beamforming scheme in Section 4.2

The retro-reflective beamforming scheme portrayed in Section 4.2 is elaborated by two experimental examples in this section. The two experimental demonstrations in this section are conducted for proof of concept. In other words, these two experiments are far from a thorough resolution to the hardware implementation of microwave power transmission to low-power mobile devices. A wide range of technical issues not covered by this section are investigated in [7–12].

The first experimental setup is shown by a photo in Figure 4.4 [13]. There is one wireless power transmitter and one wireless power receiver. The wireless power transmitter is stationary and emulates a practical wireless charger; whereas the wireless power receiver moves along the x-axis in the experiments and emulates a mobile electronic device (such as an RFID tag). In Figure 4.4, the wireless power delivery to the wireless power receiver is indicated by a light-emitting diode (LED). The wireless power transmitter consists of an array of four microstrip antenna elements and the wireless power receiver has one microstrip antenna

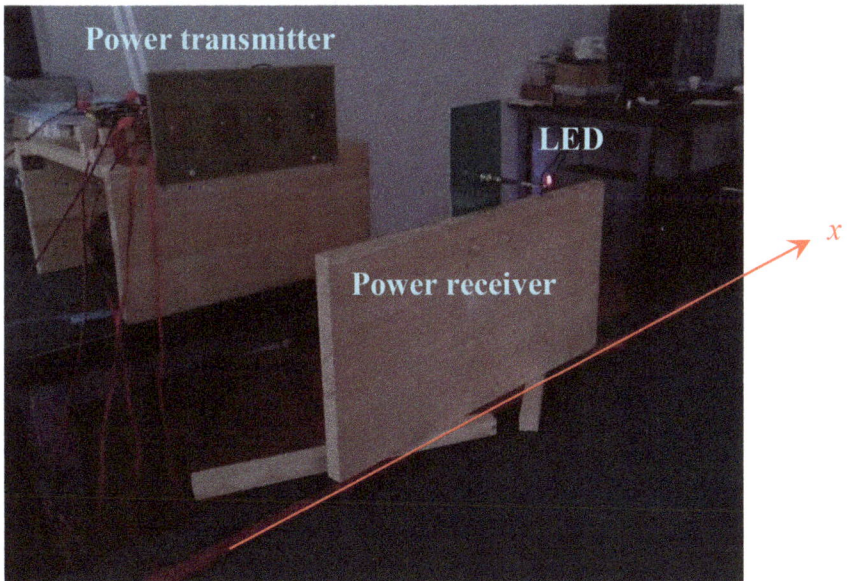

Figure 4.4   A photo of the first experimental example in Section 4.3. © [2014] IEEE. Reprinted, with permission, from [13]

element. "$x = 0$" denotes the location over the x-axis right in front of the wireless power transmitter. The distance between "$x = 0$" and the wireless power transmitter is 50 cm.

The experimental setup in Figure 4.4 is further depicted by two schematic diagrams in Figures 4.5 and 4.6. Specifically, Figures 4.5 and 4.6 illustrate the two steps of retro-reflective beamforming, respectively.

> Step (i) The wireless power receiver broadcasts a pilot signal to the wireless power transmitter, as illustrated in Figure 4.5. The wireless power receiver's antenna element is connected to a pilot signal generator. The pilot signal is received by the four antenna elements of the wireless power transmitter and then analyzed by a pilot signal analyzer.
>
> Step (ii) The wireless power transmitter transmits wireless power to the wireless power receiver, as illustrated in Figure 4.6. In this step, the wireless power transmitter's four antenna elements are fed by a microwave power generator. The power generator is configured according to the outcome of Step (i) such that a microwave power beam is constructed toward the location from which the pilot signal is emitted. The wireless power collected by the wireless power receiver's antenna element is detected by either a power meter or a rectifier.

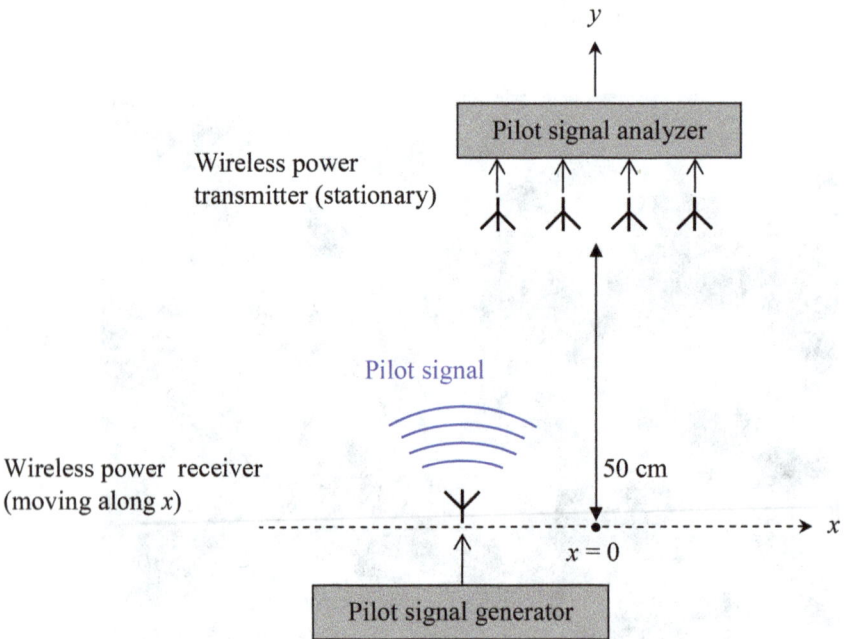

Figure 4.5   Illustration of Step (i) of the setup in Figure 4.4, in which the wireless power receiver broadcasts a pilot signal to the wireless power transmitter

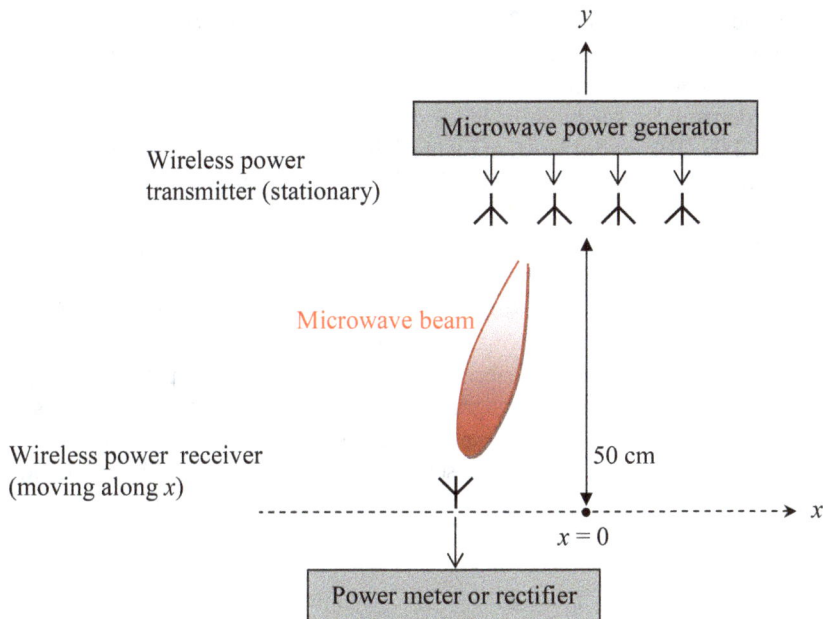

Figure 4.6   Illustration of Step (ii) of the setup in Figure 4.4, in which the wireless power transmitter generates a microwave beam to the wireless power receiver

The four antenna elements in the wireless power transmitter are identical to one another. Each is a regular rectangular microstrip patch with dimensions of 31.4 mm by 46 mm over FR4 substrate. The four antenna elements are equi-spaced with 7 cm as the distance between two adjacent elements. Each antenna element's "−10 dB return loss frequency band" is roughly from 2.055 GHz to 2.155 GHz. Each antenna element has a gain value of about 3.8 dBi along −y direction with a half-power beamwidth of 136°. When the four antenna elements are fed with a uniform amplitude and a uniform phase, the gain associated with the 4-element array is about 9.8 dBi along −y direction. The microstrip antenna element in the wireless power receiver is the same as those in the wireless power transmitter. All the antenna elements in Figure 4.4 are linearly polarized with the electric field polarized along the x direction.

A microwave oscillator could be adopted to generate the pilot signal straightforwardly over the wireless power receiver. The output of a microwave oscillator is unstable if no phase-locked loop circuits are employed. However, including phase-locked loop circuits may result in impractical size, weight, cost, or power consumption for the wireless power receiver. Thus, narrow pulses are proposed as the pilot signal in [6], which can be generated using circuits with relatively low complexity and low power consumption [14]. In the proof-of-concept experiments shown in Figure 4.4, the pilot signal is generated via amplitude-modulating a time-harmonic waveform at 2.08 GHz by periodic square pulses with a pulse width of 25 ns and a pulse repetition rate of 4 MHz. The time domain waveform of the pilot signal is illustrated in Figure 4.7(a). As the pilot signal consists of periodic pulses in the time domain, its spectrum includes discrete spectral lines with the center frequency at 2.08 GHz, as displayed in Figure 4.7(b). When the amplitude modulation is turned off, the pilot signal becomes a time-harmonic signal at 2.08 GHz.

The block diagrams of the "pilot signal analyzer" and "microwave power generator" are plotted in Figures 4.8 and 4.9, respectively. In Figure 4.8, the pilot signals received by the four antennas are amplified by four low-noise-amplifiers, down-converted via four mixers and a local oscillator at a frequency of f_{LO}, and then converted to the digital format by a four-channel analog-to-digital converter (ADC). The digital signals are stored within the memory of a personal computer and read by a signal processing program developed in the C++ language. The signal processing program calculates the phase values of the four digital signals using short-time (20 μs, to be specific) discrete Fourier transform at the intermediate frequency of f_{IF}. With the calculated phase values (denoted as α_1, α_2, α_3, and α_4), a

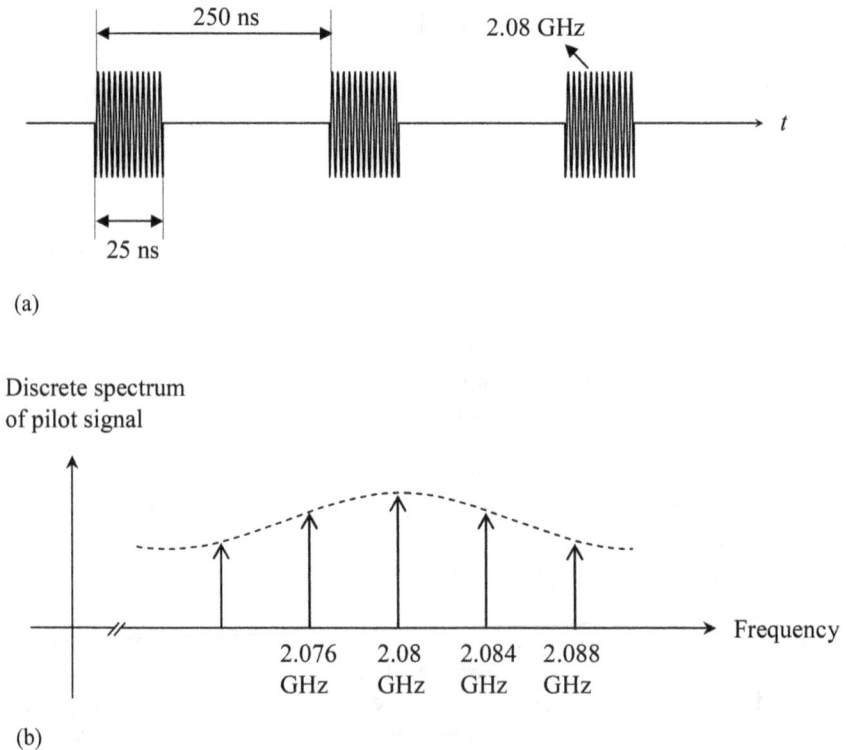

(a)

(b)

Figure 4.7   *Illustration of the pilot signal generated by the pilot signal generator in Figure 4.5. (a) Time domain waveform of pilot signal and (b) spectrum of pilot signal.*

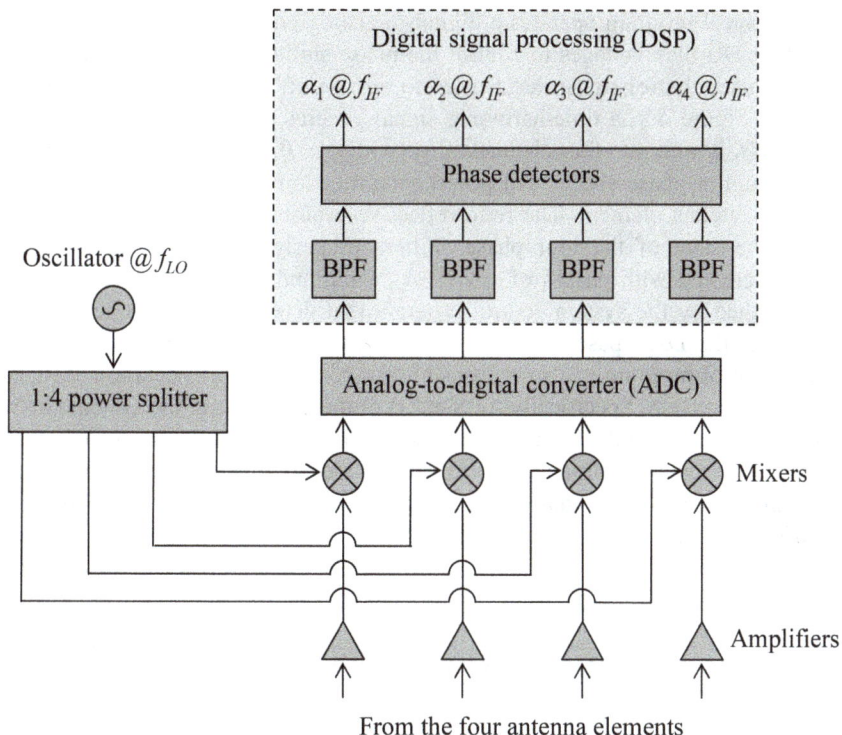

Figure 4.8   Block diagram of the pilot signal analyzer in Figure 4.5

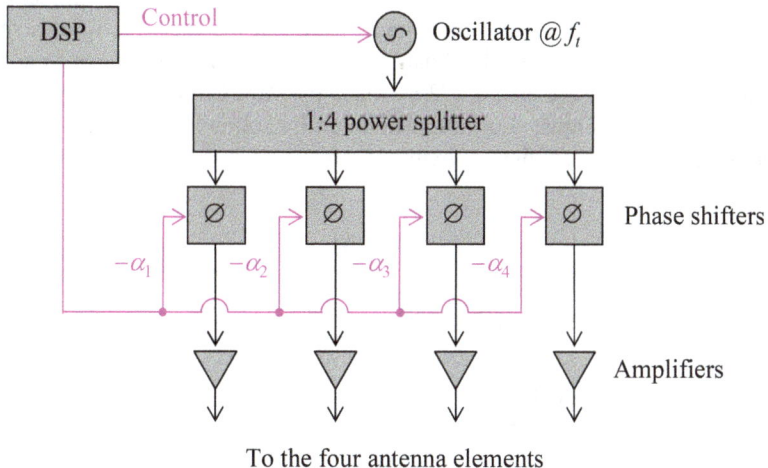

Figure 4.9   Block diagram of the microwave power generator in Figure 4.6

system control program operates a digital-to-analog converter (DAC), which provides four DC bias voltages to control the phase shifters in Figure 4.9. The DAC also outputs a control signal that is used to turn on/off the microwave power generator. In Figure 4.9, a time-harmonic signal generated by an oscillator at a frequency of f_t is split into four channels first (with $f_t = f_{LO} + f_{IF}$); next, each channel goes through a phase shifter and power amplification before reaching the corresponding antenna element. The retro-reflective beamforming is achieved by controlling the state of the four phase shifters properly so that the four antenna elements are fed with phases of $-\alpha_1$, $-\alpha_2$, $-\alpha_3$, and $-\alpha_4$, respectively. This is accomplished by the system control program that determines the output DC bias voltages of the DAC based on the characteristics of the four phase shifters. The system control program is also responsible for controlling the operation sequence of the entire system. Specifically, in Step (i) of the experiment, the system control program activates the ADC and calls the signal processing program to analyze the pilot signal. In the meantime, the oscillator in the microwave power generator is turned off by the control signal to avoid the interference of microwave power to the pilot signal analyzer. In Step (ii), the system control program deactivates the ADC and turns on the oscillator in the microwave power generator so that a microwave power beam is transmitted from the wireless power transmitter to the wireless power receiver.

When periodic pulses serve as the pilot signal, the pilot signal's spectrum is centered at 2.08 GHz and includes discrete spectral lines with a separation of 4 MHz as shown in Figure 4.7. Any discrete frequencies covered by the pilot signal could be selected as f_t, the carrier frequency of wireless power. For instance, when $f_{LO} = 2.079$ GHz and $f_{IF} = 1$ MHz, $f_t = 2.08$ GHz is used to transmit wireless power. As another example, f_t is 2.108 GHz with $f_{LO} = 2.1$ GHz and $f_{IF} = 8$ MHz.

The pilot signal analyzer and microwave power generator are implemented using commercial off-the-shelf components [13]. For instance, the four mixers in Figure 4.8 are ADL5365 made by Analog Devices Inc., and the 1:4 power splitters in Figures 4.8 and 4.9 are zn4pd-272-s+ made by Mini-Circuits.

The rectifier in Figure 4.6 is implemented by following a voltage multiplier design in [15]. The rectifier's circuit schematic is shown in Figure 4.10. In

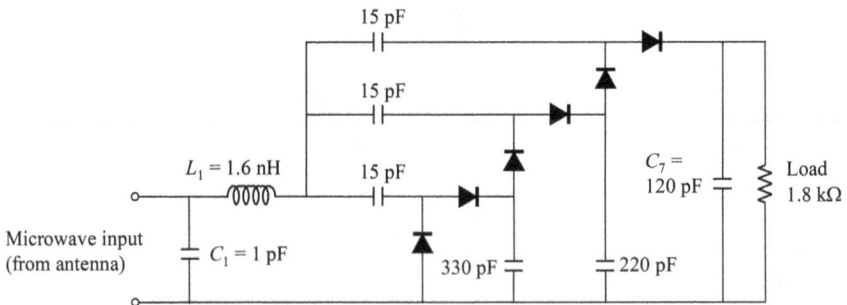

Figure 4.10   Circuit diagram of the rectifier in Figure 4.6

Figure 4.10, the six diodes are all Schottky diodes SMS7630-079LF made by Skyworks Inc.; L_1 and C_1 constitute an "L matching network"; and C_7 behaves as a single-pole low-pass-filter for the 1.8-kΩ load resistor. The rectifier is optimized around 2.1 GHz. The measured microwave-to-DC conversion efficiency results at 2.08 GHz is plotted in Figure 4.11: When the input microwave power is in the range of [10 mW, 100 mW], the microwave-to-DC conversion efficiency is between 50% and 70%.

The experiments of microwave power transmission are carried out following the two steps of retro-reflective beamforming. In the first step, a pilot signal is broadcasted by the wireless power receiver at a location denoted as x_0, and the pilot signal is received and analyzed by the wireless power transmitter. In the second step, the wireless power transmitter transmits wireless power based on the outcome of analyzing the pilot signal, and the wireless power receiver moves along the x-axis to detect the wireless power. In the second step, roughly 250 mW of microwave power is transmitted by each of the four antenna elements, and the total amount of power transmitted by the wireless power transmitter is approximately 1 Watt.

Some experimental results collected by the setup of Figure 4.4 are plotted in Figures 4.12–4.14.

In Figure 4.12, the value of microwave power measured by a power meter attached to the wireless power receiver's antenna in the second step of

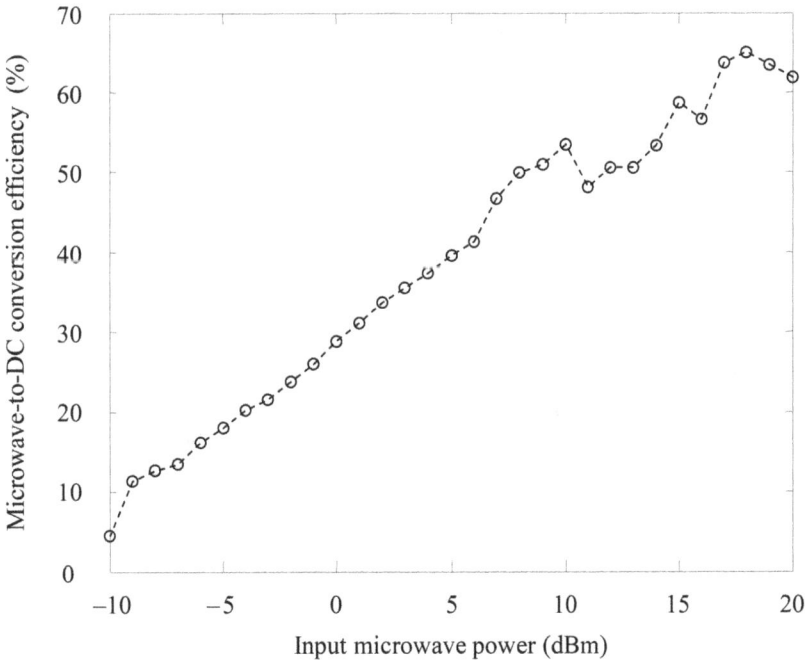

Figure 4.11   Microwave-to-DC conversion efficiency of the rectifier in Figure 4.10 measured at 2.08 GHz

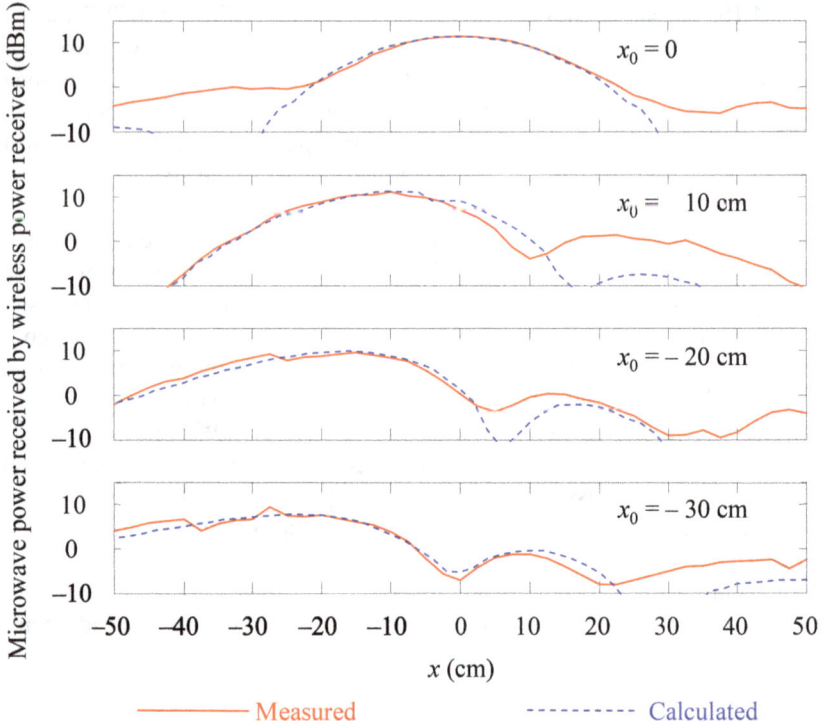

Figure 4.12   Microwave power received by the wireless power receiver in Figure 4.4 at 2.08 GHz. © [2014] IEEE. Reprinted, with permission, from [13]

retro-reflective beamforming is plotted in red color when $f_t = 2.08$ GHz. The horizontal axis in Figure 4.12 is the x coordinate of the wireless power receiver. The four subplots in Figure 4.12 correspond to "$x_0 = 0$," "$x_0 = -10$ cm," "$x_0 = -20$ cm," and "$x_0 = -30$ cm," respectively. The measured data in reaction to "$x_0 = 0$" demonstrate a power beam with the beam's center at "$x = 0$" and with 14 mW as the peak value of received microwave power. When x_0 changes to -10 cm, -20 cm, and -30 cm, the power beam is steered and the beam center follows the location of x_0. In the experiments, the beam cannot be steered beyond -30 cm due to the limitation of individual microstrip antenna elements' radiation pattern. When x_0 takes positive values, the power beam is steered and the beam center follows the location of x_0 as well; beam steering for positive x_0 values is not demonstrated as it is symmetric to the negative x_0 values.

The blue-colored curves in Figure 4.12 are calculated using the following Friis transmission equation, which was derived in Chapter 2.

$$P_r = P_t \left(\frac{\lambda_0}{4\pi d} \right)^2 G_{array} G_r. \tag{4.1}$$

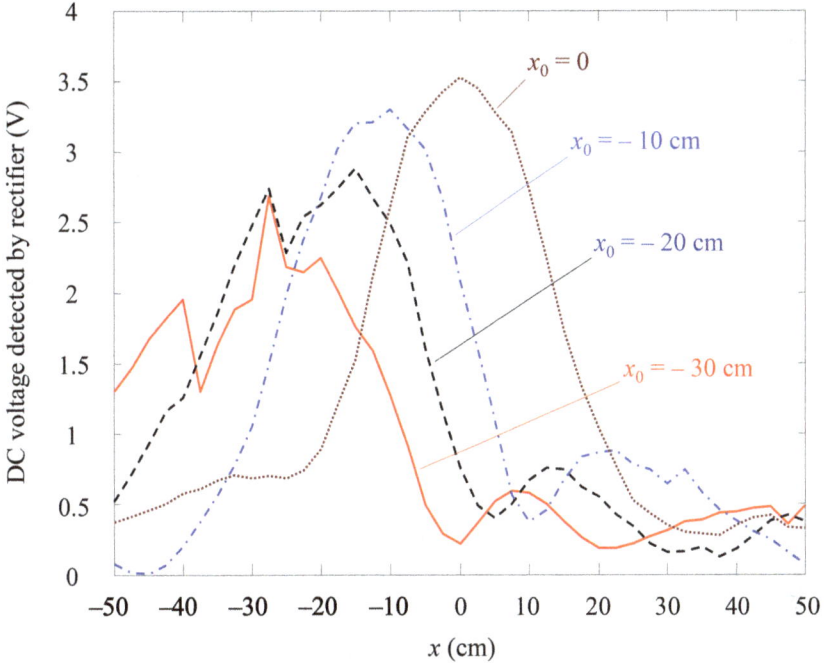

Figure 4.13   DC voltage detected by a rectifier in Figure 4.4 at 2.08 GHz. ©
[2014] IEEE. Reprinted, with permission, from [13]

As defined in Chapter 2, P_r is the received power, $P_t = 1$ Watt is the transmitted power, $\lambda_0 = c/f_t$ is the wavelength in free space at $f_t = 2.08$ GHz, c is the speed of light in free space, d is the distance between the wireless power transmitter and wireless power receiver, G_r is the gain value of receiving antenna, and G_{array} is the gain value of the 4-element array. When the wireless power transmitter only includes one microstrip antenna element fed by a power of P_0, the electric field radiated by the microstrip antenna element is denoted as

$$\mathbf{E}_0(\mathbf{r}) = \mathbf{U}_0 \frac{e^{-jk_0|\mathbf{r}-\mathbf{r}_s|}}{|\mathbf{r} - \mathbf{r}_s|}, \tag{4.2}$$

where $\mathbf{r}_s$ is the source location at ($x = 0$, $y = 50$ cm), $\mathbf{r}$ is the observation location, $k_0 = 2\pi/\lambda_0$, and $\mathbf{U}_0$ is the polarization vector associated with one microstrip antenna element. The gain value of one microstrip antenna element is

$$G_0(\mathbf{r}) = \frac{\frac{|\mathbf{E}_0|^2}{2\eta_0}}{\frac{P_0}{4\pi|\mathbf{r}-\mathbf{r}_s|^2}} = \frac{\frac{|\mathbf{U}_0|^2}{2\eta_0|\mathbf{r}-\mathbf{r}_s|^2}}{\frac{P_0}{4\pi|\mathbf{r}-\mathbf{r}_s|^2}} = \frac{4\pi|\mathbf{U}_0|^2}{2\eta_0 P_0}. \tag{4.3}$$

When the wireless power transmitter includes four microstrip antenna elements, suppose each antenna element is fed by a microwave power of P_0 with a phase

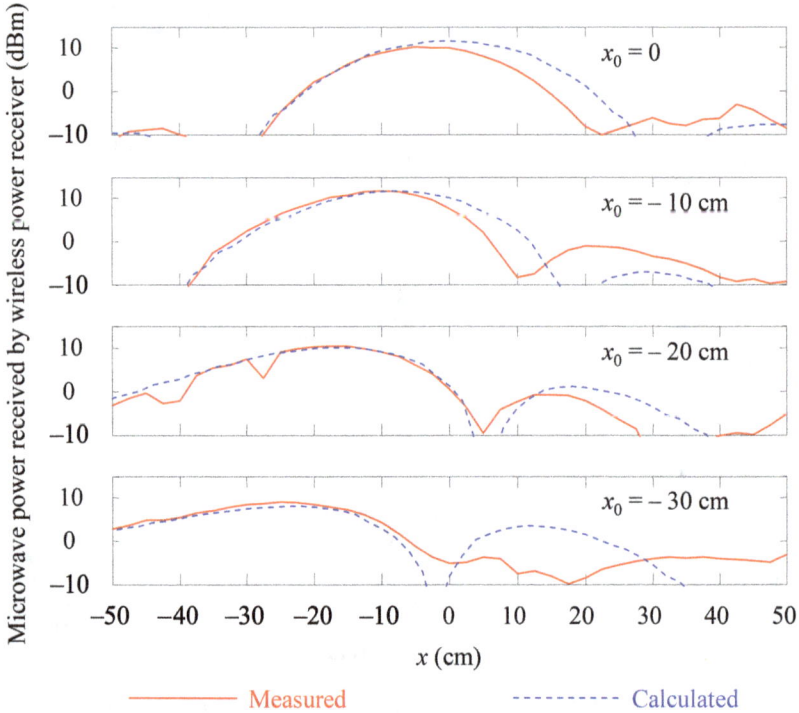

Figure 4.14   Microwave power received by the wireless power receiver in Figure 4.4 at 2.108 GHz. © [2014] IEEE. Reprinted, with permission, from [13]

value of ψ_n, $n = 1, 2, 3, 4$. The electric field radiated by the array of four microstrip antenna elements is

$$\mathbf{E}(\mathbf{r}) = \frac{\mathbf{U}_0}{|\mathbf{r} - \mathbf{r}_s|} \sum_{n=1}^{4} e^{j\psi_n} e^{-jk_0|\mathbf{r} - \mathbf{r}'_n|}, \quad (4.4)$$

where $\mathbf{r}'_n$ represents the location of the n-th antenna element. The gain value of the 4-element array is

$$G_{array}(\mathbf{r}) = \frac{\frac{|\mathbf{E}|^2}{2\eta_0}}{\frac{4P_0}{4\pi|\mathbf{r} - \mathbf{r}_s|^2}} = \frac{\frac{|\mathbf{U}_0|^2}{2\eta_0|\mathbf{r} - \mathbf{r}_s|^2} \left| \sum_{n=1}^{4} e^{j\psi_n} e^{-jk_0|\mathbf{r} - \mathbf{r}'_n|} \right|^2}{\frac{4P_0}{4\pi|\mathbf{r} - \mathbf{r}_s|^2}} = G_0(\mathbf{r}) \frac{1}{4} \left| \sum_{n=1}^{4} e^{j\psi_n} e^{-jk_0|\mathbf{r} - \mathbf{r}'_n|} \right|^2.$$

$$(4.5)$$

One sample calculation is conducted below. When $x = 0$ and $x_0 = 0$, $G_0 = G_r = 3.8$ dBi. Because of $\psi_n = k_0 |\mathbf{r} - \mathbf{r}'_n|$, the four microstrip antenna elements' contributions are in phase at $\mathbf{r}$. As a result, $G_{array} = 4 \times G_0 = 9.8$ dBi. With $G_r = 3.8$ dBi and $d = 0.5$ m, the received microwave power P_r is calculated to be 12 mW via

(4.1), which agrees with the measured power of 14 mW very well. Generally, the measured data and calculated data match each other in Figure 4.12, indicating that the Friis transmission equation is reliable in estimating the power transmission efficiency in the electrical near zone (which is one of the major conclusions of Chapter 2).

Similar sets of results are displayed in Figure 4.13, after the power meter is replaced by the rectifier in Figure 4.10. The vertical axis in Figure 4.13 represents the DC voltage measured over the 1.8-kΩ load resistor of the rectifier. Beams in reaction to four x_0 values are clearly visible. When $x_0 = 0$, the peak voltage 3.5 V corresponds to 6.8 mW of DC power.

After f_t is reconfigured to 2.108 GHz, curves similar to those in Figure 4.12 are plotted in Figure 4.14. The beamforming phenomena exhibited in Figure 4.14 are basically the same as those in Figure 4.12. With the experimental setup of Figure 4.4, retro-reflective beamforming could be clearly demonstrated in the frequency range between 2.06 GHz and 2.108 GHz.

The second experimental setup of this section is shown by a photo in Figure 4.15 [16]. As an improvement with respect to the first experiment in Figure 4.4, the wireless power transmitter includes eight antenna elements, with four of them deployed over the x-axis and the other four deployed over the y-axis. The geometrical configuration of the wireless power transmitter and wireless power receiver is illustrated in Figure 4.16. The eight antenna elements of the wireless power transmitter are identical to one another. Each is a regular rectangular

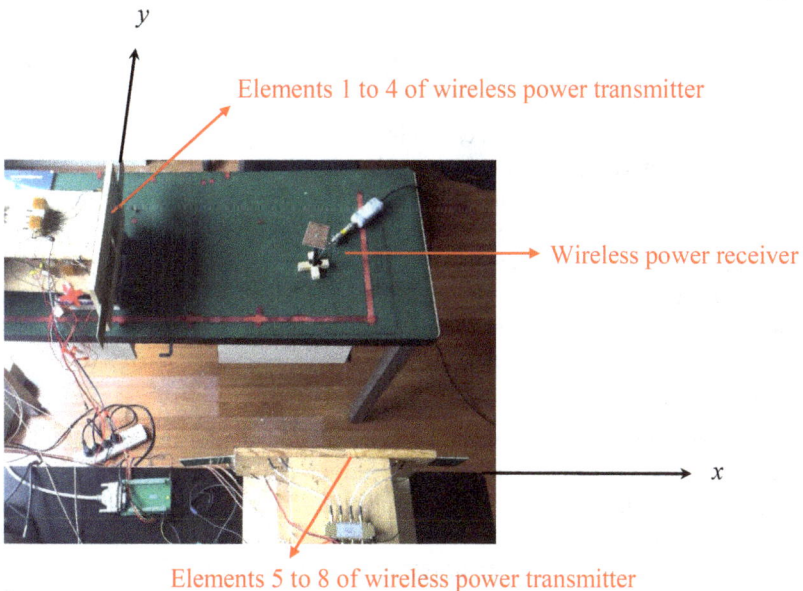

*Figure 4.15   A photo of the second experimental example in Section 4.3.
Reproduced from [16], courtesy of John Wiley and Sons*

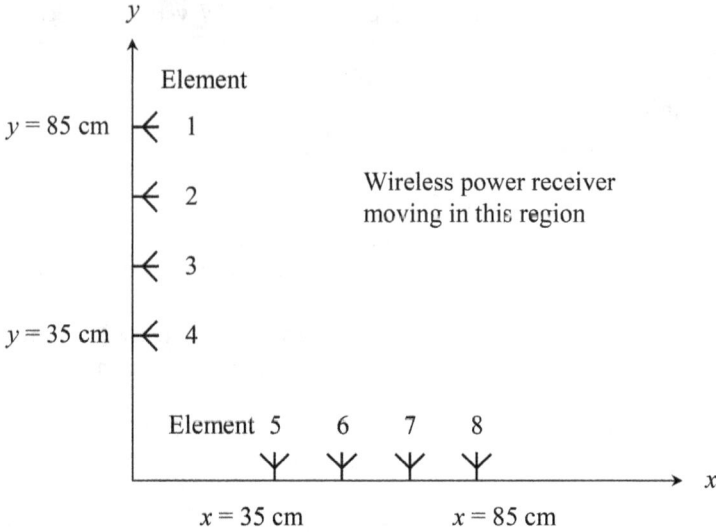

Figure 4.16   Illustration of the experimental setup in Figure 4.15

microstrip antenna polarized along the z direction. One wireless power receiver has a variable location within a certain region in the x-y plane. The antenna of the wireless power receiver is a monopole oriented along the z direction, with an omnidirectional radiation pattern in the x-y plane. The circuit modules employed in this experiment resemble those in the first experiment (shown in Figure 4.4).

The interaction between the wireless power transmitter and the wireless power receiver follows the two-step procedure of retro-reflective beamforming.

Step (i): A pilot signal is broadcasted by the wireless power receiver, which takes the form of either a time-harmonic or pulse signal. The pilot signal is detected and analyzed by the eight antenna elements of the wireless power transmitter. To be specific, the pilot signals received by the eight antenna elements are analyzed to obtain their phases at a frequency of f_t, denoted as $\alpha_1, \alpha_2, \ldots, \alpha_8$, respectively.

Step (ii): The eight antenna elements of the wireless power transmitter transmit wireless power at the frequency of f_t with a uniform amplitude. The excitation phase values to the eight antenna elements are adjusted to be $-\alpha_1, -\alpha_2, \ldots, -\alpha_8$, respectively.

In the experiments of wireless power transmission, a pilot signal is broadcasted by the wireless power receiver at a certain location (x_0, y_0) in the first step of retro-reflective beamforming. Then in the second step of retro-reflective beamforming, the wireless power receiver moves within a certain region in the x-y plane to detect wireless power. In the second step of retro-reflective beamforming, the microwave power transmitted at a frequency of $f_t = 2.125$ GHz by each antenna element of the wireless power transmitter is roughly 175 mW, with a total of 175 mW $\times$ 8 = 1.4 W. The microwave power value measured by a power meter connected to the wireless

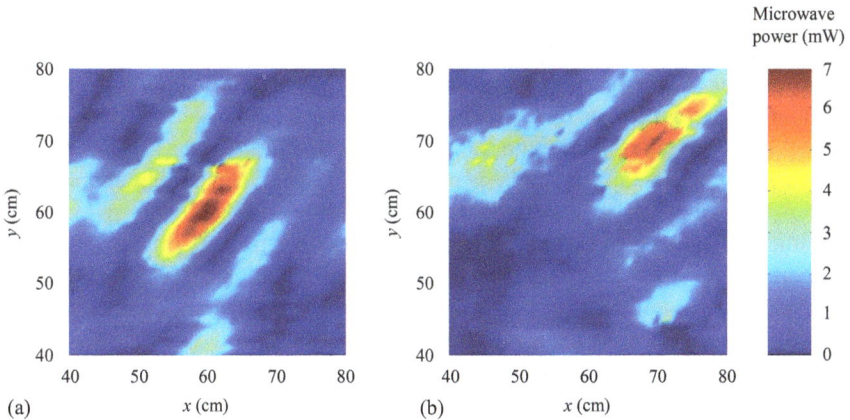

Figure 4.17   Microwave power distribution measured by the wireless power
receiver in the experiments of Figure 4.15. (a) Pilot signal from
(x_0 = 60 cm, y_0 = 60 cm) and (b) pilot signal from (x_0 = 70 cm,
y_0 = 70 cm). Reproduced from [16], courtesy of John Wiley and Sons

power receiver's antenna in the second step of retro-reflective beamforming is plotted in Figure 4.17. The two plots in Figure 4.17(a) and (b) are obtained when the pilot signal is broadcasted from (x_0 = 60 cm, y_0 = 60 cm) and (x_0 = 70 cm, y_0 = 70 cm), respectively. The plots in Figure 4.17 demonstrate that the wireless power is focused on the location from which the pilot signal is broadcasted. It is also observed from Figure 4.17 that the wireless power distribution exhibits undesired focal points; for instance in Figure 4.17(a), the power value measured at (x = 51 cm, y = 65 cm) is 4 mW while the peak power value measured at (x = 60 cm, y = 60 cm) is 7 mW. Theoretically, the undesired focal points in Figure 4.17 resemble the grating lobes of a phased array (which are discussed in Section 2.1 of this book).

Overall, the two experimental examples of this section prove the feasibility of delivering wireless power on the order of milli-Watts to low-power mobile devices (like RFID tags) using inexpensive hardware.

4.4   Theoretical study of retro-reflective beamforming for wireless power transmission to multiple targets (with "targets" standing for "wireless power receivers")

In the two experimental examples of the previous section (i.e., Section 4.3), only one target is present. In practical applications of the Internet of Things, it is quite probable that multiple low-power mobile devices require wireless power simultaneously. Actually, once a retro-reflective beamforming scheme is proved successful in generating one microwave power beam in reaction to one target's pilot signal, incorporating multiple targets seems straightforward. When multiple targets request wireless power simultaneously, each of them broadcasts its individual pilot signal to the wireless power transmitter. If the wireless power transmitter configures wireless power by analyzing the sum of the multiple pilot signals (that is, without separating the multiple pilot signals from one another), multiple microwave power beams would be generated with each beam aiming at one target, as shown in Figure 4.18. Therefore seemingly, the retro-reflective wireless power transmitter does not have to discriminate the multiple targets' pilot signals from each other. Nevertheless, in practice, various serious complications would emerge if the pilot signals of multiple targets are not differentiated from each other properly, which motivates the theoretical analysis of this section.

The theoretical analysis of this section starts with the basic configuration of retro-reflective beamforming involving only one target. As shown at the top of Figure 4.19, a wireless power transmitter includes a linear array of four antenna elements deployed along the x direction. The distance between two adjacent antenna elements in the array is 5 cm, which makes the array aperture along the x direction similar to the antenna array's size along the same direction in the experiments of the next section (i.e., Section 4.5). The position vectors of the four antenna elements are $\mathbf{r}_n$, $n = 1, 2, 3, 4$. There is one target with one antenna element at $\mathbf{r}_{\text{target}}$ ($x = -20$ cm, $y = 0$, $z = 0$). All the antenna elements over the wireless power transmitter and target are Hertzian dipoles located in the $z = 0$ plane and oriented along the z direction with a length of l_0 (the specific value of l_0 is unimportant in this section). Other than the five Hertzian dipoles, there is free space in the entire space. Hertzian dipoles are selected as the antenna elements such that the analytical formulations of Hertzian dipoles could be taken advantage of in this section. Since the underlying principle of retro-reflective beamforming does not rely on the type of antenna elements, the selection of Hertzian dipoles does not limit the generality of the conclusions drawn in this section.

The target receives wireless power from the wireless power transmitter via the following two steps. First, the target broadcasts a time-harmonic pilot signal at

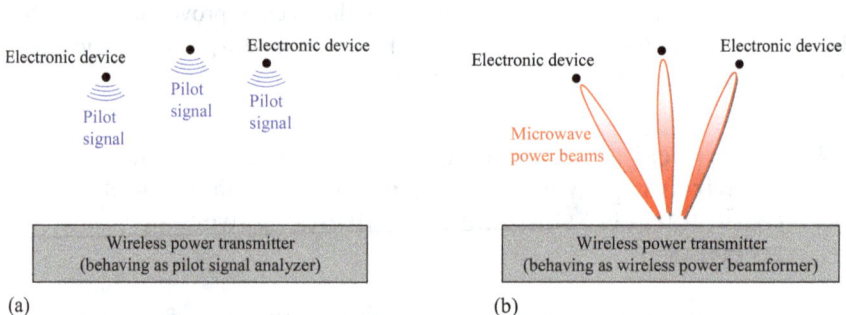

Figure 4.18   Illustration of retro-reflective beamforming for wireless power transmission to multiple targets (i.e., multiple mobile electronic devices). (a) The first step: pilot signals are broadcasted by devices and (b) the second step: wireless power beams are generated toward devices.

Target @ $x = -20$ cm
Pilot signal @ 5.75 GHz

y

x

1 meter

Array of 4 *z*-oriented
Hertzian dipoles

Wireless power
transmitter

Index of Hertzian dipole	1	2	3	4
Normalized amplitude of detected pilot signal	1.0	0.98	0.97	0.96

G$_t$ (in linear scale)

x (cm)

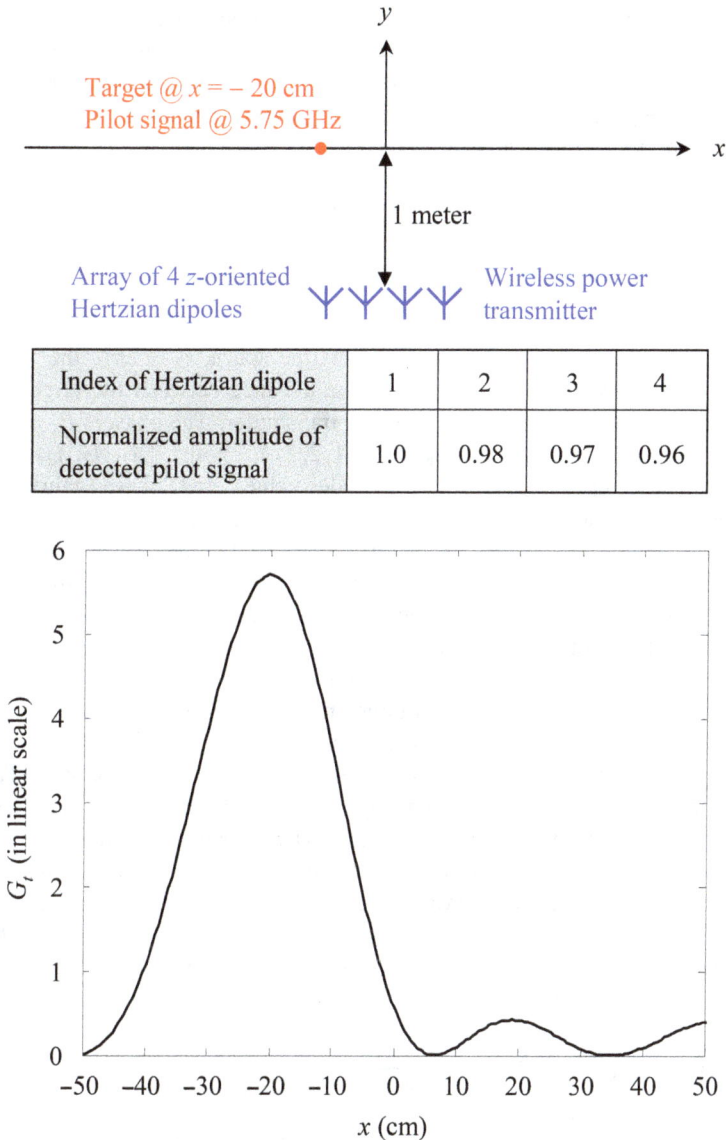

Figure 4.19 *Simulation results with one target broadcasting a time-harmonic pilot signal; gain of the 4-element array, G$_t$, in reaction to the pilot signal, is plotted when y = 0 and z = 0. © [2021] IEEE. Reprinted, with permission, from [17]*

5.75 GHz. Second, in response to the pilot signal, the wireless power transmitter generates a microwave power beam at 5.75 GHz toward the target's location. These two steps are carried out by numerical simulations next.

In the first step of retro-reflective beamforming, a pilot signal is generated by the Hertzian dipole of the target with $I_0 = 1$, where I_0 is the electric current over the dipole. The pilot signal is detected by the four antenna elements of the wireless power transmitter. Specifically, the pilot signals detected by the four Hertzian dipoles of the wireless power transmitter are denoted as phasors P_n, $n = 1, 2, 3, 4$, which are the z component of the electric field generated by the target's Hertzian dipole. Following the analytical expression of the electric field radiated by a time-harmonic Hertzian dipole [18], P_n is calculated as

$$P_n = -\sqrt{\frac{\mu_0}{\varepsilon_0}} \frac{jk_0}{4\pi} \frac{e^{-jk_0 R_{nt}}}{R_{nt}} (I_0 l_0) \left[1 + \frac{1}{jk_0 R_{nt}} - \frac{1}{(k_0 R_{nt})^2} \right] \tag{4.6}$$

where $k_0 = \omega\sqrt{\varepsilon_0\mu_0}$ is the wavenumber in free space, $\omega = 2\pi f$, the frequency $f = 5.75$ GHz, and $R_{nt} = |\mathbf{r}_n - \mathbf{r}_{\text{target}}|$ is the distance between the target and the n-th antenna element of the wireless power transmitter. The phasors P_n are further denoted as $P_n = A_n e^{j\alpha_n} = A_n \angle\alpha_n$, $n = 1, 2, 3, 4$, with $\{A_1, A_2, A_3, A_4\}$ and $\{\alpha_1, \alpha_2, \alpha_3, \alpha_4\}$ representing the amplitude values and phase values of the four pilot signals detected by the wireless power transmitter, respectively. The normalized values of $\{A_1, A_2, A_3, A_4\}$ (normalized by the largest value among $\{A_1, A_2, A_3, A_4\}$) are displayed in a table of Figure 4.19. Apparently, the four amplitudes are almost the same as each other. The distance between the target and the wireless power transmitter is roughly 1 m, much larger than the physical size of the 4-element antenna array. Thus, the pilot signals detected by the four antenna elements are not distinguishable from one another in terms of amplitude.

In the second step of retro-reflective beamforming, wireless power is transmitted by the 4-element antenna array of the wireless power transmitter. The n-th Hertzian dipole of the array is excited by $I_n = \gamma A_n e^{-j\alpha_n}$, $n = 1, 2, 3, 4$, where I_n is the electric current over the n-th dipole and γ is a real-valued constant. The value of γ is assumed to be 1 in this section, as it does not impact the numerical result graph in Figure 4.19. The excitation profile in the second step is the complex conjugate version of the pilot signal profile detected in the first step of retro-reflective beamforming. The z component of the electric field produced by the 4-element array and observed at location $\mathbf{r}_o$ is calculated as

$$E(\mathbf{r}_o) = -\sqrt{\frac{\mu_0}{\varepsilon_0}} \frac{jk_0}{4\pi} \sum_{n=1}^{4} (I_n l_0) \frac{e^{-jk_0 R_{no}}}{R_{no}} \left[1 + \frac{1}{jk_0 R_{no}} - \frac{1}{(k_0 R_{no})^2} \right] \tag{4.7}$$

where $R_{no} = |\mathbf{r}_n - \mathbf{r}_o|$ is the distance between the observation point and the n-th antenna element. It is not difficult to show that, when $\mathbf{r}_o = \mathbf{r}_{\text{target}}$, the fields produced by the four antenna elements are in phase, i.e., share the same phase. As a result, the total field produced by the four antenna elements demonstrates a focal point at the target's location.

In order to appreciate the numerical results better, the antenna gain of the 4-element array, G_t, is found as [18]

$$G_t(\mathbf{r}_o) = \frac{4\pi |\mathbf{r}_o|^2}{\Omega_{rad}} \sqrt{\frac{\varepsilon_0}{\mu_0}} \frac{|E(\mathbf{r}_o)|^2}{2} \tag{4.8}$$

In (4.8), Ω_{rad} is the total amount of power radiated by the 4-element antenna array evaluated over a fictitious spherical surface centered at the spatial origin and with infinite radius:

$$\Omega_{rad} = \sqrt{\frac{\mu_0}{\varepsilon_0}} \frac{(k_0)^2}{32\pi^2} \int_0^{\pi} d\theta \sin^3(\theta) \int_0^{2\pi} d\phi \left| \sum_{n=1}^{4} (I_n I_0) e^{jk_0 \widehat{\mathbf{k}} \cdot \mathbf{r}_n} \right|^2 \quad (4.9)$$

where $\widehat{\mathbf{k}} = \widehat{\mathbf{x}} \sin(\theta)\cos(\phi) + \widehat{\mathbf{y}} \sin(\theta)\sin(\phi) + \widehat{\mathbf{z}} \cos(\theta)$ is the outgoing direction over the fictitious spherical surface.

The simulated G_t results are plotted in Figure 4.19 with respect to the x coordinate of $\mathbf{r}_o$ when the y coordinate and z coordinate of $\mathbf{r}_o$ are zero. The spatial distribution of wireless power in Figure 4.19 exhibits a beam centered at $x = -20$ cm, the target's location. It is known that, when a Hertzian dipole is located in the $z = 0$ plane and oriented along the z direction, its gain value in the $z = 0$ plane is 1.5 [18]. In the numerical example of Figure 4.19, the four Hertzian dipoles of the wireless power transmitter are excited with a uniform amplitude. When their radiations are completely in phase at the target, the gain value would be enhanced by a factor of 4 if there is no mutual coupling among the four antenna elements. As shown in Figure 4.19, the peak G_t value at the target's location is quite close to $1.5 \times 4 = 6$, indicating that the wireless power beam in Figure 4.19 approaches the optimal spatial focusing performance of a 4-element antenna array whose elements are excited with a uniform amplitude.

The scenario in Figure 4.19, with one target requesting wireless power via a time-harmonic pilot signal, was investigated extensively in the previous section (i.e., Section 4.3). The numerical example of Figure 4.19 serves as a benchmark for the subsequent examples.

When multiple targets (rather than one target) request wireless power simultaneously, they can be differentiated from each other by various means. For example, the time-division scheme in Figure 4.3 can be extended to multiple targets straightforwardly. In Figure 4.20, the time-division scheme with two targets is illustrated. It is worth noting that "time-division" has two meanings in Figure 4.20: One is the time-division between pilot signal propagation and wireless power propagation, and the other one is the time-division between two targets. Incorporating more targets into the time-division scheme is fairly straightforward as demonstrated in [19]. A code-division scheme is designed in [20] for multiple robots to exchange power wirelessly among each other. It is also possible to differentiate multiple targets from each other via frequency-division. When multiple targets generate time-harmonic pilot signals individually, it is impossible for the pilot signals to exactly reside at the same frequency even if they share the same nominal frequency, because it is difficult and unnecessary for the multiple pilot signal generators (over multiple targets, respectively) to be synchronized among one another. Therefore in practice, each target's time-harmonic pilot signal can be assumed to be carried by a unique frequency. If the wireless power transmitter has the capability of explicitly resolving the multiple pilot signals' frequencies, it would be able to prepare one power beam in response to each pilot signal accordingly.

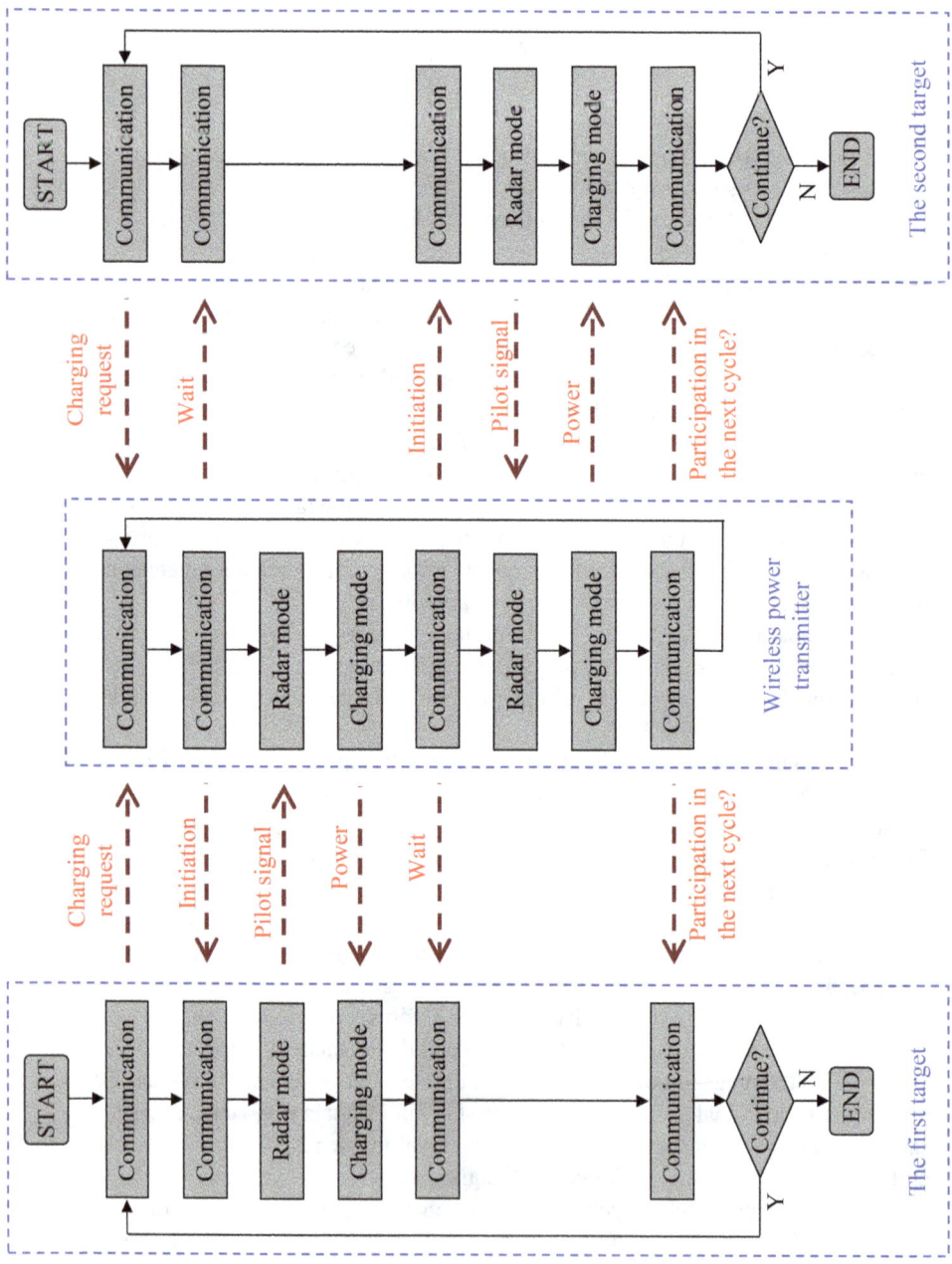

Figure 4.20 Timing sequence of retro-reflective beamforming for wireless power transmission to two targets (i.e., two wireless power receivers)

In the numerical example of Figure 4.21, two targets, namely "Target a" and "Target b," reside in front of the wireless power transmitter. The two targets' locations are denoted by $\mathbf{r}_{target}^{(a)}$ and $\mathbf{r}_{target}^{(b)}$, respectively. The targets interact with the wireless power transmitter via two time-harmonic frequencies, respectively.

In the first step of retro-reflective beamforming, the two targets broadcast time-harmonic pilot signals at $f^{(a)} = 5.745$ GHz and $f^{(b)} = 5.755$ GHz, respectively. The pilot signal from Target a is generated by the Hertzian dipole of Target a with current $I_a = 1$, and the pilot signal from Target b is generated by the Hertzian

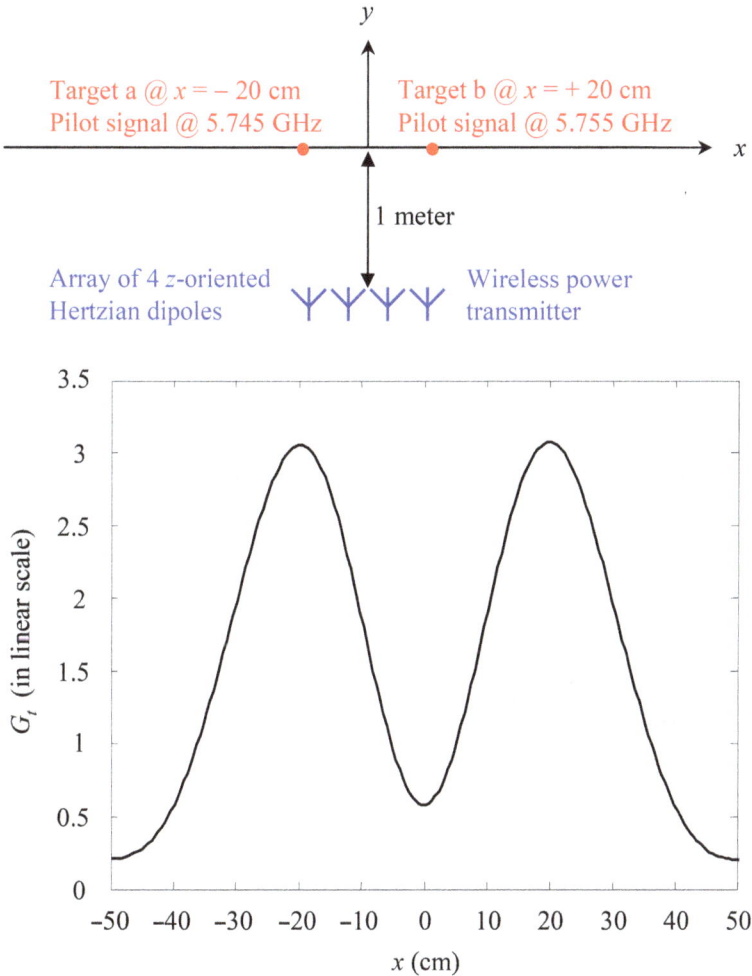

Figure 4.21   *Simulation results with two targets broadcasting time-harmonic pilot signals at 5.745 GHz and 5.755 GHz, respectively; gain of the 4-element array, G_t, in reaction to the pilot signals is plotted when $y = 0$ and $z = 0$. © [2021] IEEE. Reprinted, with permission, from [17].*

dipole of Target b with current $I_b = 1$. I_a and I_b are phasors associated with $f^{(a)}$ and $f^{(b)}$, respectively. The pilot signals detected by the four antenna elements of the wireless power transmitter, p_n, include two time-harmonic tones in the time (t) domain.

$$p_n(t) = A_n^{(a)} \cos \left\{ 2\pi f^{(a)} t + \alpha_n^{(a)} \right\} + A_n^{(b)} \cos \left\{ 2\pi f^{(b)} t + \alpha_n^{(b)} \right\}, n = 1, 2, 3, 4.$$

(4.10)

Because the propagation and detection of pilot signals can be modeled as a linear operator, the pilot signal propagation carried by $f^{(a)}$ and the pilot signal propagation carried by $f^{(b)}$ can be simulated separately. Specifically, $P_n^{(a)} = A_n^{(a)} \angle \alpha_n^{(a)}$ is found through (4.6) with the frequency of $f^{(a)}$, with I_0 replaced by I_a, and with the target at $\mathbf{r}_{\text{target}}^{(a)}$; $P_n^{(b)} = A_n^{(b)} \angle \alpha_n^{(b)}$ is found through (4.6) with the frequency of $f^{(b)}$, with I_0 replaced by I_b, and with the target at $\mathbf{r}_{\text{target}}^{(b)}$. $\{A_1^{(a)}, A_2^{(a)}, A_3^{(a)}, A_4^{(a)}\}$ are almost the same among one another, and $\{A_1^{(b)}, A_2^{(b)}, A_3^{(b)}, A_4^{(b)}\}$ are almost the same among one another too, which is similar to the phenomenon in Figure 4.19.

Next in the second step of retro-reflective beamforming, wireless power is transmitted by exciting the four Hertzian dipoles of the wireless power transmitter with

$$i_n(t) = \gamma^{(a)} A_n^{(a)} \cos \left\{ 2\pi f^{(a)} t - \alpha_n^{(a)} \right\} + \gamma^{(b)} A_n^{(b)} \cos \left\{ 2\pi f^{(b)} t - \alpha_n^{(b)} \right\}, n = 1, 2, 3, 4,$$

(4.11)

where $i_n(t)$ is the electric current over the n-th Hertzian dipole in the time domain. In this section, $\gamma^{(a)} = \gamma^{(b)} = 1$. Whereas in practice, the power budgets allocated to the two targets could be adjusted via these two parameters. The time-average excitation power at the n-th antenna element is $\left(\gamma^{(a)} A_n^{(a)} \right)^2 / 2 + \left(\gamma^{(b)} A_n^{(b)} \right)^2 / 2$, which is uniform among the four antenna elements. The wireless power propagation carried by $f^{(a)}$ and the wireless power propagation carried by $f^{(b)}$ can be simulated separately. Specifically, $E^{(a)}$ is found through (4.7) at frequency $f^{(a)}$ with $I_n = A_n^{(a)} e^{-j\alpha_n^{(a)}}$, $E^{(b)}$ is found through (4.7) at frequency $f^{(b)}$ with $I_n = A_n^{(b)} e^{-j\alpha_n^{(b)}}$, $\Omega_{rad}^{(a)}$ is found through (4.9) at frequency $f^{(a)}$ with $I_n = A_n^{(a)} e^{-j\alpha_n^{(a)}}$, and $\Omega_{rad}^{(b)}$ is found through (4.9) at frequency $f^{(b)}$ with $I_n = A_n^{(b)} e^{-j\alpha_n^{(b)}}$. The gain value of the 4-element array, G_t, is calculated by

$$G_t(\mathbf{r}_o) = \frac{4\pi |\mathbf{r}_o|^2}{\Omega_{rad}^{(a)} + \Omega_{rad}^{(b)}} \sqrt{\frac{\varepsilon_0}{\mu_0}} \frac{|E^{(a)}(\mathbf{r}_o)|^2 + |E^{(b)}(\mathbf{r}_o)|^2}{2}$$

(4.12)

The simulated G_t results are plotted in Figure 4.21 with respect to the x coordinate of $\mathbf{r}_o$ when the y coordinate and z coordinate of $\mathbf{r}_o$ are zero. Obviously, two wireless power beams are generated, centered at the two targets' locations, respectively. In Figure 4.19, there is one target, and the peak G_t value at the target's location is about 6. In Figure 4.21, the peak G_t values at the two targets' locations are about 3. This means that the power beams in Figure 4.21 approach the optimal spatial focusing performance of a 4-element antenna array whose elements are excited with a uniform amplitude. Moreover, the two power beams are independent of each other, as long as the two pilot signals can be explicitly differentiated from each other by the wireless power transmitter; to be specific, the power beam toward Target a is dictated by the pilot signal from Target a completely, and does not rely on the pilot signal from Target b. The wireless power transmitter may fail to differentiate the two pilot signals from each other in practice, especially when the two pilot signals' frequencies are close to each other, which is the scenario studied next.

The configuration illustrated at the top of Figure 4.22 is almost unchanged compared with the configuration of Figure 4.21. The only difference is that the two targets transmit time-harmonic pilot signals at the same frequency 5.75 GHz in Figure 4.22. As discussed above, even if the two targets generate pilot signals with the same nominal frequency, it is impossible for the two pilot signals to exactly reside at the same frequency in practice. However, if the wireless power transmitter is unable to differentiate the two time-harmonic pilot signals from each other, they would be interpreted as being at the same frequency by the wireless power transmitter.

The two targets are assumed to broadcast time-harmonic pilot signals at 5.75 GHz in the first step of retro-reflective beamforming. The Hertzian dipole of Target a is excited by the current $I_a = 1$, and the Hertzian dipole of Target b is excited by the current $I_b = e^{j\psi}$. I_a and I_b are phasors associated with the same frequency of $f = 5.75$ GHz. In Figure 4.22, $\psi = 0$. In other words, the two pilot signals are excited with the same amplitude and the same phase in Figure 4.22. The pilot signals detected by the four antenna elements of the wireless power transmitter are evaluated as

$$
\begin{aligned}
P_n = A_n e^{j\alpha_n} = &-\sqrt{\frac{\mu_0 j k_0}{\varepsilon_0 \, 4\pi}} l_0 \frac{e^{-jk_0 R_{na}}}{R_{na}} \left[1 + \frac{1}{jk_0 R_{na}} - \frac{1}{(k_0 R_{na})^2} \right] \\
&-\sqrt{\frac{\mu_0 j k_0}{\varepsilon_0 \, 4\pi}} l_0 e^{j\psi} \frac{e^{-jk_0 R_{nb}}}{R_{nb}} \left[1 + \frac{1}{jk_0 R_{nb}} - \frac{1}{(k_0 R_{nb})^2} \right]
\end{aligned}
\tag{4.13}
$$

where $R_{na} = |\mathbf{r}_n - \mathbf{r}_{\text{target}}^{(a)}|$ and $R_{nb} = |\mathbf{r}_n - \mathbf{r}_{\text{target}}^{(b)}|$. The normalized values of $\{A_1, A_2, A_3, A_4\}$ (normalized by the largest value among $\{A_1, A_2, A_3, A_4\}$) are displayed in a table of Figure 4.22. The nonuniformity among the four amplitude values is due to the interference pattern produced by the two targets' pilot signals in space. Particularly in Figure 4.22, when the two pilot signals reach the location of Antenna Element 1 or the location of Antenna Element 4, their phase difference is quite close to $180°$, and as a result, the sum of the two pilot signals has a weak amplitude.

In the second step of retro-reflective beamforming, wireless power is transmitted by exciting the four Hertzian dipoles of the wireless power transmitter by current $I_n = \gamma A_n e^{-j\alpha_n}$, $n = 1, 2, 3, 4$ with $\gamma = 1$. The wireless power transmission is simulated by (4.7), (4.8), and (4.9). The simulated results of G_t are plotted in Figure 4.22 with respect to x when the y coordinate of $\mathbf{r}_o$ is zero and the

y

Target a @ x = − 20 cm
Pilot signal @ 5.75 GHz
Phase of pilot signal 0

Target b @ x = + 20 cm
Pilot signal @ 5.75 GHz
Phase of pilot signal 0

x

1 meter

Array of 4 z-oriented
Hertzian dipoles

Wireless power
transmitter

Index of Hertzian dipole	1	2	3	4
Normalized amplitude of detected pilot signal	0.24	1.0	1.0	0.24

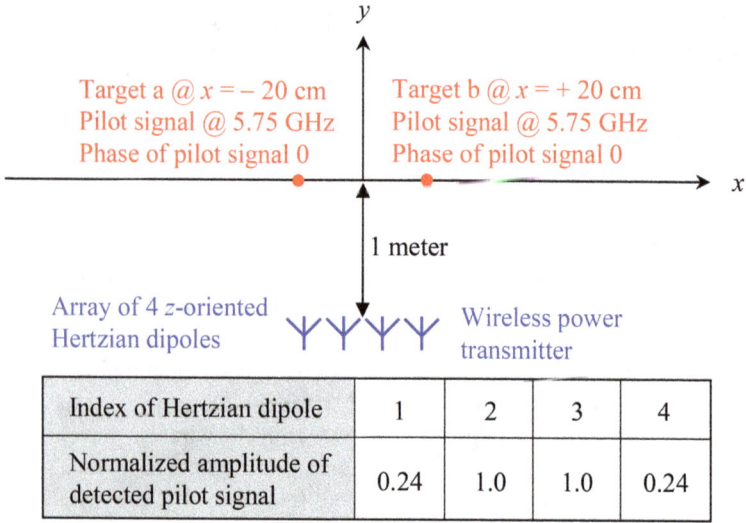

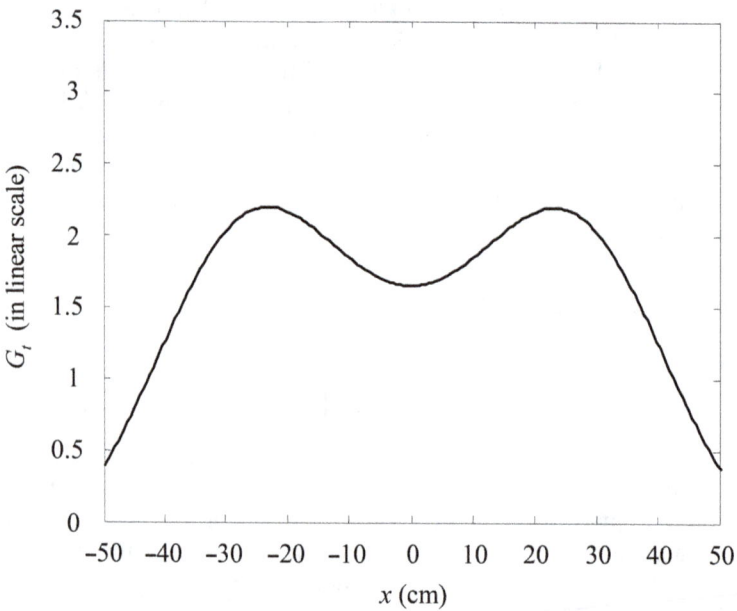

Figure 4.22   Simulation results with two targets broadcasting time-harmonic pilot
signals at the same frequency 5.75 GHz; phase difference between
the two pilot signal excitations is 0; gain of the 4-element array,
G_t, in reaction to the pilot signals is plotted when $y = 0$ and $z = 0$.
© [2021] IEEE. Reprinted, with permission, from [17]

z coordinate of $\mathbf{r}_o$ is zero. As expected, two power beams are exhibited in Figure 4.22, toward the two targets, respectively. Equation (4.13) consists of two terms, each generated by one target. If wireless power is triggered by each of these two terms individually, one power beam would appear. Since the right-hand side of (4.7) is a linear operator, the pilot signal in (4.13) leads to two power beams due to the principle of linearity. Meanwhile, it is observed that the peak G_t values of the two power beams in Figure 4.22 (which are 2.3 approximately) are considerably lower than those in Figure 4.21 (which are 3 approximately). The smaller G_t values in Figure 4.22 represent a poorer focusing performance in space, and in turn, a lower power transmission efficiency. The fact that the retro-reflective beamforming performance in Figure 4.22 is inferior to that in Figure 4.21 can be explained by the theory of antenna engineering. In retro-reflective beamforming, the excitation profile in the second step is the complex conjugate version of the pilot signal profile in the first step. Because the amplitude profile of pilot signals is non uniform as displayed in the table of Figure 4.22, the amplitude profile of wireless power excitations is nonuniform correspondingly. Specifically, in the wireless power transmission step, Antenna Element 1 and Antenna Element 4 are excited with much weaker power (weaker by 12 dB, to be specific), compared with Antenna Element 2 and Antenna Element 3. As a result, Antenna Element 2 and Antenna Element 3 play dominant roles in the 4-element array, and the other two antenna elements are almost negligible. In other words, the 4-element antenna array can be effectively considered as a 2-element antenna array in the second step of retro-reflective beamforming. In contrast, all the four antenna elements play equal roles in Figure 4.21, as they are excited by a uniform amplitude in the second step of retro-reflective beamforming. The 2-element antenna array (composed of Antenna Element 2 and Antenna Element 3) has a smaller physical aperture than the 4-element antenna array. According to the theory in Chapter 2, it is reasonable for an antenna with a smaller physical aperture to demonstrate a smaller gain value.

Another numerical example is presented in Figure 4.23. There is only one difference between Figures 4.22 and 4.23: The phase difference between the two pilot signal excitations, ψ, is zero in Figure 4.22 and 120° in Figure 4.23. The normalized values of $\{A_1, A_2, A_3, A_4\}$ are displayed in a table of Figure 4.23. Compared with Figure 4.22, a different interference pattern is formed in the space. In Figure 4.23, when the two pilot signals reach the location of Antenna Element 2, they almost cancel each other completely. In other words, Antenna Element 2 is located at a "dark spot" of the interference pattern produced by the two pilot signals. When the wireless power is tailored by the pilot signals in (4.13), the simulated G_t results are plotted in Figure 4.23 with respect to x when the y coordinate of $\mathbf{r}_o$ is zero and the z coordinate of $\mathbf{r}_o$ is zero. Two power beams appear in Figure 4.23, toward the two targets respectively, as expected. The peak G_t values of the two power beams in Figure 4.23 are greater than the peak G_t values in Figure 4.21. The fact that the spatial focusing performance in Figure 4.23 is superior to that in Figure 4.21 is not surprising. It is well known that a nonuniform amplitude excitation to the array elements (which is the scenario in Figure 4.23) might yield better performance than the uniform amplitude excitation (which is the scenario in Figure 4.21) [21].

Target a @ *x* = − 20 cm
Pilot signal @ 5.75 GHz
Phase of pilot signal 0

Target b @ *x* = + 20 cm
Pilot signal @ 5.75 GHz
Phase of pilot signal 120°

1 meter

Array of 4 *z*-oriented
Hertzian dipoles

Wireless power
transmitter

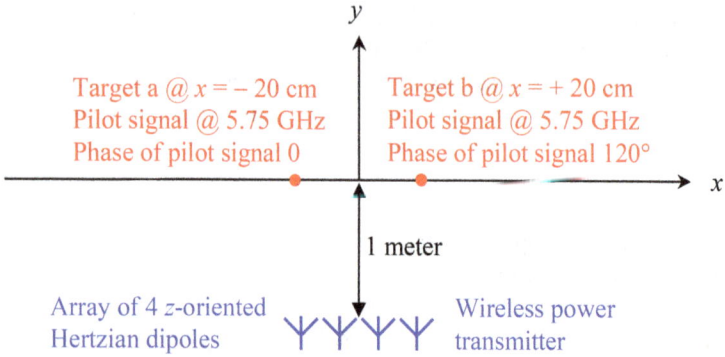

Index of Hertzian dipole	1	2	3	4
Normalized amplitude of detected pilot signal	1.0	0.07	0.95	0.79

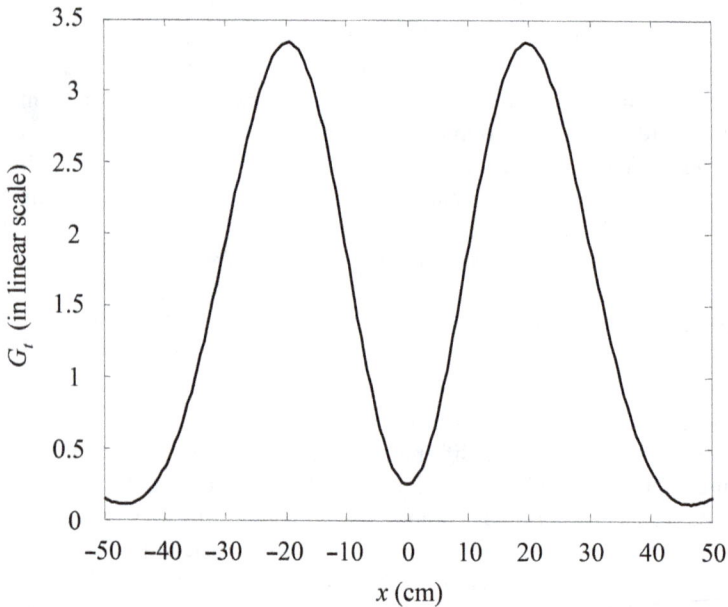

Figure 4.23   *Simulation results with two targets broadcasting time-harmonic pilot signals at the same frequency 5.75 GHz; phase difference between the two pilot signal excitations is 120°; gain of the 4-element array, G_t, in reaction to the pilot signals is plotted when y = 0 and z = 0.*
© *[2021] IEEE. Reprinted, with permission, from [17]*

A comparison between Figures 4.21–4.23 reveals several inspiring phenomena. If two targets' pilot signals could be differentiated from each other in frequency, the wireless power transmitter prepares wireless power transmission to the two targets independently, as shown in Figure 4.21. Two power beams are generated by the wireless power transmitter toward the two targets respectively. The two power beams are independent of each other in terms of the spatial focusing performance and power transmission efficiency. Moreover, the excitation power values of the antenna array elements in the wireless power transmission step are uniform. If two targets' pilot signals are considered to be carried by the same frequency, in contrast, the performance of retro-reflective beamforming depends on the interaction between the two pilot signals, as demonstrated in Figures 4.22 and 4.23. Two power beams are generated by the wireless power transmitter toward the two targets respectively, which seems similar to Figure 4.21. Nevertheless, the spatial focusing performance and power transmission efficiency associated with the two power beams heavily depend on the phase difference between the two pilot signals, because the phase difference determines the amplitude profile of the excitation to the antenna array elements in the wireless power transmission step. Compared with the uniform amplitude profile in Figure 4.21, the nonuniform amplitude profile sometimes leads to a poorer spatial focusing performance and sometimes leads to a better spatial focusing performance.

As a conclusion, it is essential to explicitly discriminate the pilot signals of multiple targets from each other. If the pilot signals of multiple targets are not differentiated from each other properly, a range of practical complications will emerge, as elaborated below.

- When two targets transmit pilot signals at the same frequency and when they are at similar distances from the wireless power transmitter, their pilot signals may create a strong interference pattern over the wireless power transmitter's aperture. A variety of interference patterns may appear, depending on the interaction between the two pilot signals; two examples are demonstrated in Figures 4.22 and 4.23. The interference pattern further determines the excitation profile of wireless power transmission. It is visualized from Figures 4.22 and 4.23 that the normalized A_2 (that is, the excitation amplitude to Antenna Element 2) is 1.0 in Figure 4.22 but becomes as small as 0.07 in Figure 4.23. If the total amount of power transmitted by the wireless power transmitter has a fixed value, the power transmitted by Antenna Element 2 changes drastically between Figures 4.22 and 4.23 (the ratio is about 24 dB, to be specific). As a result, the circuits associated with Antenna Element 2 must be able to accommodate a wide dynamic range in terms of power level. In Figure 4.21, in contrast, when the two pilot signals are differentiated from each other in frequency, the total amount of transmitted power is always distributed uniformly among the antenna elements, which greatly eases the difficulties associated with the power dynamic range in practical circuit implementations.
- When two targets transmit pilot signals at the same frequency and when one target is farther away from the wireless power transmitter than the other target,

the pilot signal from the farther target may be much weaker than the closer target's pilot signal. Consequently, the targets' pilot signals do not create a strong interference pattern over the aperture of the wireless power transmitter. However, the pilot signal from the farther target might be overwhelmed by the pilot signal from the closer target, and as a result, it is possible that the wireless power transmitter could not perceive the farther target. Suppose the weaker pilot signal is not overwhelmed and the pilot signals from both targets are detected precisely by the wireless power transmitter. The wireless power transmitter would generate a strong power beam toward the closer target in reaction to its strong pilot signal, but generate a weak power beam toward the farther target in reaction to its weak pilot signal. Such a power dispatching plan appears to be far from optimal. When the two pilot signals are told apart from each other in frequency (as demonstrated in Figure 4.21), the two wireless power beams' strength could be adjusted via $\gamma^{(a)}$ and $\gamma^{(b)}$, which facilitates better power management plans for the two targets.

- Obviously, the issues identified above would become more complicated and more serious if there were more than two targets and if their pilot signals could not be differentiated from each other by the wireless power transmitter.

Generating multiple microwave power beams toward multiple targets can be considered as an analogy of the space-division multiple access technique in wireless communication [22]. In wireless communication, multiple beams are desired to achieve the best diversity among multiple targets. The top concern of wireless power transmission applications, however, is the optimal power transmission efficiency rather than diversity. The analysis outcomes of this section indicate that, in association with space-division, it is imperative for another multiple access method (such as frequency-division, time-division, or code-division) to be enforced for the sake of achieving the optimal power transmission efficiency.

4.5   Experimental study of retro-reflective beamforming for wireless power transmission to multiple targets (with "targets" standing for "wireless power receivers")

In this section, experiments are carried out to verify the theoretical analysis of the previous section (i.e., Section 4.4).

An experimental system consistent with the geometrical configurations in Section 4.4 is constructed, as portrayed in Figure 4.24. The wireless power transmitter includes an array of four antenna elements deployed along the x direction. The geometrical center of the 4-element array is at ($x = 0$, $y = -1$ m, $z = 0$). One or more than one target is located along the x-axis.

Microstrip antenna elements are employed in the experimental system of Figure 4.24. The front view of the 4-element array in the wireless power transmitter is shown in Figure 4.25. The microstrip patches are printed over a printed circuit board manufactured by Rogers Corporation with the model number RO4003C,

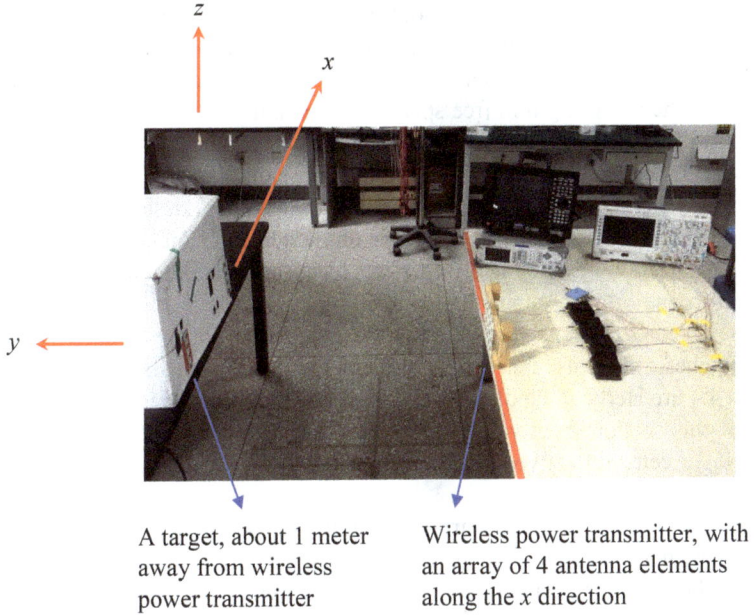

A target, about 1 meter away from wireless power transmitter

Wireless power transmitter, with an array of 4 antenna elements along the x direction

Figure 4.24 *A photo of the experimental study in Section 4.5. Reproduced courtesy of Elsevier.*

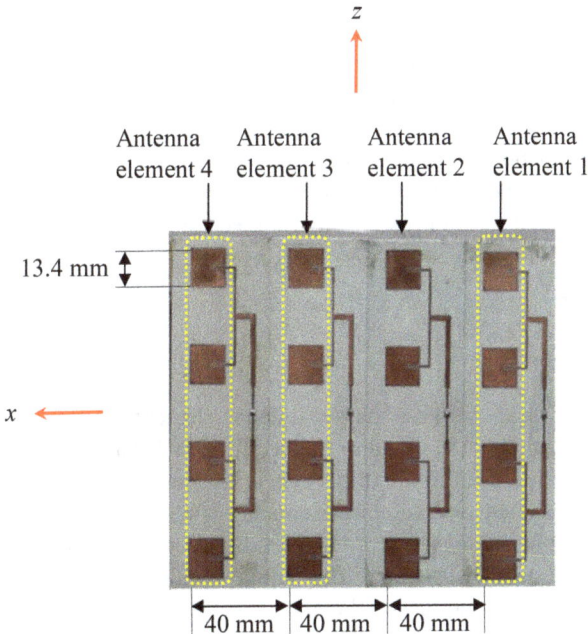

Figure 4.25 *Front view of the 4-element antenna array in Figure 4.24. Reproduced courtesy of Elsevier.*

dielectric constant of 3.55, thickness of 0.813 mm, and loss tangent of 0.0027. Each antenna element is composed of four microstrip patches. Each microstrip patch has the physical dimensions of 13.4 mm by 13.4 mm, with a resonant frequency of 5.75 GHz. The wavelength in free space corresponding to frequency 5.75 GHz, λ_0, is 52 mm. The distance between two adjacent antenna elements is about $0.77\lambda_0$. The feeding network (made of microstrip transmission lines) for the four microstrip patches in each element, which is visible in Figure 4.25, is designed to ensure that the four patches are excited with the same phase when the antenna element is used as a transmitting antenna. Each target includes only one antenna element that is identical to one of the antenna elements in Figure 4.25.

The antenna elements employed in the experiments of this section are microstrip patches, whereas the antenna elements in the theoretical analysis of Section 4.4 are Hertzian dipoles. This difference does not prevent the experimental study of this section from verifying the conclusions drawn in Section 4.4, as the principle of retro-reflective beamforming does not rely on the type of antenna elements.

The aperture of the antenna array in Figure 4.25 has the physical dimensions of $D \times D$, with $D = 150$ mm approximately. According to the far-field criterion (discussed in Chapter 2), a wireless power transmitter and a wireless power receiver reside in each other's far zone if the distance between them is greater than $2D^2/\lambda_0$ = 0.86 m. In the experimental setup of Figure 4.24, the wireless power receiver is 1 m away from the wireless power transmitter roughly, with the far-field condition barely satisfied.

The experimental characterization of one antenna element is presented in Figure 4.26. One transmitting antenna element is located at ($x = 0$, $y = -1$ m, $z = 0$); it is connected to a time-harmonic power source with a frequency of 5.75 GHz and the supplied power is P_t. One receiving antenna element moves along the x-axis; it is connected to a power meter and the power measured by the power meter is denoted as received power P_r. The power transmission efficiency (PTE) is defined as P_r/P_t. According to the Friis transmission equation (derived in Chapter 2), the power transmission efficiency is

$$\text{PTE} = \frac{P_r}{P_t} = (G_0)^2 \left(\frac{\lambda_0}{4\pi d}\right)^2 \tag{4.14}$$

where d is the distance between the two antenna elements and G_0 is the gain value of one antenna element. The G_0 data obtained from (4.14) are plotted in Figure 4.26 with respect to the x coordinate of the receiving antenna element. It is observed from Figure 4.26 that one antenna element has a fairly wide radiation beam. Each patch's gain along the broadside direction (i.e., the y direction) is approximately 5 dBi. Because the four patches included in one antenna element are excited with the same phase, the antenna element's gain at "$x = 0$" is about 11 dBi.

The experiments of retro-reflective beamforming follow the two-step procedure in Figure 4.18. In the first step, one or more than one target broadcasts pilot

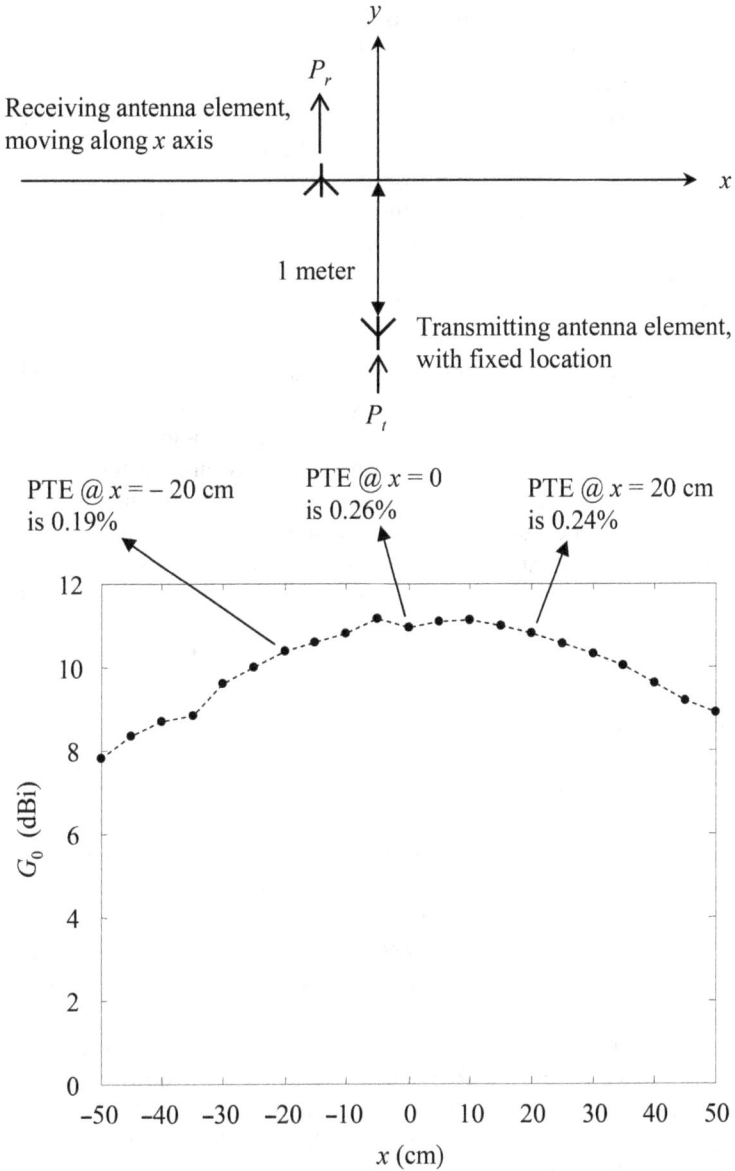

Figure 4.26 G_0, the gain of one antenna element with respect to the x coordinate of the receiving antenna element; "PTE" stands for the power transmission efficiency. Reproduced courtesy of Elsevier.

signals, and the wireless power transmitter analyzes the pilot signals. In the second step, the wireless power transmitter generates power beams toward the targets based on the outcome of analyzing the pilot signals.

The wireless power transmitter is constructed as a "digital retro-reflective beamformer." The pilot signals detected by the four antenna elements in the first step of retro-reflective beamforming are amplified by four low noise amplifiers respectively, down-converted to intermediate-frequency band with the aid of four mixers, and then converted to the digital format using analog-to-digital converters. The four low-noise amplifiers are manufactured by Analog Devices with model number ADL5611. The four mixers are purchased from Mini-Circuits with model number SIM-73L+. A digital oscilloscope manufactured by Teledyne LeCroy with four channels is employed as the analog-to-digital converter. Pilot signals' information, including frequency, amplitude, and phase, is obtained via processing the outputs of analog-to-digital converters. In the second step (i.e., wireless power transmission step), the waveforms fed to the four antenna elements, which consist of one or multiple time-harmonic tones, are generated by a digital signal processing unit, converted to the analog format using digital-to-analog converters, up-converted to the 5.75 GHz frequency band, and pass through four power amplifiers before reaching the four antenna elements. The four digital-to-analog converters are implemented by two RIGOL DG4202 arbitrary waveform generators (each with two output channels). The frequency up-conversion is accomplished by four mixers with image rejection composed of 90° hybrid couplers JSPQ-65W+ (manufactured by Mini-Circuits) along with I/Q mixers HMC8193 (manufactured by Analog Devices). The power amplifiers are HS5805Z1 signal boosters with a maximum output power of 5 Watts.

The experimental setup illustrated at the top of Figure 4.27 corresponds to the theoretical setup in Figure 4.19, with one target requesting wireless power using a time-harmonic pilot signal. The experiment associated with Figure 4.27 includes two steps, as elaborated next.

In the first step, one target broadcasts a time-harmonic pilot signal of 5.75 GHz from the location ($x = -20$ cm, $y = 0$, $z = 0$). The wireless power transmitter receives and analyzes the pilot signal. The pilot signals received by the four antenna elements of the wireless power transmitter are denoted as $p_n(t) = A_n \cos(2\pi ft + \alpha_n)$, $n = 1, 2, 3, 4$, where $f = 5.75$GHz. Specifically, the amplitude and phase values of the four pilot signals are $\{A_1, A_2, A_3, A_4\}$ and $\{\alpha_1, \alpha_2, \alpha_3, \alpha_4\}$ respectively. With the aid of one local oscillator at 5.735 GHz, pilot signals detected by the four antenna elements are down-converted to an intermediate frequency $f_{IF} = 15$ MHz. Four amplitudes $\{A_1, A_2, A_3, A_4\}$ and four phases $\{\alpha_1, \alpha_2, \alpha_3, \alpha_4\}$ are obtained at 15 MHz through Fourier transformation, as the outcome of the first step. The normalized amplitude of $\{A_1, A_2, A_3, A_4\}$ (normalized by the largest value among $\{A_1, A_2, A_3, A_4\}$) is displayed in a table of Figure 4.27. It is observed that the four amplitudes are quite uniform among one another.

In the second step of retro-reflective beamforming, a digital signal processing unit generates four waveforms $A_n \cos\{2\pi f_{IF}t - \alpha_n\}$, $n = 1, 2, 3, 4$. After these four waveforms are up-converted to 5.75 GHz using four mixers with image rejection, they are amplified and fed to the four antenna elements. Specifically, the four antenna elements are excited by $\gamma A_n \cos\{2\pi ft - \alpha_n\}$, $n = 1, 2, 3, 4$, which is the

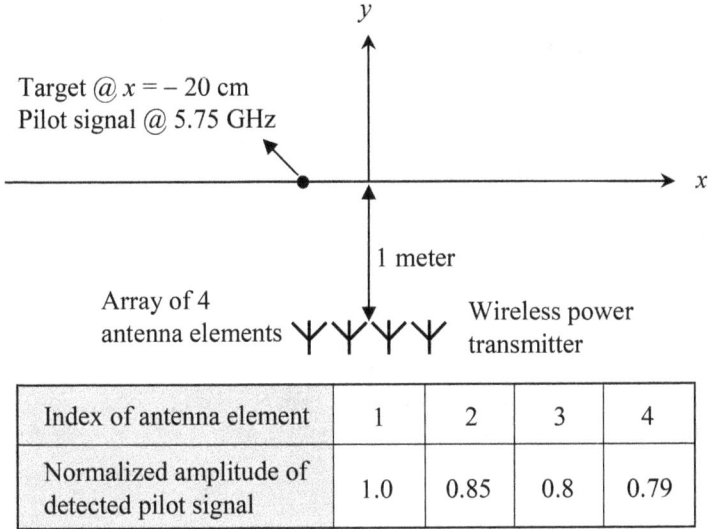

Index of antenna element	1	2	3	4
Normalized amplitude of detected pilot signal	1.0	0.85	0.8	0.79

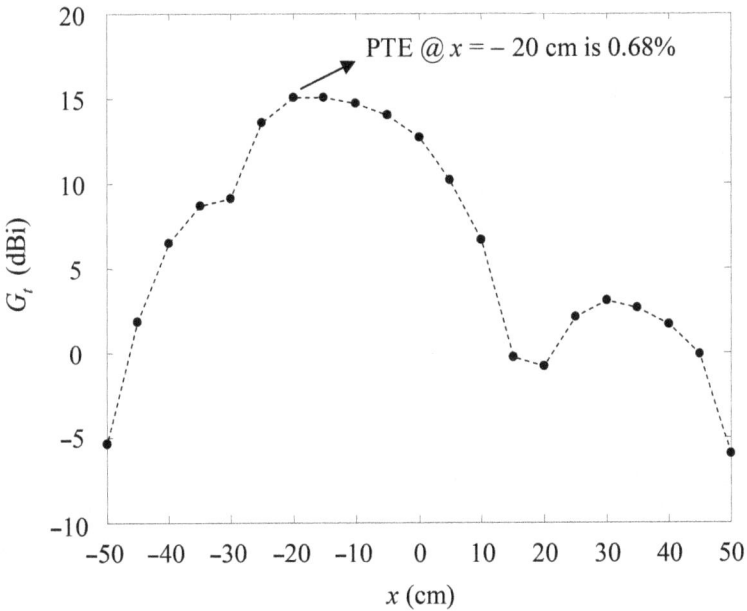

Figure 4.27 Experimental results with one target broadcasting a time-harmonic pilot signal of 5.75 GHz; G_t is the gain of the 4-element array in reaction to the pilot signal; "PTE" stands for the power transmission efficiency. Reproduced courtesy of Elsevier.

conjugate version of the pilot signal profile detected in the first step. The conjugate relationship between pilot signal reception and wireless power excitation ensures that the wireless power radiated by the four antenna elements is in phase at ($x = -20$ cm, $y = 0$, $z = 0$). The value of γA_n is determined by the output power level of four power amplifiers. A power meter is used to measure the power fed to each antenna element. The total amount of power fed to the four antenna elements is denoted as P_t. The target moves along the x-axis as the power detector, terminated by a power meter. The power detected by the power meter is denoted as P_r. The wireless power transmission efficiency (PTE) is defined as the ratio between P_r and P_t.

Some experimental results are plotted in Figure 4.27. The horizontal axis of Figure 4.27 is the x-coordinate of the power detector. The vertical axis of Figure 4.27 is G_t, the antenna gain of the 4-element array. G_t is calculated through the Friis transmission equation:

$$\text{PTE} = \frac{P_r}{P_t} = G_t G_0 \left(\frac{\lambda_0}{4\pi d} \right)^2 \tag{4.15}$$

When (4.15) is used to find G_t, G_0 is the gain value of one antenna element shown in Figure 4.26, λ_0 is the wavelength in free space corresponding to 5.75 GHz, and d is the distance between the power detector and wireless power transmitter. One power beam is clearly visible in Figure 4.27. The center of the power beam is at $x = -20$ cm, from which the pilot signal is broadcasted. The power transmission efficiency value at $x = -20$ cm is 0.68%, approximately four times greater than the efficiency value at $x = -20$ cm in Figure 4.26. The phenomenon in Figure 4.27 agrees with that in Figure 4.19 very well.

The experimental setup illustrated at the top of Figure 4.28 corresponds to the numerical setup in Figure 4.21. The experiments associated with Figure 4.28 include two steps, as elaborated next.

In the first step of retro-reflective beamforming, two targets, namely "Target a" and "Target b," broadcast pilot signals. Target a is located at ($x = -20$ cm, $y = 0$, $z = 0$); its time-harmonic pilot signal is at $f^{(a)} = 5.745$ GHz. Target b is located at ($x = 20$ cm, $y = 0$, $z = 0$); its time-harmonic pilot signal is at $f^{(b)} = 5.755$ GHz. The two pilot signals are generated by two oscillators individually, and the two oscillators have similar output power. The pilot signals are received and analyzed by the wireless power transmitter. To be specific, the pilot signals received by the four antenna elements of the wireless power transmitter are

$$p_n(t) = A_n^{(a)} \cos\left\{ 2\pi f^{(a)} t + \alpha_n^{(a)} \right\} + A_n^{(b)} \cos\left\{ 2\pi f^{(b)} t + \alpha_n^{(b)} \right\}, n = 1, 2, 3, 4. \tag{4.16}$$

With the aid of one local oscillator at 5.735 GHz, two intermediate frequency components at $f_{IF}^{(a)} = 10$ MHz and $f_{IF}^{(b)} = 20$ MHz are outstanding after down-conversion. $A_n^{(a)}$, $\alpha_n^{(a)}$, $A_n^{(b)}$, $\alpha_n^{(b)}$, $n = 1, 2, 3, 4$ are obtained through Fourier transforming the intermediate frequency components. Unsurprisingly, $\{A_1^{(a)}, A_2^{(a)}, A_3^{(a)}, A_4^{(a)}\}$ are quite

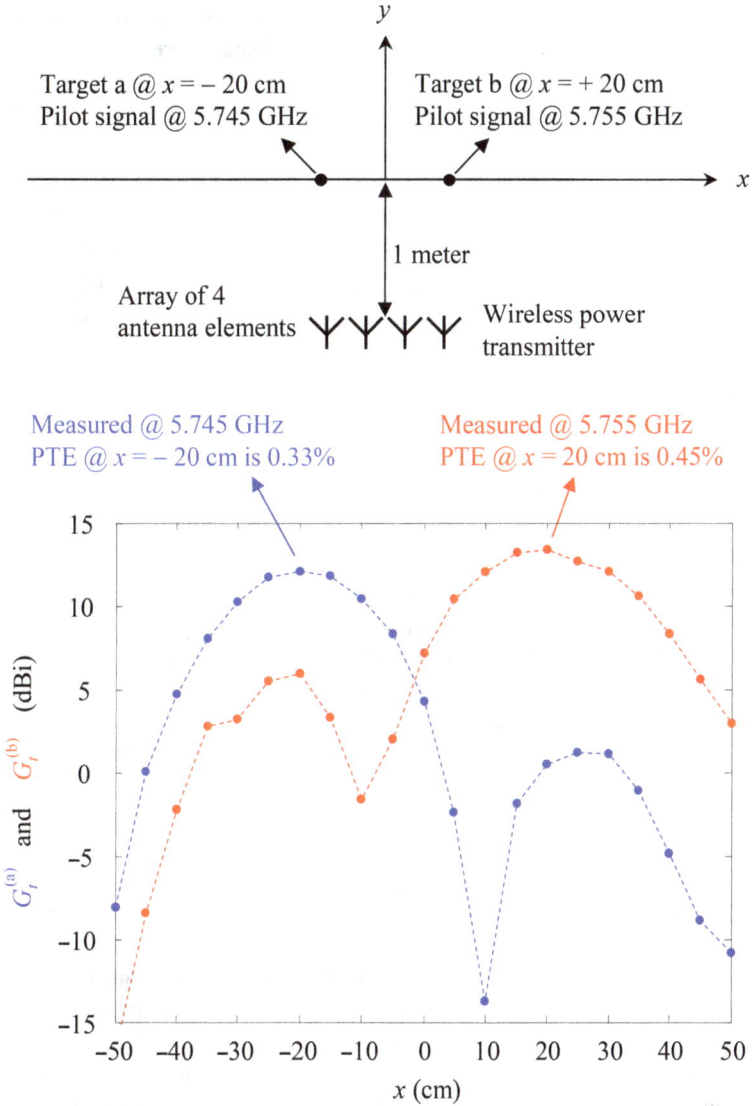

Figure 4.28   *Experimental results with two targets broadcasting time-harmonic pilot signals of 5.745 GHz and 5.755 GHz respectively; G_t is the gain of the 4-element array in reaction to the pilot signals; "PTE" stands for the power transmission efficiency. Reproduced courtesy of Elsevier*

uniform among one another, and $\{A_1^{(b)}, A_2^{(b)}, A_3^{(b)}, A_4^{(b)}\}$ are quite uniform among one another as well.

In the second step of retro-reflective beamforming, four waveforms are generated as $A_n^{(a)}\cos\left\{2\pi f_{IF}^{(a)}t - \alpha_n^{(a)}\right\} + A_n^{(b)}\cos\left\{2\pi f_{IF}^{(b)}t - \alpha_n^{(b)}\right\}$, $n = 1, 2, 3, 4$ by a digital signal processing unit. After these four waveforms are up-converted and amplified, the waveforms fed to the four antenna elements are

$$q_n(t) = \gamma^{(a)} A_n^{(a)} \cos\left\{2\pi f^{(a)} t - \alpha_n^{(a)}\right\} + \gamma^{(b)} A_n^{(b)} \cos\left\{2\pi f^{(b)} t - \alpha_n^{(b)}\right\}, n = 1, 2, 3, 4.$$

$$(4.17)$$

In the experiments of this section, $\gamma^{(a)} = \gamma^{(b)}$. Whereas in practice, the power budgets allocated to the two targets could be adjusted via $\gamma^{(a)}$ and $\gamma^{(b)}$. The time-average excitation power at the n-th antenna element is $\left(\gamma^{(a)} A_n^{(a)}\right)^2 / 2 + \left(\gamma^{(b)} A_n^{(b)}\right)^2 / 2$, which is uniform among the four antenna elements. The transmitted power P_t is the total amount of power fed to the four antenna elements. One target moves along the x-axis as the power detector, terminated by a spectrum analyzer. A spectrum analyzer is used instead of a power meter, such that the power values at $f^{(a)}$ and $f^{(b)}$ could be measured separately. The power measured at $f^{(a)}$ is denoted as $P_r^{(a)}$ and the power measured at $f^{(b)}$ is denoted as $P_r^{(b)}$.

Two curves are plotted in Figure 4.28. The vertical axis associated with the blue curve is $G_t^{(a)}$, calculated through (4.15) with P_r replaced by $P_r^{(a)}$. The vertical axis associated with the red curve in Figure 4.28 is $G_t^{(b)}$, calculated through (4.15) with P_r replaced by $P_r^{(b)}$. λ_0 corresponding to 5.75 GHz is used as $f^{(a)}$ and $f^{(b)}$ are both very close to 5.75 GHz. The horizontal axis of Figure 4.28 is the x-coordinate of the power detector. Obviously, two power beams are generated by the wireless power transmitter. The power transmission efficiency values at $x = -20$ cm and $x = 20$ cm in Figure 4.28 are approximately twice as much as those in Figure 4.26, approaching the optimal performance of a 4-element antenna array whose elements are excited with uniform power. The two power beams in Figure 4.28 are independent of each other, once the pilot signals of two targets are explicitly differentiated from each other by the wireless power transmitter. To be specific, the power beam toward Target a is dictated by the pilot signal from Target a completely, and does not rely on the pilot signal from Target b. Overall, the phenomena observed in Figure 4.28 are consistent with those observed in Figure 4.21.

The experimental configuration illustrated at the top of Figures 4.29 and 4.30 is the same as that in Figure 4.28 except that the two targets broadcast time-harmonic pilot signals at the same frequency of 5.75 GHz in Figures 4.29 and 4.30. The experiments associated with Figures 4.29 and 4.30 intend to verify the numerical results in Figures 4.22 and 4.23.

In the experiments of Figures 4.29 and 4.30, a time-harmonic signal at 5.75 GHz is generated by one oscillator and is then divided into two signals using a 1:2 power divider. The two output signals of the power divider are supplied to the two targets' antennas as the excitation of pilot signals, respectively. A phase shifter is applied before one of the two excitation signals reaches Target b's antenna. Consequently, the two excitation signals share the same frequency of 5.75 GHz, but the phase difference between them is tunable. Eighteen (18) sets of experiments are carried out; specifically, the phase delay created by the phase shifter is adjusted from 0 to 360° with 20° as the step size. Each set of experiments follows the

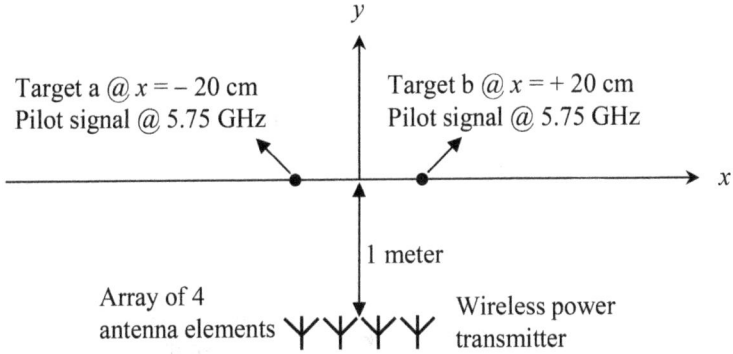

Index of antenna element	1	2	3	4
Normalized amplitude of detected pilot signal	0.14	1.0	0.96	0.34

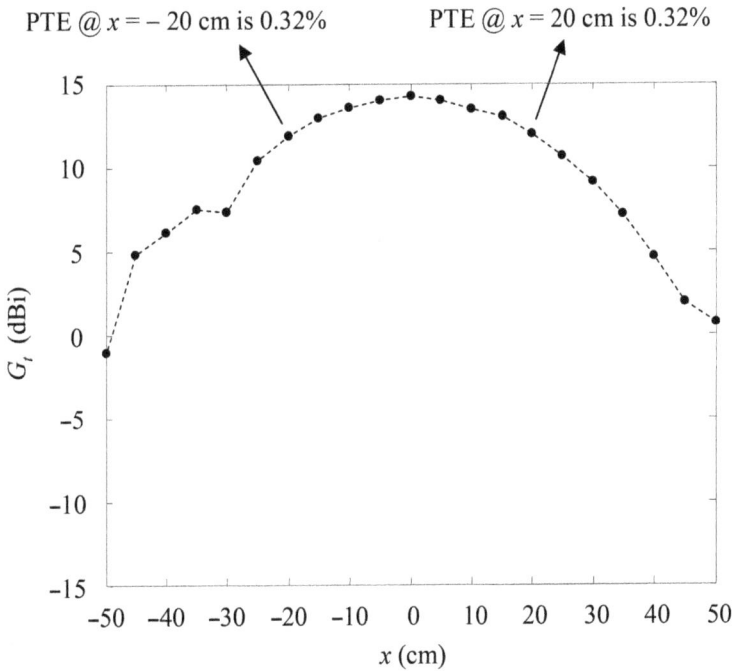

Figure 4.29   Experimental results with two targets broadcasting time-harmonic pilot signals of the same frequency 5.75 GHz; the phase difference between the excitations of two pilot signals is about 0°; G_t is the gain of the 4-element array in reaction to the pilot signals; "PTE" stands for the power transmission efficiency. Reproduced courtesy of Elsevier.

Index of antenna element	1	2	3	4
Normalized amplitude of detected pilot signal	1.0	0.49	1.0	0.98

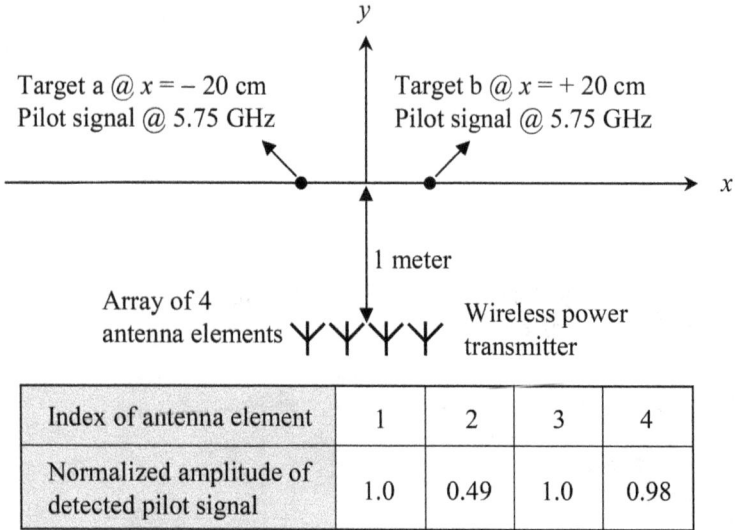

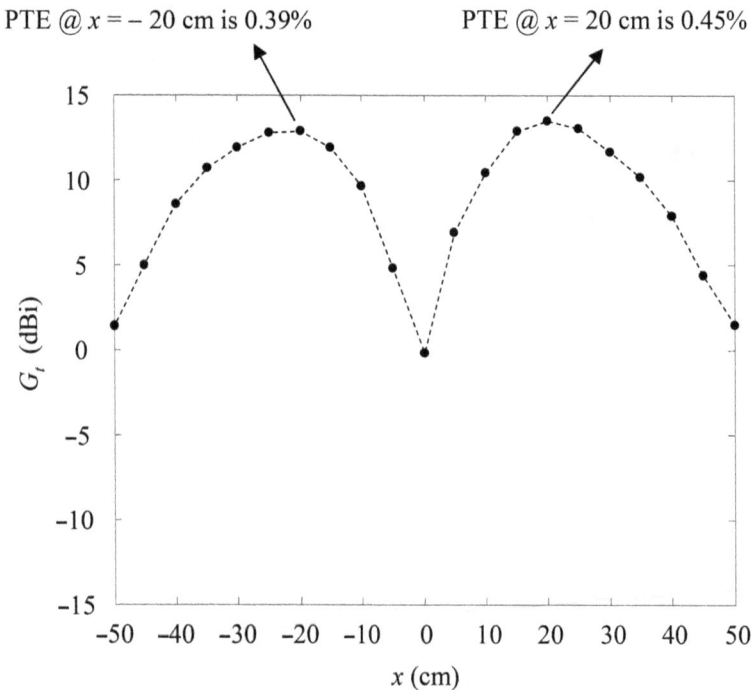

Figure 4.30 *Experimental results with two targets broadcasting time-harmonic pilot signals of the same frequency 5.75 GHz; the phase difference between the excitations of two pilot signals is about 120°; G_t is the gain of 4-element array in reaction to the pilot signals; "PTE" stands for the power transmission efficiency. Reproduced courtesy of Elsevier*

procedure laid out in Figure 4.27. Significant variations of the beamforming performance are observed among the 18 sets of experimental data, which indicate that the phase difference between the two pilot signals generates a strong impact on the performance of retro-reflective beamforming. Out of the 18 sets, two sets' results are presented in Figures 4.29 and 4.30 to demonstrate the variations of beamforming performance. The phase difference between the excitations of two pilot signals is about $0°$ in Figure 4.29 and about $120°$ in Figure 4.30.

In the first step of retro-reflective beamforming, the pilot signals detected by the four antenna elements of the wireless power transmitter are denoted as $p_n(t) = A_n \cos(2\pi ft + \alpha_n)$, $n = 1, 2, 3, 4$, with $f = 5.75\text{GHz}$. The interaction between the pilot signals from two targets creates an interference pattern over the aperture of the 4-element antenna array of wireless power transmitter. Consequently, the four amplitude values $\{A_1, A_2, A_3, A_4\}$ are not uniform among one another. In each of Figures 4.29 and 4.30, the normalized values of $\{A_1, A_2, A_3, A_4\}$ (normalized by the largest value among $\{A_1, A_2, A_3, A_4\}$) are displayed in a table. In Figure 4.29, when the pilot signals from the two targets reach the location of Antenna Element 1 or the location of Antenna Element 4, their phase difference is quite close to $180°$, and as a result, the sum of the two pilot signals has a weak amplitude. Compared with Figure 4.29, a different interference pattern is observed in Figure 4.30. Specifically in Figure 4.30, when the pilot signals from the two targets reach the location of Antenna Element 2, they significantly cancel each other.

In the second step of retro-reflective beamforming, the four antenna elements of the wireless power transmitter are excited by $\gamma A_n \cos\{2\pi ft - \alpha_n\}$, $n = 1, 2, 3, 4$. The antenna gain of the 4-element array, G_t, obtained from the measurement data is plotted in Figures 4.29 and 4.30. In Figure 4.29, little wireless power is radiated by Antenna Element 1 or Antenna Element 4 in the second step of retro-reflective beamforming, since A_1 and A_4 are much smaller than A_2 and A_3. As a result, the power transmission efficiency values at $x = -20$ cm and $x = 20$ cm in Figure 4.29 are lower than those in Figure 4.28, which is in agreement with the observations from Figure 4.22. The wireless power transmitter actually attempts to generate two beams, one toward $x = -20$ cm and one toward $x = 20$ cm. However, because of their large beamwidth, the two beams merge into one beam in Figure 4.29. In Figure 4.30, the fact that Antenna Element 2 is located at a "dark spot" of the interference pattern does not change the aperture size of the 4-element array. As predicted by the numerical results in Figure 4.23, the G_t values of the two power beams in Figure 4.30 are higher than those in Figure 4.28.

Overall, the experimental results in this section excellently verify the numerical analysis in the previous section. It is concluded from Sections 4.4 and 4.5 that, when multiple wireless power beams are generated toward multiple targets in retro-reflective beamforming, certain multiple access techniques (based on frequency-division, time-division, or code-division) must be enforced among the targets' pilot signals.

Based on the experimental setup of Figure 4.28, another target is included as depicted at the top of Figure 4.31. In Figure 4.31, three targets broadcast

Target c @ $x = 0$
Pilot signal @ 5.75 GHz

Target b @ $x = + 20$ cm
Pilot signal @ 5.755 GHz

Target a @ $x = - 20$ cm
Pilot signal @ 5.745 GHz

1 meter

Wireless power transmitter

Measured @ 5.75 GHz
PTE @ $x = 0$ is 0.43%

Measured @ 5.755 GHz
PTE @ $x = 20$ cm is 0.32%

Measured @ 5.745 GHz
PTE @ $x = - 20$ cm is 0.24%

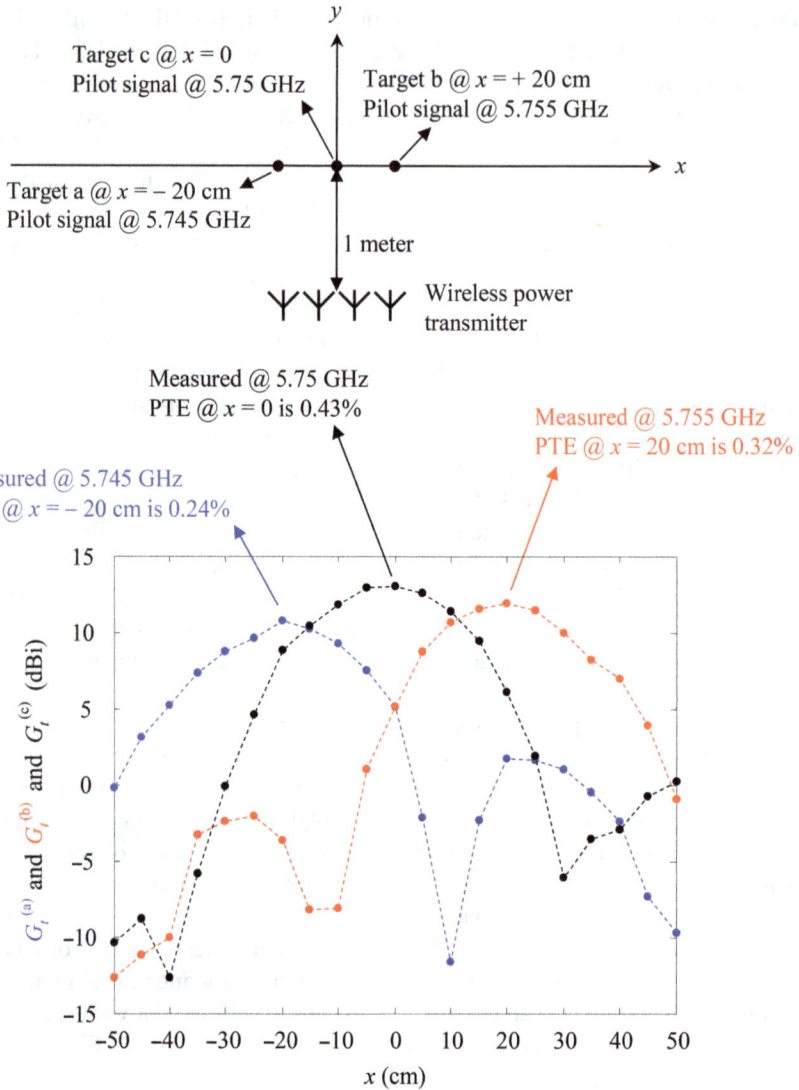

Figure 4.31   Experimental results with three targets broadcasting time-harmonic pilot signals of 5.745 GHz, 5.75 GHz, and 5.755 GHz, respectively; G_t is the gain of the 4-element array in reaction to the pilot signals; "PTE" stands for the power transmission efficiency. Reproduced courtesy of Elsevier.

time-harmonic pilot signals at three different frequencies respectively. The scheme described in Figure 4.28 can be extended to the scenario of three targets straight-forwardly. As long as the wireless power transmitter is able to explicitly resolve the three frequencies of the pilot signals, three power beams are generated by the wireless power transmitter in reaction to the three targets' pilot signals as observed from the experimental results in Figure 4.31. Each power beam in Figure 4.31 is solely dependent on one target's pilot signal; in other words, the interaction among the three targets' pilot signals does not impact the performance of retro-reflective beamforming.

In Figure 4.28, there are two targets, and their pilot signals' frequencies are separated by 10 MHz. In Figure 4.31, the three targets' pilot signals are separated from one another by 5 MHz. In all the experiments of this section, the temporal data length of Fourier transformation is as short as 200 ns. The temporal window of 200 ns corresponds to a spectral resolution of 5 MHz, which is sufficient to resolve the three pilot signals' frequencies in this section. Obviously, if more than three targets are allocated within the 10 MHz bandwidth, the temporal length of the Fourier transformation must be larger to refine the spectral resolution, which will increase the complexity of digital signal processing. In practice, numerous other factors (such as the drift of pilot signals' frequencies) must also be taken into account. Thus, when many targets are densely populated within a certain frequency band, it would be probable for a wireless power transmitter to fail to discriminate their pilot signals from each other. It should be noted that, if the wireless power transmitter fails to discriminate multiple pilot signals from each other, all the complications identified in Section 4.4 will ensue. Therefore, it does not appear to be the optimal resolution to merely rely on discriminating multiple targets' pilot signals in frequency.

References

[1]   Atzori L., Iera A., Morabito G. 'The Internet of Things: A survey'. *Computer Networks*. 2010;54:2787–805.

[2]   Kopetz H. Internet of Things. *Real-Time Systems: Design Principles for Distributed Embedded Applications*. 2nd ed.: Berlin: Springer; 2011.

[3]   Disruptive civil technologies – six technologies with potential impacts on US interests out to 2025. National Intelligence Council, Report No.: Conference Report CR 2008-07, 2008.

[4]   PowerStream ultrathin rechargeable Lithium polymer batteries. [cited 2023]; Available from: https://www.powerstream.com/thin-lithium-ion.htm.

[5]   Molex thin-film battery. [cited 2023]; Available from: https://www.molex.com/en-us/products/printed-circuit-solutions/printed-electronics/thin-film-battery.

[6]   Zhai H., Pan H.K., Lu M. 'A practical wireless charging system based on ultra-wideband retro-reflective beamforming'. *Presented at IEEE Antennas and Propagation Symposium*; Toronto, Canada. 2010.

[7]   Khang S.T., Lee D.J., Hwang I.J., Yeo T.D., Yu J.W. 'Microwave power transfer with optimal number of rectenna arrays for midrange applications'. *IEEE Antennas and Wireless Propagation Letters*. 2018;17(1):155–9.

[8]   Cho Y.S. 'Development of transmitter/receiver front-end module with automatic Tx/Rx switching scheme for retro-reflective beamforming'. *Journal of Information and Communication Convergence Engineering*. 2019;17(3):221–6.

[9]   Hilario Re P.D., Podilchak S.K., Rotenberg S.A., Goussetis G., Lee J. 'Circularly polarized retrodirective antenna array for wireless power transmission'. *IEEE Transactions on Antennas and Propagation*. 2020;68(4):2743–52.

[10]   Koo H., Bae J., Choi W., *et al.* 'Retroreflective transceiver array using a novel calibration method based on optimum phase searching'. *IEEE Transactions on Industrial Electronics*. 2020;68(3):2510–20.

[11]   Fairouz M., Saed M.A. 'A complete system of wireless power transfer using a circularly polarized retrodirective array'. *Journal of Electromagnetic Engineering and Science*. 2020;20(2):139–44.

[12]   Kang Y., Lin X.Q., Li Y., Wang B. 'Dual-frequency retrodirective antenna array with wide dynamic range for wireless power transfer'. *IEEE Antennas and Wireless Propagation Letters*. 2023;22(2):427–31.

[13]   Wang X., Sha S., He J., Guo L., Lu M. 'Wireless power delivery to low-power mobile devices based on retro-reflective beamforming'. *IEEE Antennas and Wireless Propagation Letters*. 2014;13:919–22.

[14]   Fontana R., Ameti A., Richley E., Beard L., Guy D. 'Recent advances in ultra wideband communications systems'. *Presented at IEEE Conference on Ultra Wideband Systems and Technologies*; Baltimore, MD. 2002.

[15]   Lamantia A., Maranesi P.G., Radrizzani L. 'Small-signal model of the Cockcroft-Walton voltage multiplier'. *IEEE Transactions on Power Electronics*. 1994;9(1):18–25.

[16]   He J., Wang X., Guo L., Shen S., Lu M. 'A distributed retro-reflective beamformer for wireless power transmission'. *Microwave and Optical Technology Letters*. 2015;57(8):1873–6.

[17]   Liu M., Wang X., Zhang S., Lu M. 'Theoretical analysis of retro-reflective beamforming schemes for wireless power transmission to multiple mobile targets'. *Presented at IEEE Wireless Power Transfer Conference*; Virtual. 2021.

[18]   Balanis C.A. *Antenna Theory: Analysis and Design*. 3rd ed. New York: Wiley-Interscience; 2005.

[19]   Pabbisetty G., Murata K., Taniguchi K., Mitomo T., Mori H. 'Evaluation of space time beamforming algorithm to realize maintenance-free IoT sensors with wireless power transfer system in 5.7-GHz band'. *IEEE Transactions on Microwave Theory and Techniques*. 2019;67(12):5228–34.

[20]   Sarin A., Avestruz A.-T. 'Code division multiple access wireless power transfer for energy sharing in heterogenous robot swarms'. *IEEE Access*. 2020;8:132121–33.

[21]   Massa A., Oliveri G., Viani F., Rocca P. 'Array designs for long-distance wireless power transmission: State-of-the-art and innovative solutions'. *Proceedings of the IEEE*. 2013;101(6):1464–81.

[22]   Hong W., Jiang Z.H., Yu C., *et al.* 'Multibeam antenna technologies for 5G wireless communications'. *IEEE Transactions on Antennas and Propagation*. 2017;65(12):6231–49.

Chapter 5

Retro-reflective beamforming technique for microwave power transmission in space solar power applications

The concept of space solar power was proposed in the 1960s as a potential resolution to the global energy crisis. If successfully implemented, solar power on the order of Giga-Watts would be harvested by artificial satellites over the Earth's geostationary orbit and then be delivered to the Earth wirelessly. Wireless power transmission is a critical element of the space solar power concept. This chapter studies applying the retro-reflective beamforming technique to achieve efficient wireless power transmission from a geostationary satellite to the Earth. In Section 5.1, the basic concepts pertinent to space solar power applications are narrated. A satellite-borne retro-reflective antenna array is modeled in Section 5.2 for the wireless power transmission from a geostationary satellite to the Earth. In Sections 5.3 and 5.4, the formulations in Section 2.1 (which are under the condition of electrical far zone) and the formulations in Section 2.4 (which are under the condition of electrical near zone) are employed to analyze the retro-reflective beamforming antenna array modeled in Section 5.2, respectively. One bench-scale experimental example is presented in Section 5.5 to demonstrate the fundamental scheme of wireless power transmission from a geostationary satellite to the Earth. The topic of wireless power reception over a ground station on the Earth is discussed briefly in Section 5.6.

5.1 Basic concepts of space solar power

The concept of space solar power satellites (SSPS) was proposed in 1968 [1]. It aims to harvest solar power using satellites over the Earth's geostationary orbit and then deliver the harvested power to the Earth wirelessly. If successfully implemented, SSPS would supply power on the order of Giga-Watts to the Earth steadily and continuously, and thus is anticipated to be a resolution to the global energy crisis mankind is facing. Compared with solar power harvesting on the Earth, solar power harvesting in the outer space offers tremendous benefits. Solar power is not available at night on the Earth. Consequently, solar power harvesting on the Earth must be coupled with either another energy source or an energy storage system in practice. When the location of a space solar power satellite is selected appropriately, in contrast, the satellite is almost always under the sunshine. Moreover, solar power harvesting in the outer space is not impacted by the Earth's atmospheric conditions whereas the performance of solar power harvesting on the Earth heavily depends on the weather conditions. In summary, space-based solar power harvesting does not suffer from the shortcomings of discontinuity or instability as most of the renewable energy sources do.

As an extremely complex engineering system, the feasibility of SSPS is still under assessment/discussion to date. The readers are referred to [2] for an overview of the engineering/technical topics covered by SSPS.

A variety of system designs have been proposed for the practical implementation of SSPS; one of them is depicted in Figure 5.1 [3]. A space solar power satellite in Figure 5.1 includes one planar solar panel and one planar antenna panel (needless to say, the satellite includes numerous other components, albeit the solar panel and antenna panel are the two most bulky ones). The solar power harvested by the solar panel is delivered to the antenna panel and then radiated toward the Earth in the form of a microwave beam by the antenna panel. While the Earth rotates (with one round per day) and the Earth orbits around the sun (with one round per year), the solar panel must keep facing the sun and the antenna panel must keep facing the Earth as shown in Figure 5.2. The two requirements above could not be satisfied simultaneously if there is no relative motion between the solar panel and antenna panel. As a result, rotary joints are proposed to fulfill the mechanical connection between the solar panel and antenna panel in Figures 5.1 and 5.2. The aperture of the solar panel is on the order of 10 $(km)^2$ in order to collect solar power of Giga-Watt level, and the aperture of the antenna panel is on the order of 1 $(km)^2$ in order to efficiently transmit the power to the Earth. Due to their large physical aperture, both the solar panel and antenna panel must be assembled in the outer

Figure 5.1   One design of a space solar power satellite, including a planar solar panel and a planar antenna panel. Reproduced from [3], courtesy of Aerospace China.

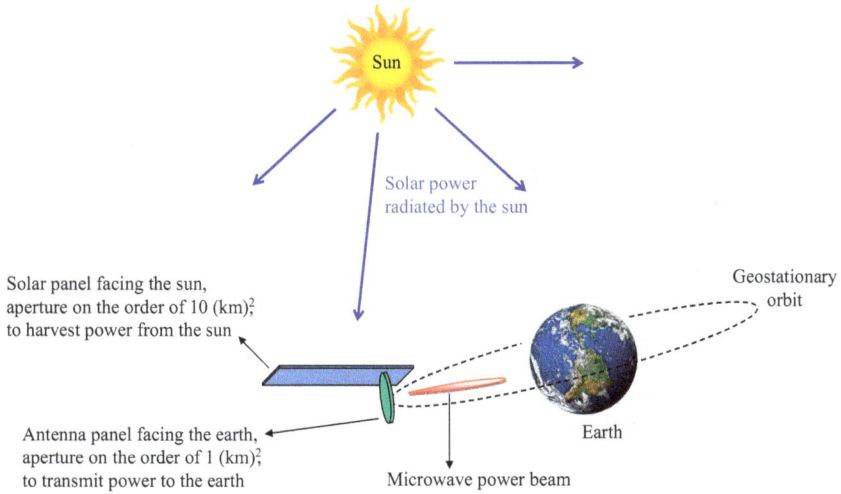

Sun

Solar power
radiated by the sun

Solar panel facing the sun,
aperture on the order of 10 (km)²,
to harvest power from the sun

Geostationary
orbit

Antenna panel facing the earth,
aperture on the order of 1 (km)²,
to transmit power to the earth

Earth

Microwave power beam

Figure 5.2   Further illustration of the space solar power satellite design in
Figure 5.1

space after parts are transported from the Earth via multiple space launches. Obviously, the assembly process would be much easier when the solar panel and antenna panel have planar configurations rather than nonplanar configurations. Nevertheless, the efficient transmission of electrical power on the order of Giga-Watts through rotary joints involves a range of unprecedented technical problems. Therefore, rotary joints between the solar power collection part and the antenna part are avoided intentionally in some other designs of space solar power satellites [4]. For instance, nonplanar antenna configurations are proposed for SSPS in [5] such that the mechanical connections between the solar power collection part and the antenna part are rigid.

Comparative studies among various designs of space solar power satellites are beyond the scope of this book. Chapter 5 of this book focuses on one technical aspect of SSPS: Wireless power transmission from a geostationary satellite to the Earth. Specifically, this chapter aims to investigate applying the retro-reflective beamforming technique to achieve efficient wireless power transmission in the context of SSPS applications. The antenna panel is assumed to have a planar configuration in this chapter, although it is possible that the planar configuration is not optimal from the perspective of system design as argued in the previous paragraph.

5.2   Theoretical model of wireless power transmission from a geostationary satellite to the Earth based on retro-reflective beamforming technique

As depicted in Figures 5.1 and 5.2, the solar power harvested by a satellite over the Earth's geostationary orbit is transmitted to the Earth by wireless means, with the distance of wireless power transmission being as large as approximately 36,000 km. During the past few decades, enormous research efforts have been reported on the wireless power transmission from a space solar power satellite to the Earth. Reference [6] (which was published in 2013) provides a comprehensive description of the research endeavors before 2013. The recent and ongoing developments pertinent to wireless power transmission in SSPS applications were reviewed in [7–9]. The microwave frequency range between 1 GHz and 10 GHz is found suitable for carrying the wireless power in SSPS applications after a large number of practical factors are taken into account [4,6]. Two Industrial, Scientific, and Medical (ISM) frequency bands, which are around 2.45 GHz and 5.8 GHz respectively, are believed to be particularly excellent candidates.

To transmit wireless power efficiently over 36,000 km, the antenna panel over the geostationary satellite must have a large physical aperture (on the order of 1 (km)^2, to be specific), as analyzed in Section 5.3. For the Earth, the physical position and attitude/orientation of the antenna panel are not completely fixed. Rather, the antenna panel's physical position and attitude/orientation are maintained around the desired status by a sophisticated mechanical control system over the satellite. To mitigate power loss and possible hazards, it is imperative to ensure that most of the wireless power transmitted by the satellite is collected by a designated ground station on the Earth when the antenna panel's physical condition is under constant change. Therefore, the wireless power transmission from a geostationary satellite to the Earth must meet two technical requirements. First, a narrow beam needs to be generated by the antenna panel as the carrier of wireless power. Second, the beam needs to be steered toward the designated ground station precisely in response to the slight change in the antenna panel's position or attitude.

Indeed, the Internet of Things applications (which are discussed in Chapter 4) impose two similar requirements for wireless power transmission. Specifically in the Internet of Things, a wireless power transmitter is required to deliver wireless power to a mobile wireless power receiver (e.g., an RFID tag) through a narrow beam, and the narrow beam must be reconfigured to follow the location of the mobile wireless power receiver in real-time. Now that the retro-reflective beamforming technique has the potential to address the two requirements imposed by the Internet of Things, it should apply to SSPS applications without fundamental alterations. Following the fundamental two-step procedure of retro-reflective beamforming, the wireless power transmission between a satellite and the Earth could be accomplished as follows.

> *Step (i)* A pilot signal is broadcasted from the Earth to a satellite, and the pilot signal is detected and analyzed by the satellite.
> *Step (ii)* Based on the outcome of analyzing the pilot signal, a microwave power beam is constructed by the satellite to the Earth, from which the pilot signal originates.

Of course, there are numerous differences when the retro-reflective beamforming technique is applied to the two types of applications, i.e., Internet of Things applications and SSPS applications. Some of the differences are articulated in Table 5.1. Obviously, the wireless power transmission in SSPS application is a lot

*Table 5.1   Differences between "wireless power transmission for Internet of
Things applications" and "wireless power transmission for space solar
power applications"*

	Wireless power transmission for Internet of Things applications	Wireless power transmission for space solar power applications
Distance of wireless power transmission	< 10 meters	~ 36,000,000 meters
Power level transmitted by wireless power transmitter	~ 1 Watt	> 1 Giga-Watt
Power level delivered to wireless power receiver	~ 10 milli-Watts	> 1 Giga-Watt
Electrical far zone?	No	No
Geometrical far zone?	No	Yes
Duplexing between pilot signal and wireless power	Time-division	Frequency-division

more complex/difficult, and it calls for vast investments. Consequently, the SSPS
project would be considered a failure if its ultimate power output does not reach the
level of Giga-Watt. As discussed in the previous chapters, the transmitted power level
in Internet of Things applications is severely restricted by a range of established
policies/regulations. It is thus envisioned that customized policies/regulations would be
prescribed and designated for SSPS applications in the future. In the Internet of Things
applications, the physical size of wireless power transmitters and wireless power
receivers is limited, and as a result, high power transmission efficiency is quite unlikely. A power transmission efficiency as poor as 1% may be acceptable in the Internet
of Things since the associated power loss is a few Watts at most. In SSPS applications,
however, a decrease in power transmission efficiency by 1% is equivalent to a power
loss on the order of ten million Watts. Therefore, the targeted power transmission
efficiency of SSPS must be as high as possible, which in turn requires the wireless
power transmitter and wireless power receiver to have large physical dimensions. The
last three rows of Table 5.1 are elucidated in Sections 5.3, 5.4, and 5.5.

The ideal scenario of wireless power transmission based on retro-reflective
beamforming in SSPS applications is modeled in Figure 5.3. The antenna panel
over a space solar power satellite includes a planar retro-reflective antenna array. A
pilot signal is broadcasted from a ground station on the Earth. After the pilot signal
is detected and analyzed by the retro-reflective antenna array, a microwave power
beam is constructed by the retro-reflective antenna array toward the ground station,
from which the pilot signal originates. The retro-reflective antenna array resides in
the x-z plane, as depicted in Figure 5.3. There are $N \times N$ antenna elements deployed
along the x direction and z direction. The antenna elements in the array are identical
to one another. They are assumed to be either microstrip antennas or slot antennas
[10] printed over a large metallic plate such that they have little radiation in the

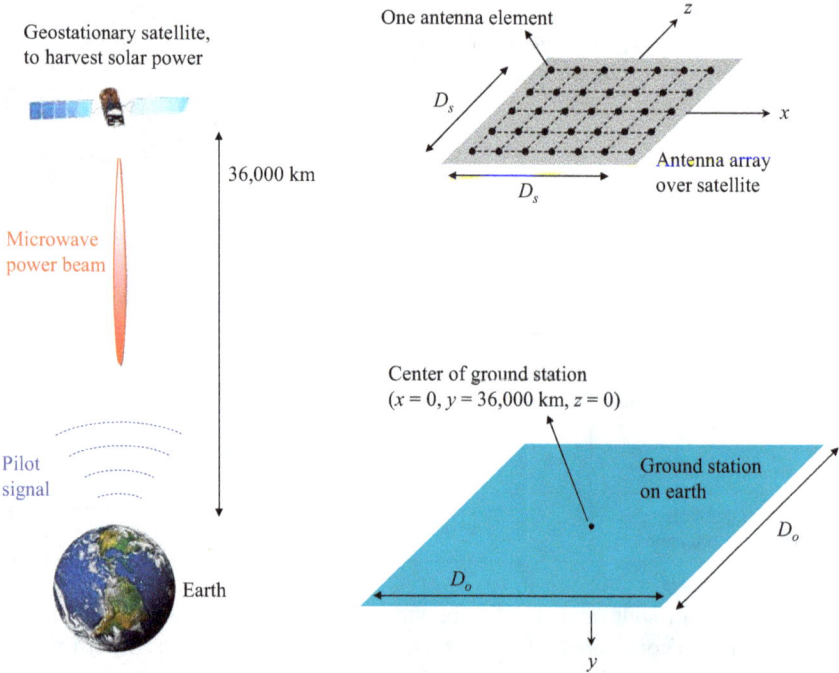

Figure 5.3   Ideal scenario of wireless power transmission from a geostationary satellite to the Earth based on retro-reflective beamforming technique (not to scale)

half-space of $y < 0$. The spacing between two adjacent antenna elements is s_x along the x direction and s_z along the z direction. In this chapter, it is assumed that $s_x = s_z = s_0$. The physical dimensions of the retro-reflective antenna array are $D_s \times D_s$, with $D_s = N \times s_0$ approximately. A two-dimensional linear array configuration is assumed in this chapter, such that the analysis of this chapter is consistent with the previous few chapters. It is worth noting that the retro-reflective beamforming technique, which is the theme of this book, applies to other array configurations such as the nonplanar configuration in [5]. The geometrical center of the antenna array is designated as the spatial origin. The position vector of each antenna element is denoted by

$$\mathbf{r}'_{mn} = \hat{\mathbf{x}}\left(m - \frac{N+1}{2}\right)s_0 + \hat{\mathbf{z}}\left(n - \frac{N+1}{2}\right)s_0, \tag{5.1}$$

$m = 1, 2, \ldots, N$, $n = 1, 2, \ldots, N$. The ground station is centered at $(x = 0, \ y = d, \ z = 0)$, where $d = 36,000$ km is the distance between the geostationary satellite and the Earth. The ground station is assumed to have a square aperture parallel to the x-z plane with a side length of D_o. In this chapter, D_o is assumed to be as large as several kilometers such that most of the wireless power radiated by the retro-reflective antenna array is collected by the ground station. The pilot signal is broadcasted from the center of the ground station.

The practical scenario of wireless power transmission in SSPS applications would deviate from the ideal model in Figure 5.3, due to the following two facts.

First, the antenna array's position and attitude are not fixed with respect to the ground station. Specifically, the antenna array's position and attitude are maintained around the desired status by a mechanical control system over the space solar power satellite. Figure 5.4 illustrates the practical scenario with the antenna array's position and attitude deviating from the ideal status. The geometrical center of the antenna array is the spatial origin. The ground station's aperture is parallel to the x-z plane centered at $(x = 0, \ y = d, \ z = 0)$ where the value of d varies slightly around 36,000 km. The attitude deviation is characterized by two parameters: $\Delta\theta$ and $\Delta\phi$. The direction normal to the antenna array's aperture is

$$\hat{\mathbf{n}} = \hat{\mathbf{x}} \sin\left(90^\circ + \Delta\theta\right)\cos\left(90^\circ + \Delta\phi\right) + \hat{\mathbf{y}} \sin\left(90^\circ + \Delta\theta\right)\sin\left(90^\circ + \Delta\phi\right)$$

$$+ \hat{\mathbf{z}} \cos\left(90^\circ + \Delta\theta\right).$$

$$(5.2)$$

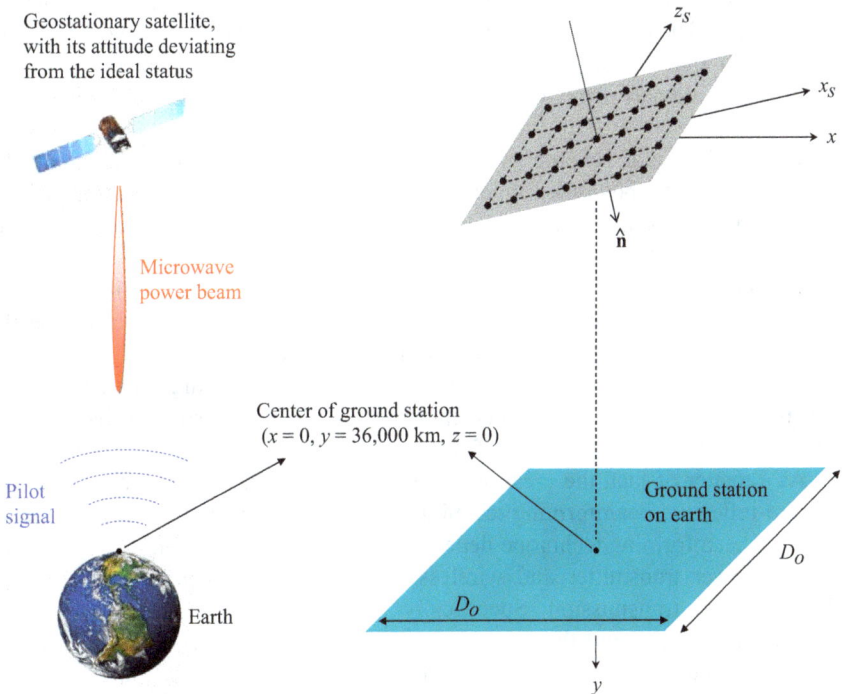

Figure 5.4   *Practical scenario of wireless power transmission from a geostationary satellite to the Earth with the antenna array's position and attitude deviating from the ideal status (not to scale)*

The direction normal to the antenna array's aperture, $\hat{\mathbf{n}}$, reverts to $\hat{\mathbf{y}}$ (i.e., the ideal scenario) when $\Delta\theta$ and $\Delta\phi$ are zero. The antenna elements are deployed along $\hat{\mathbf{x}}_s$ and $\hat{\mathbf{z}}_s$, with

$$\hat{\mathbf{x}}_s = \hat{\mathbf{x}} \sin(90° + \Delta\phi) - \hat{\mathbf{y}} \cos(90° + \Delta\phi), \tag{5.3}$$

$$\hat{\mathbf{z}}_s = -\hat{\mathbf{x}} \cos(90° + \Delta\theta)\cos(90° + \Delta\phi) - \hat{\mathbf{y}} \cos(90° + \Delta\theta)\sin(90° + \Delta\phi)$$

$$+ \hat{\mathbf{z}} \sin(90° + \Delta\theta).$$

$$\tag{5.4}$$

The location of antenna elements in Figure 5.4 is

$$\mathbf{r}'_{mn} = \hat{\mathbf{x}}_s \left(m - \frac{N+1}{2} \right) s_0 + \hat{\mathbf{z}}_s \left(n - \frac{N+1}{2} \right) s_0, \tag{5.5}$$

$m = 1, 2, 3, \ldots, N, n = 1, 2, 3, \ldots, N.$

Second, only a very limited region over the Earth's geostationary orbit is suitable for the location of a space solar power satellite. The selection of the satellite's location must ensure that the satellite is always under the sunshine, in other words, there is no night at the satellite's location. The ground station is assumed to be located at $(x = 0, y = d, z = 0)$ in Figure 5.5(a). However, in practice, it is unlikely that a ground station would be built over the Earth's equator right underneath a space solar power satellite. In other words, very probably a ground station would deviate from the $(x = 0, y = d, z = 0)$ location, as depicted by Figure 5.5(b).

The fundamental scheme of the retro-reflective beamforming technique is insensitive to the above-mentioned deviations, including the deviation of the antenna array's position, the deviation of the antenna array's attitude, and the deviation of the ground station's location. As discussed in the previous few chapters, the underlying physical principle of the retro-reflective beamforming technique is channel reciprocity. The aforementioned deviations do not invalidate the channel reciprocity between the satellite and the ground station. As a result, the retro-reflective beamforming technique is capable of generating a narrow microwave power beam toward the ground station as long as a pilot signal is broadcasted from the ground station in Figures 5.3–5.5.

As a matter of fact, the location of the ground station is unimportant as far as the retro-reflective beamforming technique is concerned. This is because the retro-reflective beamforming technique depends on the (reciprocal) channel between the wireless power transmitter and wireless power receiver when it is applied for wireless power transmission. Specifically, the relative geometrical relationship between the satellite and the ground station is critical for retro-reflective beamforming in Figure 5.5, while the absolute location of the ground station is not. In other words, the two plots in Figure 5.5 have no fundamental differences between each other, as long as the satellite and the ground station have line-of-sight interaction. Thus in the rest of this chapter, the ground station is assumed to be located at $(x = 0, y = d, z = 0)$, which facilitates the theoretical analysis.

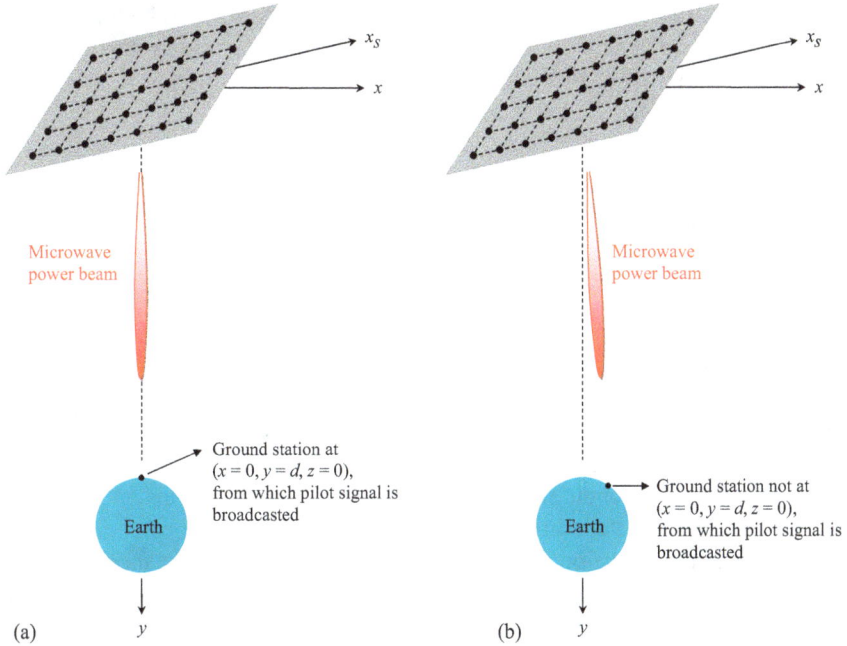

Figure 5.5 *Wireless power transmission from a geostationary satellite to the Earth, with the ground station's location (a) at (x = 0, y = d, z = 0) and (b) not at (x = 0, y = d, z = 0)*

In SSPS applications, the variation of value d is very small with respect to 36,000 km. Thus in this chapter, d is assumed to have a constant value of 36,000 km. It is not difficult to show that a little deviation of value d away from 36,000 km does not generate any significant impacts on retro-reflective beamforming.

Compared with the deviation of the antenna array's location and the deviation of the ground station's location, the deviation of the antenna array's attitude may have a strong influence on the performance of wireless power transmission, and thus it is analyzed carefully in the rest of this chapter. It is assumed that $\Delta\theta$ varies in the range of $[-(\Delta\theta)_{max}, (\Delta\theta)_{max}]$ and $\Delta\phi$ varies in the range of $[-(\Delta\phi)_{max}, (\Delta\phi)_{max}]$. In practice, $(\Delta\theta)_{max}$ and $(\Delta\phi)_{max}$ are determined by the mechanical control system that maintains the satellite's physical attitude around the desired status. It is typical that $(\Delta\theta)_{max}$ and $(\Delta\phi)_{max}$ are (much) smaller than $1°$ for a regular satellite with today's sophisticated satellite control technology [11]. Nevertheless, controlling a physical structure with size on the order of kilometers in outer space is an unprecedented engineering problem, and thus $(\Delta\theta)_{max}$ and $(\Delta\phi)_{max}$ associated with a space solar power satellite are currently hard to predict. In some of the numerical examples of this chapter, $(\Delta\theta)_{max}$ and $(\Delta\phi)_{max}$ are as large as $20°$, which is likely an over-estimation.

The medium between the satellite and the Earth is assumed to be free space in this chapter. Though not rigorous, this assumption tremendously facilitates the theoretical analysis of this chapter. The validity of this assumption is a complex/difficult research topic [2] and is beyond the scope of this book.

5.3   Theoretical analysis of retro-reflective antenna array in SSPS applications using the formulations of Section 2.1 under the condition of electrical far zone

In this section and the next section (i.e., Sections 5.3 and 5.4), the retro-reflective antenna array modeled in Figures 5.3 and 5.4 is analyzed theoretically to estimate the performance of wireless power transmission from a geostationary satellite to the Earth. Sections 5.3 and 5.4 are based on the formulations of Section 2.1 (which are under the condition of electrical far zone) and the formulations of Section 2.4 (which are under the condition of electrical near zone), respectively.

In Chapter 2, two "conditions of far zone" are defined: The condition of geometrical far zone and the condition of electrical far zone. The distance between a geostationary satellite and the Earth is as large as 36,000 km. Thus, the condition of geometrical far zone is undoubtedly satisfied in SSPS applications. Nevertheless, the condition of electrical far zone is not satisfied if a high power transmission efficiency (greater than 50%, for instance) is desired. This section and the next section (i.e., Sections 5.3 and 5.4) intend to analyze the performance of a satellite-borne retro-reflective antenna array using the formulations of the electrical far zone as well as the formulations of the electrical near zone in the context of SSPS applications.

Consider the ideal configuration in Figure 5.3. Suppose each antenna element of the antenna array is excited by a time-harmonic voltage with phasor $X_{mn}e^{j\psi_{mn}}$, $m = 1, 2, 3, \ldots, N$, $n = 1, 2, 3, \ldots, N$. The electric field radiated by the antenna array is observed at an observation point $\mathbf{r}_o$ with spherical coordinates of (r_o, θ_o, ϕ_o). When the observation point is in each antenna element's far zone, the electric field observed at $\mathbf{r}_o$ is

$$\mathbf{E}(\mathbf{r}_o) = \sum_{m=1}^{N}\sum_{n=1}^{N} X_{mn}e^{j\psi_{mn}}\mathbf{U}_{mn}\frac{e^{-jk_0|\mathbf{r}_o-\mathbf{r}'_{mn}|}}{|\mathbf{r}_o - \mathbf{r}'_{mn}|}. \tag{5.6}$$

In (5.6), k_0 is the wavenumber in free space as free space is assumed to be the medium between the satellite and the Earth. Under the condition of geometrical far zone (i.e., $r_o \gg D_s$ as defined in Chapter 2), the following two relationships are approximately valid,

$$\frac{1}{|\mathbf{r}_o - \mathbf{r}'_{mn}|} = \frac{1}{r_o}, \quad \mathbf{U}_{mn} = \mathbf{U}_0,$$

where $\mathbf{U}_0$ represents the polarization direction of the electric field radiated by one antenna element when the antenna element resides at the spatial origin and is excited by a voltage of phasor value 1. Moreover, assume that $X_{mn} = X_0$, that is, the antenna elements are excited with a uniform amplitude of X_0. Then, (5.6) becomes

$$\mathbf{E}(\mathbf{r}_o) = \frac{X_0 U_0}{r_o} \sum_{m=1}^{N} \sum_{n=1}^{N} e^{j\psi_{mn}} e^{-jk_0|\mathbf{r}_o - \mathbf{r}'_{mn}|}. \tag{5.7}$$

The two-fold sum in (5.7) is usually termed the array factor. Following Chapter 2, the constructive coefficient Φ is defined based on the array factor:

$$\Phi(\mathbf{r}_o) = \frac{1}{N^2} \sum_{m=1}^{N} \sum_{n=1}^{N} e^{j\psi_{mn}} e^{-jk_0|\mathbf{r}_o - \mathbf{r}'_{mn}|}. \tag{5.8}$$

Under the condition of electrical far zone (i.e., $r_o > 2(\sqrt{2}D_s)^2/\lambda_0$), the term $|\mathbf{r}_o - \mathbf{r}'_{mn}|$ in (5.8) can be expressed as $|\mathbf{r}_o - \mathbf{r}'_{mn}| = r_o - \hat{\mathbf{r}}_o \cdot \mathbf{r}'_{nm}$, following the derivations in Section 2.1. Substituting $|\mathbf{r}_o - \mathbf{r}'_{mn}| = r_o - \hat{\mathbf{r}}_o \cdot \mathbf{r}'_{nm}$ into (5.8) leads to

$$\Phi(\mathbf{r}_o) = \frac{1}{N^2} e^{-jk_0 r_o} \sum_{m=1}^{N} \sum_{n=1}^{M} e^{j\psi_{mn}} e^{jk_0 \hat{\mathbf{r}}_o \cdot \mathbf{r}'_{mn}}, \tag{5.9}$$

where

$$\hat{\mathbf{r}}_o = \hat{\mathbf{x}} \sin(\theta_o)\cos(\phi_o) + \hat{\mathbf{y}} \sin(\theta_o)\sin(\phi_o) + \hat{\mathbf{z}} \cos(\theta_o). \tag{5.10}$$

The dot product $\hat{\mathbf{r}}_o \cdot \mathbf{r}'_{mn}$ in (5.9) is evaluated as

$$\begin{aligned}
\hat{\mathbf{r}}_o \cdot \mathbf{r}'_{mn} &= (\hat{\mathbf{r}}_o \cdot \hat{\mathbf{x}}) \left(m - \frac{N+1}{2}\right) s_0 + (\hat{\mathbf{r}}_o \cdot \hat{\mathbf{z}}) \left(n - \frac{N+1}{2}\right) s_0 \\
&= \left(m - \frac{N+1}{2}\right) s_0 \sin(\theta_o)\cos(\phi_o) + \left(n - \frac{N+1}{2}\right) s_0 \cos(\theta_o)
\end{aligned} \tag{5.11}$$

by making use of (5.1) and (5.10).

According to the derivations of Section 2.1.2, the radiation pattern of the antenna array in Figure 5.3 would exhibit a beam toward the $+y$ direction when $\psi_{mn} = 0$. With $\psi_{mn} = 0$, the constructive coefficient in (5.9) is

$$\Phi(\mathbf{r}_o) = \frac{1}{N^2} e^{-jk_0 r_o} \times \sum_{m=1}^{N} e^{jk_0 \left(m - \frac{N+1}{2}\right) s_0 \sin(\theta_o)\cos(\phi_o)} \times \sum_{n=1}^{N} e^{jk_0 \left(n - \frac{N+1}{2}\right) s_0 \cos(\theta_o)}. \tag{5.12}$$

Due to the following identity

$$\sum_{n=1}^{N} e^{jn\xi} = \frac{\sin\left\{\frac{N\xi}{2}\right\}}{\sin\left\{\frac{\xi}{2}\right\}} e^{j\frac{N+1}{2}\xi}, \tag{5.13}$$

the constructive coefficient in (5.12) can be evaluated in closed form:

$$\Phi(r_o) = \frac{1}{N^2} e^{-jk_0 r_o} \times \frac{\sin\left\{\frac{Nk_0 s_0 \sin(\theta_o)\cos(\phi_o)}{2}\right\}}{\sin\left\{\frac{k_0 s_0 \sin(\theta_o)\cos(\phi_o)}{2}\right\}} \times \frac{\sin\left\{\frac{Nk_0 s_0 \cos(\theta_o)}{2}\right\}}{\sin\left\{\frac{k_0 s_0 \cos(\theta_o)}{2}\right\}}. \qquad (5.14)$$

Some numerical results evaluated via (5.14) are plotted in Figure 5.6. In Figure 5.6, the frequency $f = 5.8$ GHz (which is believed to be one of the candidate carrier frequencies of SSPS applications [6]), the wavelength in free space $\lambda_0 = 5.17$ cm, the inter-element spacing $s_0 = 0.7 \times \lambda_0$, $N = 27,633$, and the physical dimension of the antenna array $D_s = N \times s_0$ is approximately 1,000 m. The amplitude of constructive coefficient $|\Phi|$ is plotted when (θ_o, ϕ_o) vary around $(\theta_o = 90°, \phi_o = 90°)$. Obviously, the plot in Figure 5.6 demonstrates a narrow beam. Specifically, $|\Phi| = 1$ when $\theta_o = 90°$ and $\phi_o = 90°$, indicating that the beam is toward $+y$ direction. Letting

$$\frac{Nk_0 s_0 \cos(\theta_{null})}{2} = \pi \qquad (5.15)$$

results in $\theta_{null} = 90° \pm 0.003°$. When $\theta_o = \theta_{null}$ is substituted into (5.14), Φ is 0, which can be verified by observing Figure 5.6. Similarly, a null appears when $\phi_o = \phi_{null} = 90° \pm 0.003°$. Suppose the beam is projected onto an observation plane of "$y = d = 36,000$ km." It is easy to find that $(\theta_o = 90° - 0.003°, \phi_o = 90°)$ corresponds to the x coordinate of $x_o = 0$ and z coordinate of $z_o = d \tan(0.003°) = 1.86$ km over the "$y = d$" plane. The ranges of $\theta_o \in [90° - 0.0032°, \ 90° + 0.0032°]$ and $\phi_o \in [90° - 0.0032°, \ 90° + 0.0032°]$ in Figure 5.6 are selected such that they correspond to $x_o \in [-2 \text{ km}, 2 \text{ km}]$ and $z_o \in [-2 \text{ km}, \ 2 \text{ km}]$ over the "$y = d$" plane. According to the plot in Figure 5.6, if a

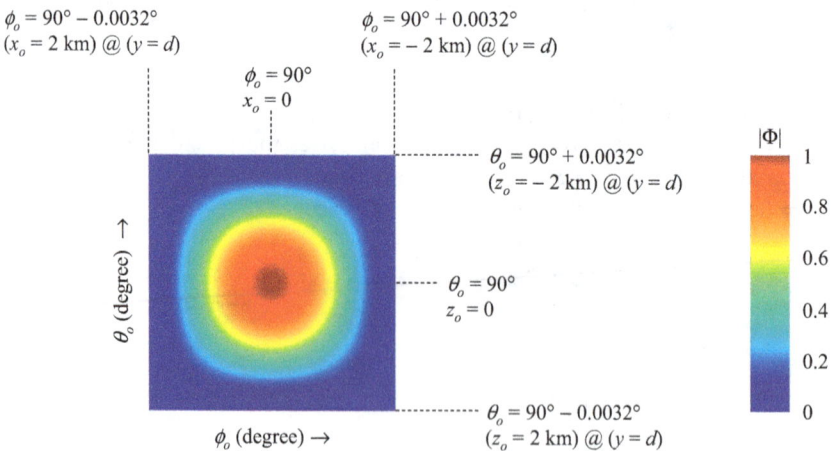

Figure 5.6   Constructive coefficient $|\Phi|$ calculated via (5.14) when (θ_o, ϕ_o) vary around $(\theta_o = 90°, \phi_o = 90°)$

ground station centered at $(x = 0, y = d, z = 0)$ has a square aperture with a side length of $D_o = 3.7$ km, it would be able to capture most of the power transmitted by the antenna array, in other words, the power transmission efficiency would approach 100%. This conclusion is consistent with the findings of [12,13]. In [12,13], it was found that the power transmission efficiency would approach 100% when $\tau = \sqrt{A_t A_r}/(\lambda_0 d)$ is greater than 2, where A_t and A_r are the surface areas of the wireless power transmitter's aperture and wireless power receiver's aperture, respectively. With $\sqrt{A_t} = D_s = 1$ km, $\sqrt{A_r} = D_o = 3.7$ km, $\lambda_0 = 5.17$ cm, and $d = 36,000$ km, the value of τ is calculated to be 1.99. Therefore, the numerical results in Figure 5.6 are in excellent agreement with the theoretical analysis of [12,13]. In addition, it is obvious that the power transmission efficiency is determined by the product of D_s (the size of the antenna array) and D_o (the size of the ground station). In the numerical example of Figure 5.6, $D_s = 1$ km, and D_o needs to be 3.7 km if a power transmission efficiency close to 100% is desired. If D_s is reduced by 10 times to 100 m, the ground station would reach an unaffordable size of 37 km by 37 km. Thus, it appears that the physical dimension of the antenna array over a space solar power satellite ought to be on the order of 1 km in order for a practically viable ground station to achieve high power transmission efficiency in SSPS applications.

Although Figure 5.6 portrays a rough picture of the wireless power transmission from a space solar power satellite to the Earth, the numerical results in Figure 5.6 are not precise as they are evaluated via (5.14). Equation (5.14) is equivalent to (5.8) under the condition of electrical far zone. When applied to Figure 5.3, the condition of electrical far zone is $d > 2(\sqrt{2}D_s)^2/\lambda_0$, which is not satisfied with $D_s = 1$ km, $\lambda_0 = 5.17$ cm, and $d = 36,000$ km. As an example of the inaccuracy of (5.14), with the parameters of Figure 5.6 (i.e., $f = 5.8$ GHz, $\lambda_0 = 5.17$ cm, $s_0 = 0.7\lambda_0$, $N = 27,633$, and $\psi_{mn} = 0$), $|\Phi|$ evaluated at an observation point $(x_o = 0, y = d, z_o = 0)$ is 0.984 through (5.8) but is 1 through (5.14). In terms of power, the difference between 1 and "$(0.984)^2 \cong 0.97$" is about 3%. Since a space solar power satellite intends to deliver power on the order of Giga-Watts to the Earth, a 3% error might result in an enormous amount of power loss in SSPS applications. Therefore, it is necessary to analyze the wireless power transmission in SSPS applications under the condition of electrical near zone.

It is obvious that D_s and D_o, which are on the order of 1 km, are much smaller than $d = 36,000$ km. As a result, the condition of geometrical far zone is satisfied in SSPS applications, as specified in Table 5.1.

The fact that a space solar power satellite and the Earth are in each other's electrical near zone can be visualized clearly by analyzing the pilot signal propagation. In the first step of retro-reflective beamforming, a pilot signal is broadcasted from the Earth to the satellite. As shown in Figure 5.7, a time-harmonic pilot signal at frequency $f = 5.8$ GHz is transmitted from $\mathbf{r}_{ps}$ with coordinates of $(x = 0, y = d, z = 0)$. The pilot signal is received by a pilot signal receiver, and the phase angle of the detected pilot signal is obtained. It is assumed that the physical sizes of the pilot signal transmitter and pilot signal receiver are both on the order of λ_0. With respect to the distance of 36,000 km, their physical sizes can both be modeled as zero. When the pilot signal receiver is located at $(x = x_s, y = 0, z = z_s)$, the phase angle of

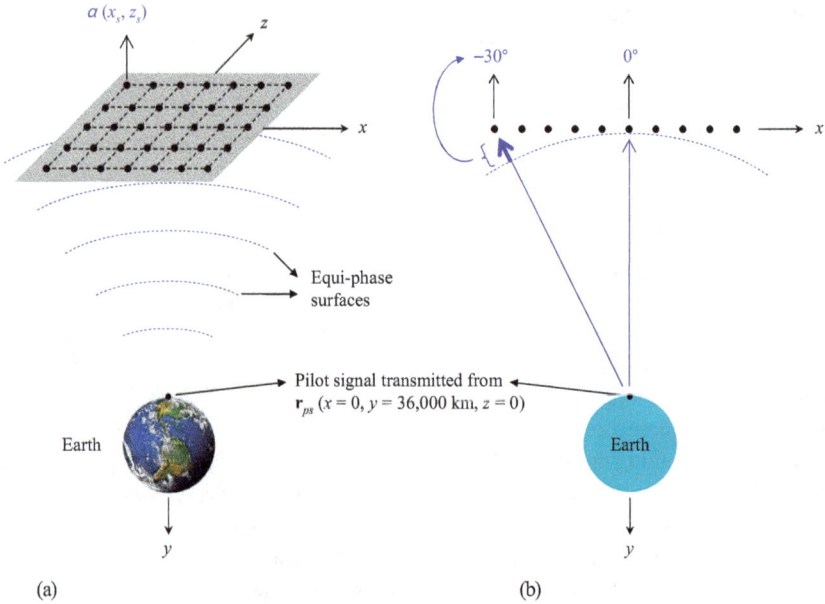

Figure 5.7   Illustration of pilot signal propagation from Earth to satellite: (a) three-dimensional view and (b) view in the x-y plane

the detected pilot signal is denoted as $a(x_s, z_s)$. The phase angle at the spatial origin, $a_0 = a(0,0)$, is used as the phase reference. When x_s and z_s are much smaller than d, the distribution of $a(x_s, y_s) - a_0$ is solely determined by the path of line-of-sight propagation:

$$a(x_s, z_s) - a_0 = -k_0 \sqrt{d^2 + (x_s)^2 + (z_s)^2} + k_0 d \xrightarrow{d^2 \gg (x_s)^2 + (z_s)^2} - k_0 \frac{(x_s)^2 + (z_s)^2}{2d} \tag{5.16}$$

In Figure 5.8, the spatial distribution of phase angle $a(x_s, z_s) - a_0$ calculated from (5.16) in the range of $x_s \in [-500 \text{ m}, 500 \text{ m}]$ and $z_s \in [-500 \text{ m}, 500 \text{ m}]$ is plotted. With respect to a_0, the phase angle at the four corners is as large as 48°. The non-uniform spatial distribution of a means that the Earth is located in the electrical near zone of the antenna array. As a result, the wavefronts (i.e., equi-phase surfaces) are curved rather than planar when the pilot signal propagation reaches the satellite, as illustrated in Figure 5.7. In Figure 5.7(b), the nonuniform phase distribution of a is explained using one wavefront. Suppose a is 0 at one location and a is $-30°$ at another location (as shown in Figure 5.7(b)), the phase difference of 30° is due to the extra propagation path highlighted by a thick line segment.

Since the Earth is located in the electrical near zone of the antenna array, applying a uniform phase to excite the antenna elements in the second step of retro-reflective beamforming is sub-optimal. As discussed in Chapter 3, the optimal

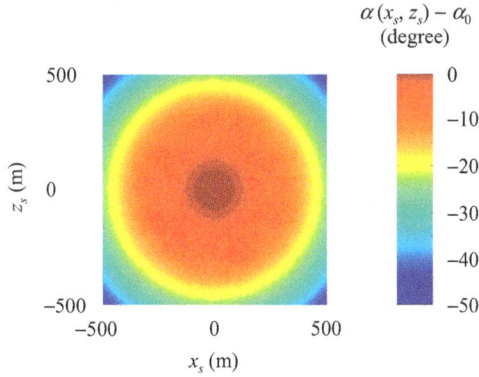

Figure 5.8 Numerical results of pilot signal's phase angle detected over y = 0 plane

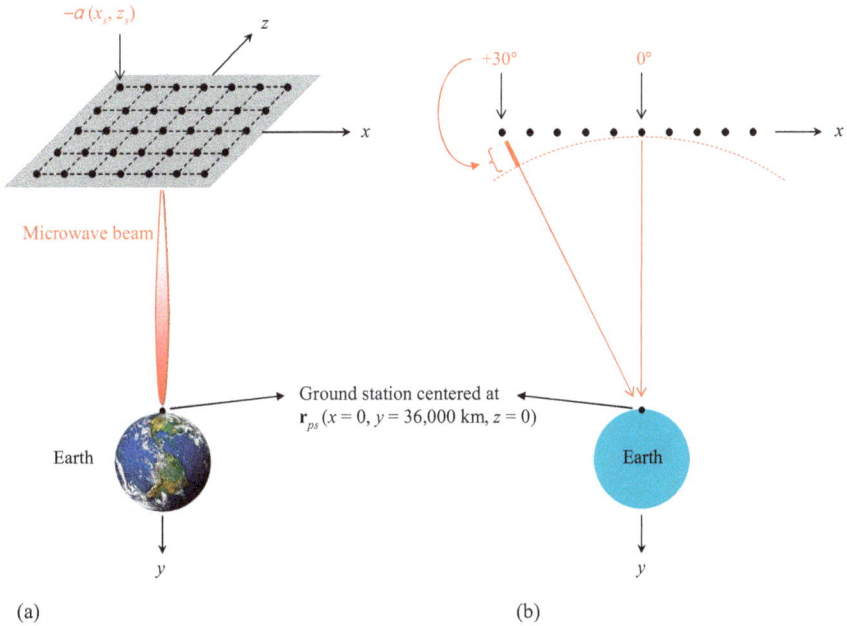

Figure 5.9 Illustration of microwave power propagation from satellite to Earth. (a) Three-dimensional view. (b) View in the x-y plane.

excitation phases are $\psi_{mn} = -\alpha_{mn}$, where α_{mn} is the phase of the pilot signal if the pilot signal is detected at $\mathbf{r}'_{mn}$, the location of antenna element with indices mn. In other words, the phase profile in the second step of retro-reflective beamforming is conjugate to the phase profile in the first step of retro-reflective beamforming. The principle of phase conjugation is justified by Figure 5.9. The geometrical configuration of Figure 5.9 is identical to that in Figure 5.7. It is known from Figure 5.7 that the propagation path highlighted by a thick line segment corresponds to the phase value of 30°. If the antenna element over the left edge is excited by 30°, its propagation would have the phase value of 0° when it reaches the wavefront in Figure 5.9(b). Meanwhile, the antenna element at the center is excited by 0°. As a result, the two antenna elements' radiations would end up with the same phase (i.e., in phase) when they reach the ground station on the Earth. Figures 5.7 and 5.9 constitute a specific example of Figure 3.2. Whereas Figure 3.2 depicts the general principle of retro-reflective beamforming, Figures 5.7 and 5.9 illustrate the application of retro-reflective beamforming technique in SSPS.

With the parameters of Figure 5.6 (i.e., $f = 5.8$ GHz, $\lambda_0 = 5.17$ cm, $s_0 = 0.7\lambda_0$, and $N = 27,633$), the amplitude of constructive coefficient $|\Phi|$ evaluated by the following two formulations are plotted in Figure 5.10.

$$\Phi(\mathbf{r}_o) = \frac{1}{N^2}\sum_{m=1}^{N}\sum_{n=1}^{N}e^{j\psi_{mn}}e^{-jk_0|\mathbf{r}_o - \mathbf{r}'_{mn}|} \xrightarrow{\psi_{mn}=0} \Phi(\mathbf{r}_o) = \frac{1}{N^2}\sum_{m=1}^{N}\sum_{n=1}^{N}e^{-jk_0|\mathbf{r}_o - \mathbf{r}'_{mn}|}$$

$$\Phi(\mathbf{r}_o) = \frac{1}{N^2}\sum_{m=1}^{N}\sum_{n=1}^{N}e^{j\psi_{mn}}e^{-jk_0|\mathbf{r}_o - \mathbf{r}'_{mn}|} \xrightarrow{\psi_{mn}=-\alpha_{mn}}$$

$$\Phi(\mathbf{r}_o) = \frac{1}{N^2}\sum_{m=1}^{N}\sum_{n=1}^{N}e^{-j\alpha_{mn}}e^{-jk_0|\mathbf{r}_o - \mathbf{r}'_{mn}|}$$

The two formulations above are both based on (5.8). The top one assumes $\psi_{mn} = 0$, that is, a uniform excitation phase. The bottom one assumes $\psi_{mn} = -\alpha_{mn}$, that is, the conjugation of the pilot signal's phase profile. The observation point $\mathbf{r}_o$ varies along

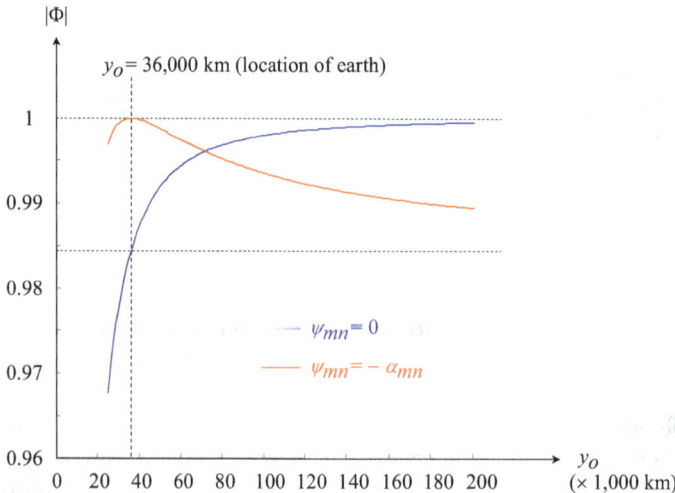

Figure 5.10   Numerical results of $|\Phi|$ with two profiles of the excitation phase

the y-axis with coordinates of $x = 0$, $y = y_o$, and $z = 0$. According to (5.16), $\alpha_{mn} = -k_0|\mathbf{r}_{ps} - \mathbf{r}'_{mn}| + \alpha_0 + k_0 d$. The results in Figure 5.10 are evaluated with $\alpha_{mn} = -k_0|\mathbf{r}_{ps} - \mathbf{r}'_{mn}|$; in other words, a constant phase $\alpha_0 + k_0 d$ is ignored as the constant phase does not change the amplitude of Φ. It is observed from Figure 5.10 that $|\Phi| = 1$ when y_o is 36,000 km with $\psi_{mn} = -\alpha_{mn}$. This means that the phase conjugation scheme achieves the optimal performance of wireless power transmission at the Earth's location. In contrast, the result of the uniform excitation phase (i.e., $\psi_{mn} = 0$) is $|\Phi| = 0.984$ when y_o is 36,000 km. With $|\Phi| = 0.984$, $|\Phi|^2 \cong 0.97$ embodies a power loss of about 3% with respect to the optimal performance. The curve of the uniform excitation phase in Figure 5.10 approaches $|\Phi| = 1$ when y_o approaches infinity, which is very reasonable as the uniform excitation phase is optimal under the far zone condition. It is interesting to notice the following two facts.

- With $\psi_{mn} = 0$ and $y_o = d$, $|\Phi| = \frac{1}{N^2}\left|\sum_{m=1}^{N}\sum_{n=1}^{N} e^{-jk_0|\mathbf{r}_{ps} - \mathbf{r}'_{mn}|}\right|$ is 0.984.

- With $\psi_{mn} = -\alpha_{mn}$ and $y_o = \infty$, $|\Phi| = \frac{1}{N^2}\left|\sum_{m=1}^{N}\sum_{n=1}^{N} e^{jk_0|\mathbf{r}_{ps} - \mathbf{r}'_{mn}|}\right|$ is also 0.984.

As discussed in Chapter 2, the electromagnetic wave propagation between a wireless transmitter and a wireless receiver can be characterized by one plane wave when the condition of electrical far zone is satisfied. The phase distribution associated with a plane wave exhibits a linear pattern in space. According to the well-known far-field condition (derived in Section 2.1), the phase distribution is considered "approximately linear" as long as its deviation from a linear pattern does not exceed 22.5°. The phase of pilot signal reception displayed in Figure 5.8 varies between 0 and $-48°$, which cannot be considered linear. However, if the 1 $(\text{km})^2$ aperture in Figure 5.8 is divided into 16 regions as illustrated by Figure 5.11, the phase distribution in each region is "approximately linear." Since the excitation phase profile in the second step of retro-reflective beamforming is conjugate to the phase profile of pilot signal reception, the excitation phase distribution over each region is "almost linear" as well. Thanks to the linear excitation phase profile, the closed-form formulations in Section 2.1 are applicable. A larger number of regions (64 or 256, for instance) would make the phase profile more linear. Nevertheless, "16 regions" seems fairly optimal in this section, as "more than 16 regions" does not improve the numerical accuracy significantly in the numerical examples of this section.

Assume that $r_o \to \infty$, that is, the observation point is located in the far zone. According to (5.8) and (5.9), the constructive coefficient in the far zone is

$$\Phi^{far}(r_o, \theta_o, \phi_o) = \frac{1}{N^2} e^{-jk_0 r_o} \sum_{m=1}^{N}\sum_{n=1}^{N} e^{j\psi_{mn}} e^{jk_0 \hat{\mathbf{r}}_o \cdot \mathbf{r}'_{mn}}. \tag{5.17}$$

With $\psi_{mn} = -\alpha_{mn}$, the constructive coefficient in the far zone becomes

$$\Phi^{far}(r_o, \theta_o, \phi_o) = \frac{1}{N^2} e^{-jk_0 r_o} \sum_{m=1}^{N}\sum_{n=1}^{N} e^{-j\alpha_{mn}} e^{jk_0 \hat{\mathbf{r}}_o \cdot \mathbf{r}'_{mn}}. \tag{5.18}$$

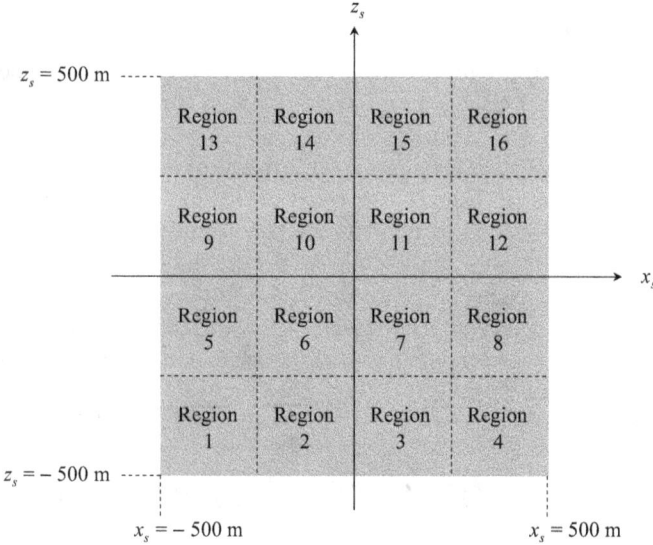

Figure 5.11   Dividing the aperture of the retro-reflective antenna array into 16 regions such that the phase distribution over each region is "approximately linear"

Following Figure 5.11, the constructive coefficient in (5.18) can be evaluated as

$$\Phi^{far}(r_o, \theta_o, \phi_o) = \frac{1}{N^2} e^{-jk_0 r_o} \sum_{m=M_{s,1}}^{M_{e,1}} \sum_{n=N_{s,1}}^{N_{e,1}} e^{-j\alpha_{mn}} e^{jk_0 \hat{r}_o \cdot r'_{mn}}$$

$$+ \frac{1}{N^2} e^{-jk_0 r_o} \sum_{m=M_{s,2}}^{M_{e,2}} \sum_{n=N_{s,2}}^{N_{e,2}} e^{-j\alpha_{mn}} e^{jk_0 \hat{r}_o \cdot r'_{mn}} \qquad (5.19)$$

$$+ \cdots + \frac{1}{N^2} e^{-jk_0 r_o} \sum_{m=M_{s,16}}^{M_{e,16}} \sum_{n=N_{s,16}}^{N_{e,16}} e^{-j\alpha_{mn}} e^{jk_0 \hat{r}_o \cdot r'_{mn}}$$

As indicated by (5.19), the $N \times N$ antenna elements are divided into 16 groups, each corresponding to one region in Figure 5.11. The 16th group, for instance, includes the antenna elements with indices in the range of $m \in [M_{s,16}, M_{e,16}]$ and $n \in [N_{s,16}, N_{e,16}]$ where $M_{s,16} = N_{s,16} = 20725$ and $M_{e,16} = N_{e,16} = 27633$; the antenna elements in the 16th group reside in the spatial region of $x_s \in [250 \text{ m}, 500 \text{ m}]$ and $z_s \in [250 \text{ m}, 500 \text{ m}]$.

The spatial distribution of $a(x_s, z_s) - a_0$ in the 16th region is plotted in Figure 5.12(a). In order to reveal how linear $a(x_s, z_s)$ is, the mathematical operation of linear regression is conducted. Specifically, c_0, c_x, and c_z are found such that $a(x_s, z_s) - a_0$ could be approximated by a linear function $c_0 + c_x x_s + c_z z_s$. In other words, the operation of linear regression aims to minimize the difference between $a(x_s, z_s) - a_0$ and $c_0 + c_x x_s + c_z z_s$. After the algorithm of linear regression in Appendix C is executed, the outcomes are $c_0 = 0.457$ (rad),

(a) x_s (m)   (b) x_s (m)

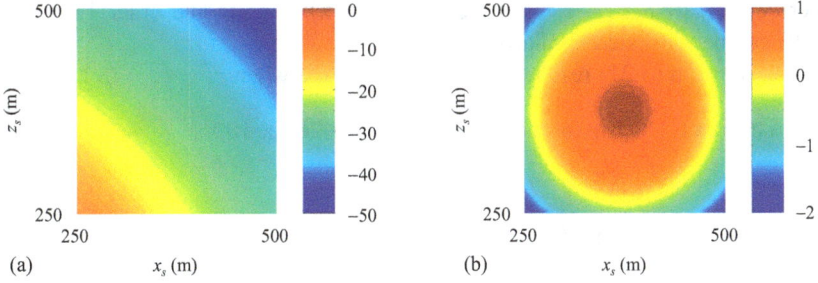

Figure 5.12   *Linear regression of phase distribution in the 16th region with $\Delta\theta = 0$ and $\Delta\phi = 0$: (a) $[\alpha\,(x_s, z_s) - \alpha_0]$, in degree and (b) $[\alpha\,(x_s, z_s) - \alpha_0] - [c_0 + c_x x_s + c_z z_s]$, in degree*

$c_x = -1.266 \times 10^{-3}$ (rad/m), and $c_z = -1.266 \times 10^{-3}$ (rad/m). The error of linear regression, defined as $[\alpha(x_s, z_s) - \alpha_0] - [c_0 + c_x x_s + c_z z_s]$, is plotted in Figure 5.12 (b). It is observed that the error values are in the range of $-2°$ and $1°$, which are typically negligibly small in the practice of microwave engineering. The error of linear regression in the other 15 regions is similarly small. Since the error due to linear regression is negligibly small, the following relationship is approximately true:

$$\alpha(x_s, z_s) - \alpha_0 = c_0 + c_x s_x + c_z s_z. \tag{5.20}$$

By making use of (5.1), (5.10), (5.13), and (5.20), the term of the 16th region in (5.19) can be evaluated analytically.

$$\frac{1}{N^2} e^{-jk_0 r_0} \sum_{m=M_{s,16}}^{M_{e,16}} \sum_{n=N_{s,16}}^{N_{e,16}} e^{-j\alpha_{mn}} e^{jk_0 \hat{\mathbf{r}}_0 \cdot \mathbf{r}'_{mn}}$$

$$= \frac{1}{N^2} e^{-jk_0 r_0} \sum_{m=M_{s,16}}^{M_{e,16}} \sum_{n=N_{s,16}}^{N_{e,16}} e^{-j\left[\alpha_0 + c_0 + c_x\left(m-\frac{N+1}{2}\right)s_0 + c_z\left(n-\frac{N+1}{2}\right)s_0\right]}$$

$$\times \; e^{jk_0\left[(\hat{\mathbf{r}}_0 \cdot \hat{\mathbf{x}})\left(m-\frac{N+1}{2}\right)s_0 + (\hat{\mathbf{r}}_0 \cdot \hat{\mathbf{z}})\left(n-\frac{N+1}{2}\right)s_0\right]}$$

$$= \frac{1}{N^2} e^{-jk_0 r_0} e^{-j(\alpha_0 + c_0)} \sum_{m=M_{s,16}}^{M_{e,16}} e^{j\left(m-\frac{N+1}{2}\right)\left[k_0(\hat{\mathbf{r}}_0 \cdot \hat{\mathbf{x}}) - c_x\right]s_0}$$

$$\times \sum_{n=N_{s,16}}^{N_{e,16}} e^{j\left(n-\frac{N+1}{2}\right)\left[k_0(\hat{\mathbf{r}}_0 \cdot \hat{\mathbf{z}}) - c_z\right]s_0} \tag{5.21}$$

$$= \frac{1}{N^2} e^{-jk_0 r_0} e^{-j(\alpha_0 + c_0)} \frac{\sin\left\{\dfrac{(M_{e,16} - M_{s,16} + 1)\varsigma_x}{2}\right\}}{\sin\left\{\dfrac{\varsigma_x}{2}\right\}} e^{j\left(\frac{M_{e,16} + M_{s,16}}{2} - \frac{N+1}{2}\right)\varsigma_x}$$

$$\times \frac{\sin\left\{\dfrac{(N_{e,16} - N_{s,16} + 1)\varsigma_z}{2}\right\}}{\sin\left\{\dfrac{\varsigma_z}{2}\right\}} e^{j\left(\frac{N_{s,16} + N_{e,16}}{2} - \frac{N+1}{2}\right)\varsigma_z}$$

where $\varsigma_x = [k_0(\hat{\mathbf{r}}_o \cdot \hat{\mathbf{x}}) - c_x]s_0$ and $\varsigma_z = [k_0(\hat{\mathbf{r}}_o \cdot \hat{\mathbf{z}}) - c_z]s_0$. The other 15 terms in (5.19) can be evaluated similarly.

The term in (5.21) reaches the peak magnitude when $\varsigma_x = \varsigma_z = 0$. Letting

$$k_0(\hat{\mathbf{r}}_b \cdot \hat{\mathbf{x}}) = k_0 \sin(\theta_b)\cos(\phi_b) = c_x = -1.266 \times 10^{-3}$$

and

$$k_0(\hat{\mathbf{r}}_b \cdot \hat{\mathbf{z}}) = k_0 \cos(\theta_b) = c_z = -1.266 \times 10^{-3}$$

results in $\theta_b = 90° + 0.0006°$ and $\phi_b = 90° + 0.0006°$. It is therefore interesting to note that the antenna elements in the 16th region jointly generate a beam toward ($\theta_b = 90° + 0.0006°$, $\phi_b = 90° + 0.0006°$) rather than toward the $+y$ direction in the second step of retro-reflective beamforming. Although an angle deviation of $0.0006°$ appears small, it actually corresponds to a large length deviation of 375 m if multiplied by 36,000 km, as $36,000,000 \times 0.0006° \times \frac{\pi}{180°} = 375$.

With the aid of (5.21), the constructive coefficient Φ^{far} in (5.19) can be efficiently evaluated. When $\theta_o = 90°$ and $\phi_o = 90°$, $|\Phi^{far}|$ is evaluated to be 0.984, which agrees with the numerical results in Figure 5.10. When θ_o and ϕ_o vary around $90°$, the numerical results of $|\Phi^{far}|$ are plotted in Figure 5.13. A main lobe centered at $\theta_o = 90°$ and $\phi_o = 90°$ is outstanding in Figure 5.13. In addition to the main lobe, side lobes are also visible in Figure 5.13. The two most visible side lobes along $\theta_o = 90°$ are located at $\phi_o = 90° + 0.0045°$ and $\phi_o = 90° - 0.0045°$, respectively; their amplitude is approximately $2/(3\pi)$. The next two side lobes along $\theta_o = 90°$ are located at $\phi_o = 90° + 0.0075°$ and $\phi_o = 90° - 0.0075°$, respectively; their amplitude is approximately $2/(5\pi)$. Roughly, the distance between two adjacent side lobes is $0.003°$, and their amplitude attenuation follows the pattern of $2/(3\pi)$, $2/(5\pi)$, $2/(7\pi)$, $2/(9\pi)$, ... away from the main lobe. The side lobes located along $\phi_o = 90°$ exhibit the same pattern. Overall, the location and

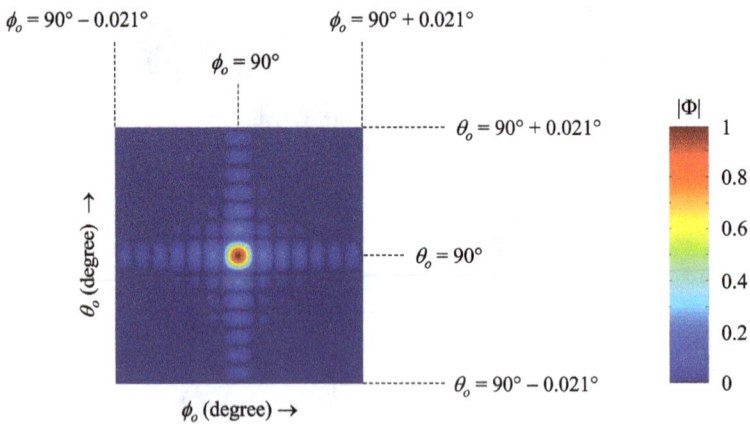

Figure 5.13   Constructive coefficient $|\Phi^{far}|$ calculated via (5.19) when (θ_o, ϕ_o) vary around ($\theta_o = 90°$, $\phi_o = 90°$), with $\Delta\theta = 0$ and $\Delta\phi = 0$

magnitude of the side lobes along $\phi_o = 90°$ and $\theta_o = 90°$ in Figure 5.13 resemble the two Dirichlet functions in (5.14) very well, which is reasonable as the profile of the excitation phase in Figure 5.8 is not far off the uniform phase profile.

Using the data of Figure 5.13, the following integral is evaluated.

$$I^{far} = \int_{\frac{\pi}{2}-\delta}^{\frac{\pi}{2}+\delta} \int_{\frac{\pi}{2}-\delta}^{\frac{\pi}{2}+\delta} |\Phi^{far}|^2 \sin(\theta_o)d\theta_o d\phi_o. \tag{5.22}$$

Because Φ^{far} is proportional to the electric field, the integral I^{far} in (5.22) embodies the amount of power radiated in a region of size $\delta \times \delta$ centered at ($\theta_o = 90°$, $\phi_o = 90°$). The values of I^{far} with $\delta = 0.003°$, $0.006°$, $0.009°$, $\ldots$, and $0.021°$ are plotted in Figure 5.14. When $\delta = 0.003°$, the integration region only includes the main lobe. When $\delta = 0.006°$, the strongest side lobes are included in the integration. All the radiation power displayed in Figure 5.13 is integrated with $\delta = 0.021°$. Since the side lobes attenuate away from ($\theta_o = 90°$, $\phi_o = 90°$), I^{far} converges to a certain value with the increase of δ. The dashed curve in Figure 5.14 predicts the convergence of I^{far} when δ is greater than $0.021°$ by assuming the attenuation of side lobes' magnitude follows the Dirichlet functions in (5.14). It is observed in Figure 5.14 that I^{far} would approach a value of $\Omega_{rad} = 2.67 \times 10^{-9}$ when δ is beyond $0.2°$ (which is 0.0035 radian). In other words, the radiation power beyond the region of $\theta \in [90° - 0.2°, 90° + 0.2°]$ and $\phi \in [90° - 0.2°, 90° + 0.2°]$ (which is $\theta \in [\pi/2 - 0.0035, \pi/2 + 0.0035]$ and $\phi \in [\pi/2 - 0.0035, \pi/2 + 0.0035]$ in terms of radian) ought to be negligibly small. It is noted that the value of I^{far} with $\delta = 0.003°$ is about $0.82 \times \Omega_{rad}$. This means that the main lobe only carries 82% of the power radiated by the antenna array. Since it is almost impossible for the ground station to collect the other 18% of the radiation power, it appears very challenging to reach a power transmission efficiency above 80% in SSPS applications. The power transmission efficiency may be improved by suppressing the

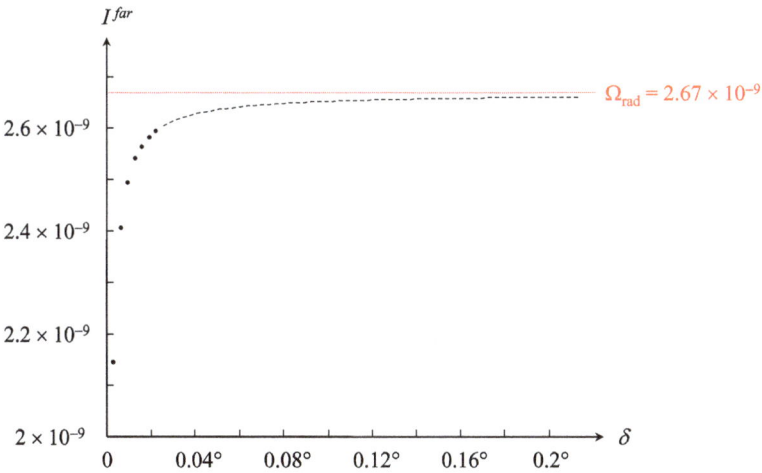

Figure 5.14   Integral I^{far} with $\Delta\theta = 0$ and $\Delta\phi = 0$

side lobes [14,15]. However, suppressing the side lobes is at the cost of adjusting the excitation amplitude profile delicately (throughout this book, it is always assumed that the antenna elements in a retro-reflective beamforming array are excited with a uniform amplitude, which facilitates the theoretical analysis as well as practical implementation tremendously).

As pointed out above, the integral I^{far} in (5.22) embodies the amount of power radiated by the antenna array. The relationship between I^{far} and P_{rad}, which denotes the total amount of power radiated by the antenna array, is derived next. P_{rad} can be found by integrating the Poynting vector over a fictitious spherical surface centered at the spatial origin and with $r_o \to \infty$ as the radius:

$$P_{rad} = \int_0^{2\pi} \int_0^{\pi} \frac{|\mathbf{E}|^2}{2\eta_0} (r_o)^2 \sin(\theta_o) d\theta_o d\phi_o, \tag{5.23}$$

where η_0 is the intrinsic impedance of free space. By making using of (5.7), (5.8), and (5.17), it is not difficult to arrive at

$$P_{rad} = \frac{|X_0|^2}{2\eta_0} N^4 \int_0^{2\pi} \int_0^{\pi} |\mathbf{U}_0(\theta_o, \phi_o)\Phi^{far}(r_o, \theta_o, \phi_o)|^2 \sin(\theta_o) d\theta_o d\phi_o. \tag{5.24}$$

Because most of the radiation power resides in the region of $\theta \in [\pi/2 - 0.0035, \pi/2 + 0.0035]$ and $\phi \in [\pi/2 - 0.0035, \pi/2 + 0.0035]$, P_{rad} can be evaluated effectively by

$$P_{rad} = \frac{|X_0|^2}{2\eta_0} N^4 \int_{\frac{\pi}{2}-0.0035}^{\frac{\pi}{2}+0.0035} \int_{\frac{\pi}{2}-0.0035}^{\frac{\pi}{2}+0.0035} |\mathbf{U}_0(\theta_o, \phi_o)\Phi^{far}(r_o, \theta_o, \phi_o)|^2 \sin(\theta_o) d\theta_o d\phi_o. \tag{5.25}$$

Apparently, $\mathbf{U}_0$, which is associated with one antenna element, stays as a constant in the small integration region of (5.25). Thus, (5.25) becomes

$$P_{rad} = \frac{|X_0|^2}{2\eta_0} N^4 \left| \mathbf{U}_0\left(\theta_o = \frac{\pi}{2}, \phi_o = \frac{\pi}{2}\right) \right|^2 \int_{\frac{\pi}{2}-0.0035}^{\frac{\pi}{2}+0.0035} \int_{\frac{\pi}{2}-0.0035}^{\frac{\pi}{2}+0.0035} |\Phi^{far}|^2 \sin(\theta_o) d\theta_o d\phi_o$$

$$= \frac{|X_0|^2}{2\eta_0} N^4 \left| \mathbf{U}_0\left(\theta_o = \frac{\pi}{2}, \phi_o = \frac{\pi}{2}\right) \right|^2 \Omega_{rad} \tag{5.26}$$

When the expression of $\mathbf{r}'_{mn}$ in (5.5) is utilized, the formulations in (5.19) and (5.21) can be extended to the configuration of Figure 5.4 with the antenna array's attitude deviating from the ideal status.

The plots in Figure 5.15 are similar to those in Figure 5.12. In Figure 5.12, $\Delta\theta = 0$ and $\Delta\phi = 0$; whereas in Figure 5.15, $\Delta\theta = 10°$ and $\Delta\phi = 10°$. Although the phase values of the pilot signal in Figure 5.15(a) span a large range, they fit a linear function very well. Specifically, the phase distribution in Figure 5.15(a) can be approximated by $c_0 + c_x x_s + c_z z_s$ with $c_0 = 0.43$ (rad), $c_x = 21.1$ (rad/m), and $c_z = 20.78$ (rad/m), after the linear regression algorithm in Appendix C is applied. The error of linear regression is shown in Figure 5.15(b). It is interesting to note

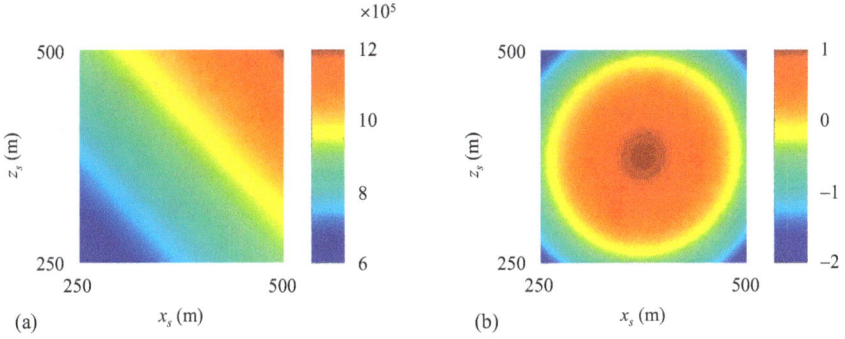

Figure 5.15   *Linear regression of phase distribution in the 16th region with $\Delta\theta = 10°$ and $\Delta\phi = 10°$: (a) $[\alpha\,(x_s,\,z_s) - \alpha_0]$, in degree and (b) $[\alpha\,(x_s,\,z_s) - \alpha_0] - [c_0 + c_x x_s + c_z z_s]$, in degree*

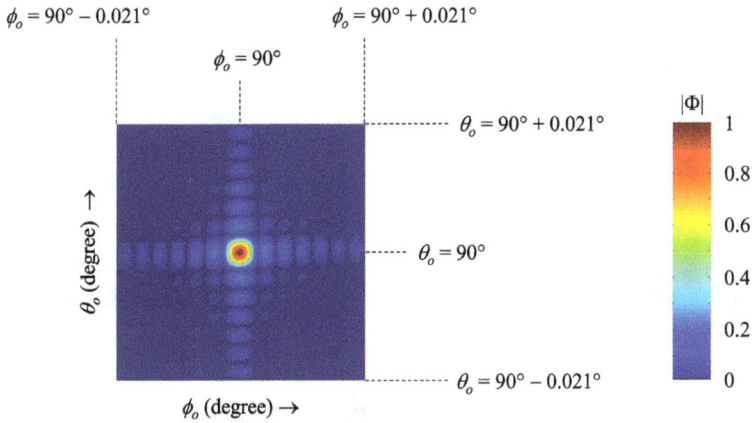

Figure 5.16   *Constructive coefficient $|\Phi^{far}|$ calculated via (5.19) when $(\theta_o,\,\phi_o)$ vary around $(\theta_o = 90°,\,\phi_o = 90°)$, with $\Delta\theta = 10°$ and $\Delta\phi = 10°$*

that Figure 5.15(b) is almost the same as Figure 5.12(b), albeit Figures 5.15(a) and 5.12(a) are drastically different from each other.

The numerical results of $|\Phi^{far}|$ evaluated via (5.19) and (5.21) are plotted in Figure 5.16 with $\Delta\theta = 10°$ and $\Delta\phi = 10°$. Compared with Figure 5.13, the main lobe and side lobes have little change.

The plot in Figure 5.17 does not appear very different from that in Figure 5.14. In Figure 5.14, I^{far} is plotted with $\Delta\theta = 0$ and $\Delta\phi = 0$; whereas in Figure 5.17, I^{far} is plotted with $\Delta\theta = 10°$ and $\Delta\phi = 10°$. It is observed from Figure 5.17 that I^{far} approaches $\Omega_{rad} = 2.76 \times 10^{-9}$ when δ is beyond $0.2°$.

The numerical results in Figures 5.18, 5.19, and 5.20 are obtained with $\Delta\theta = 20°$ and $\Delta\phi = 20°$. When $\Delta\theta = 20°$ and $\Delta\phi = 20°$, the phase distribution of

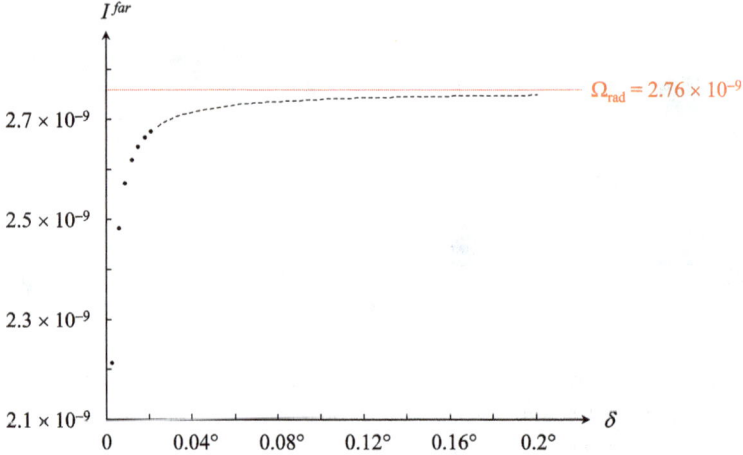

Figure 5.17   Integral I^{far} with $\Delta\theta = 10°$ and $\Delta\phi = 10°$

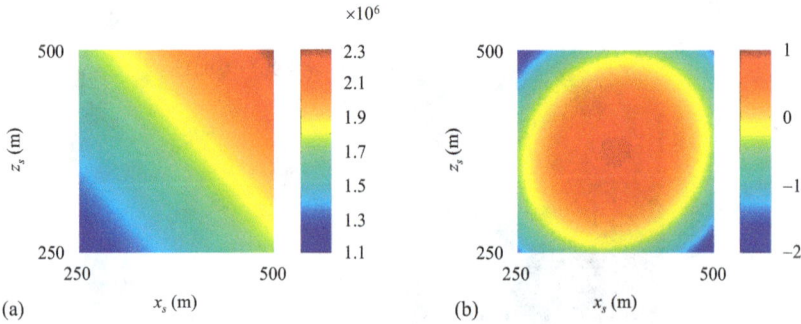

Figure 5.18   Linear regression of phase distribution in the 16th region with $\Delta\theta = 20°$ and $\Delta\phi = 20°$: (a) $[\alpha (x_s, z_s) - \alpha_0]$, in degree and (b) $[\alpha (x_s, z_s) - \alpha_0] - [c_0 + c_x x_s + c_z z_s]$, in degree

$\alpha(x_s, y_s) - \alpha_0$ in Figure 5.18(a) can be approximated very well by $c_0 + c_x x_s + c_z z_s$ with $c_0 = 0.355$ (rad), $c_x = 41.56$ (rad/m), and $c_z = 39.06$ (rad/m). The difference between $\alpha(x_s, y_s) - \alpha_0$ and $c_0 + c_x x_s + c_z z_s$ does not exceed 2° as visualized in Figure 5.18(b). Numerical results of $|\Phi^{far}|$ evaluated via (5.19) and (5.21) are plotted in Figure 5.19 with $\Delta\theta = 20°$ and $\Delta\phi = 20°$. Compared with Figures 5.13 and 5.16, the main lobe appears "distorted" a bit but the distortion seems insignificant. Figure 5.20 shows a plot of I^{far} versus δ with $\Delta\theta = 20°$ and $\Delta\phi = 20°$. When δ is greater than 0.2°, I^{far} approaches $\Omega_{rad} = 3.02 \times 10^{-9}$.

The results in Figures 5.12–5.20 demonstrate that the radiation characteristics of the retro-reflective antenna array illustrated in Figure 5.4 are stable when the attitude change of a space solar power satellite is as large as 20°.

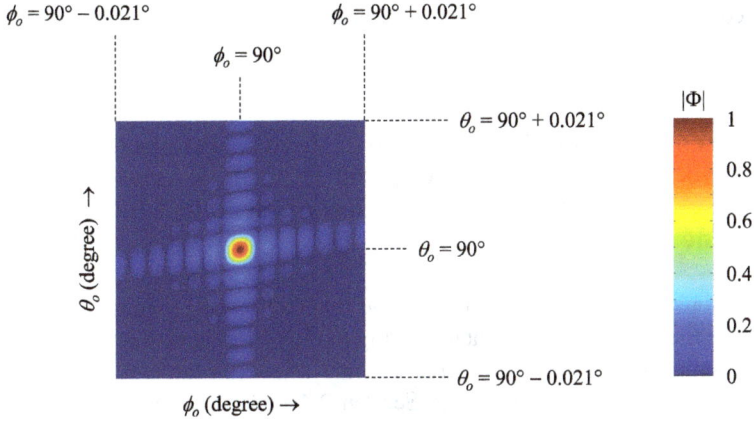

Figure 5.19 *Constructive coefficient $|\Phi^{far}|$ calculated via (5.19) when (θ_o, ϕ_o) vary around $(\theta_o = 90°, \phi_o = 90°)$, with $\Delta\theta = 20°$ and $\Delta\phi = 20°$*

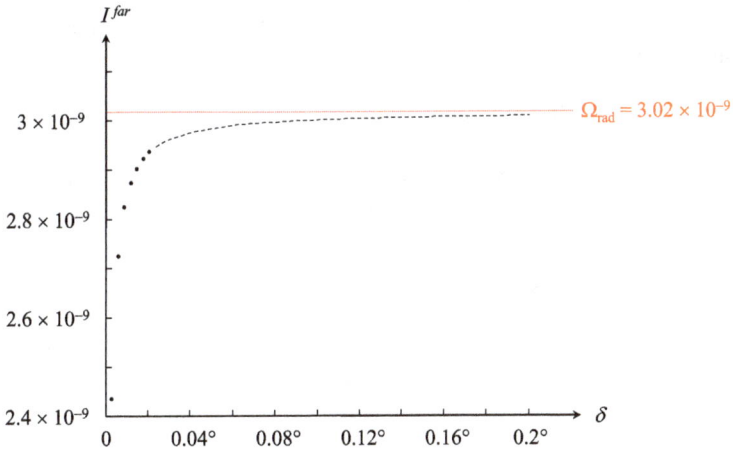

Figure 5.20 *Integral I^{far} with $\Delta\theta = 20°$ and $\Delta\phi = 20°$*

The analysis and numerical results of this section are based on the classic theory of phased array, which is reviewed in Section 2.1. Most of the research on antenna arrays for transmitting wireless power from a space solar power satellite to the Earth is based on the classic theory of phased array as well [15–19]. The classic theory of phased array is under the condition of electrical far zone, whereas it is very probable that the Earth is located in the electrical near zone of a satellite-borne retro-reflective antenna array (as concluded earlier in this section). Therefore in the next section (i.e., Section 5.4), the satellite-borne retro-reflective antenna array is analyzed under the condition of electrical near zone by employing the formulations of Section 2.4.

5.4 Theoretical analysis of retro-reflective antenna array in SSPS applications using the formulations of Section 2.4 under the condition of electrical near zone

The previous section (i.e., Section 5.3) and this section (i.e., Section 5.4) both aim to analyze the retro-reflective antenna array modeled in Figure 5.4. Whereas the far zone radiation of the retro-reflective antenna array is evaluated in the previous section, the formulations derived in Section 2.4 under the condition of electrical near zone are applied in this section.

In the first step of retro-reflective beamforming, suppose a time-harmonic pilot signal with frequency $f = 5.8$ GHz is broadcasted from $\mathbf{r}_{ps}$ ($x = 0$, $y = d$, $z = 0$) in the model of Figure 5.4. The pilot signal is detected by each antenna element of the retro-reflective array. The phase angle of the pilot signal detected by the antenna element with indices mn is denoted as α_{mn}. If an antenna element is located at the spatial origin, the phase angle of the detected pilot signal is denoted as α_0. According to (5.16),

$$\alpha_{mn} - \alpha_0 = -k_0 |\mathbf{r}_{ps} - \mathbf{r}'_{mn}| + k_0 d. \tag{5.27}$$

In the second step of retro-reflective beamforming, the excitation phase of wireless power to the antenna element with indices mn is selected to be conjugate to α_{mn}. According to (5.7) and (5.8), the electric field observed at an observation point $\mathbf{r}_o$ is

$$\mathbf{E}(\mathbf{r}_o) = \frac{X_0 \mathbf{U}_0}{r_o} N^2 \Phi(\mathbf{r}_o) \tag{5.28}$$

and the constructive coefficient is

$$\Phi(\mathbf{r}_o) = \frac{1}{N^2} \sum_{m=1}^{N} \sum_{n=1}^{N} e^{j\psi_{mn}} e^{-jk_0|\mathbf{r}_o - \mathbf{r}'_{mn}|} = \frac{1}{N^2} \sum_{m=1}^{N} \sum_{n=1}^{N} e^{-j\alpha_{mn}} e^{-jk_0|\mathbf{r}_o - \mathbf{r}'_{mn}|}. \tag{5.29}$$

Substituting (5.27) into (5.29) yields

$$\Phi(\mathbf{r}_o) = \frac{1}{N^2} e^{-jk_0 d} e^{-j\alpha_0} \sum_{m=1}^{N} \sum_{n=1}^{N} e^{jk_0|\mathbf{r}_{ps} - \mathbf{r}'_{mn}|} e^{-jk_0|\mathbf{r}_o - \mathbf{r}'_{mn}|}. \tag{5.30}$$

In this section, the observation point $\mathbf{r}_o$ is assumed to have coordinates of ($x = x_o$, $y = d$, $z = z_o$). In other words, the observation point is located over the aperture of the ground station as depicted in Figure 5.4. Since the ground station resides in the electrical near zone (rather than the electrical far zone) of the retro-reflective antenna array, the analysis of this section follows the derivations in Section 2.4. Specifically, $|\mathbf{r}_{ps} - \mathbf{r}'_{mn}|$ and $|\mathbf{r}_o - \mathbf{r}'_{mn}|$ in (5.30) are expressed as

$$
\begin{aligned}
|\mathbf{r}_{ps} - \mathbf{r}'_{mn}| &= \sqrt{(\mathbf{r}_{ps} - \mathbf{r}'_{mn}) \cdot (\mathbf{r}_{ps} - \mathbf{r}'_{mn})} \\
&= d\sqrt{1 - \frac{2\mathbf{r}_{ps} \cdot \mathbf{r}'_{mn}}{d^2} + \frac{|\mathbf{r}'_{mn}|^2}{d^2}} \\
&= d - \hat{\mathbf{y}} \cdot \mathbf{r}'_{mn} + \frac{|\mathbf{r}'_{mn}|^2}{2d} \\
&\quad - \frac{d}{8}\left[-\frac{2\hat{\mathbf{y}} \cdot \mathbf{r}'_{mn}}{d} + \frac{|\mathbf{r}'_{mn}|^2}{d^2}\right]^2 \\
&\quad + \frac{d}{16}\left[-\frac{2\hat{\mathbf{y}} \cdot \mathbf{r}'_{mn}}{d} + \frac{|\mathbf{r}'_{mn}|^2}{d^2}\right]^3 + \cdots
\end{aligned}
\tag{5.31}
$$

and

$$
\begin{aligned}
|\mathbf{r}_o - \mathbf{r}'_{mn}| &= \sqrt{[\mathbf{r}_{ps} + (\mathbf{r}_o - \mathbf{r}_{ps}) - \mathbf{r}'_{mn}] \cdot [\mathbf{r}_{ps} + (\mathbf{r}_o - \mathbf{r}_{ps}) - \mathbf{r}'_{mn}]} \\
&= d\sqrt{1 + \frac{2\mathbf{r}_{ps} \cdot (\mathbf{r}_o - \mathbf{r}_{ps} - \mathbf{r}'_{mn})}{d^2} + \frac{|(\mathbf{r}_o - \mathbf{r}_{ps}) - \mathbf{r}'_{mn}|^2}{d^2}} \\
&= d + \hat{\mathbf{y}} \cdot (\mathbf{r}_o - \mathbf{r}_{ps}) - \hat{\mathbf{y}} \cdot \mathbf{r}'_{mn} + \frac{|(\mathbf{r}_o - \mathbf{r}_{ps}) - \mathbf{r}'_{mn}|^2}{2d} \\
&\quad - \frac{d}{8}\left[\frac{2\hat{\mathbf{y}} \cdot [(\mathbf{r}_o - \mathbf{r}_{ps}) - \mathbf{r}'_{mn}]}{d} + \frac{|(\mathbf{r}_o - \mathbf{r}_{ps}) - \mathbf{r}'_{mn}|^2}{d^2}\right]^2 \\
&\quad + \frac{d}{16}\left[\frac{2\hat{\mathbf{y}} \cdot [(\mathbf{r}_o - \mathbf{r}_{ps}) - \mathbf{r}'_{mn}]}{d} + \frac{|(\mathbf{r}_o - \mathbf{r}_{ps}) - \mathbf{r}'_{mn}|^2}{d^2}\right]^3 + \cdots
\end{aligned}
\tag{5.32}
$$

The difference between (5.31) and (5.32) is

$$
\begin{aligned}
|\mathbf{r}_{ps} - \mathbf{r}'_{mn}| - |\mathbf{r}_o - \mathbf{r}'_{mn}| &= -\hat{\mathbf{y}} \cdot (\mathbf{r}_o - \mathbf{r}_{ps}) - \frac{|(\mathbf{r}_o - \mathbf{r}_{ps}) - \mathbf{r}'_{mn}|^2}{2d} + \frac{|\mathbf{r}'_{mn}|^2}{2d} \\
&\quad + \left\{\frac{d}{8}\left[\frac{2\hat{\mathbf{y}} \cdot [(\mathbf{r}_o - \mathbf{r}_{ps}) - \mathbf{r}'_{mn}]}{d} + \frac{|(\mathbf{r}_o - \mathbf{r}_{ps}) - \mathbf{r}'_{mn}|^2}{d^2}\right]^2 - \frac{d}{8}\left[-\frac{2\hat{\mathbf{y}} \cdot \mathbf{r}'_{mn}}{d} + \frac{|\mathbf{r}'_{mn}|^2}{d^2}\right]^2\right\} \\
&\quad + \left\{-\frac{d}{16}\left[\frac{2\hat{\mathbf{y}} \cdot [(\mathbf{r}_o - \mathbf{r}_{ps}) - \mathbf{r}'_{mn}]}{d} + \frac{|(\mathbf{r}_o - \mathbf{r}_{ps}) - \mathbf{r}'_{mn}|^2}{d^2}\right]^3 + \frac{d}{16}\left[-\frac{2\hat{\mathbf{y}} \cdot \mathbf{r}'_{mn}}{d} + \frac{|\mathbf{r}'_{mn}|^2}{d^2}\right]^3\right\} \\
&\quad + \cdots
\end{aligned}
\tag{5.33}
$$

Only the first three terms on the right-hand side of (5.33) are retained. Consequently,

$$
|\mathbf{r}_{ps} - \mathbf{r}'_{mn}| - |\mathbf{r}_o - \mathbf{r}'_{mn}| = -\hat{\mathbf{y}} \cdot (\mathbf{r}_o - \mathbf{r}_{ps}) - \frac{|(\mathbf{r}_o - \mathbf{r}_{ps}) - \mathbf{r}'_{mn}|^2}{2d} + \frac{|\mathbf{r}'_{mn}|^2}{2d}.
\tag{5.34}
$$

Because $\hat{\mathbf{y}} \cdot (\mathbf{r}_o - \mathbf{r}_{ps}) = 0$ in the setup of Figure 5.4 and because

$$
\begin{aligned}
|(\mathbf{r}_o - \mathbf{r}_{ps}) - \mathbf{r}'_{mn}|^2 &= [(\mathbf{r}_o - \mathbf{r}_{ps}) - \mathbf{r}'_{mn}] \cdot [(\mathbf{r}_o - \mathbf{r}_{ps}) - \mathbf{r}'_{mn}] \\
&= |\mathbf{r}_o - \mathbf{r}_{ps}|^2 - 2(\mathbf{r}_o - \mathbf{r}_{ps}) \cdot \mathbf{r}'_{mn} + |\mathbf{r}'_{mn}|^2
\end{aligned}
\tag{5.35}
$$

Equation (5.34) can be re-arranged to be

$$
|\mathbf{r}_{ps} - \mathbf{r}'_{mn}| - |\mathbf{r}_o - \mathbf{r}'_{mn}| = \frac{(\mathbf{r}_o - \mathbf{r}_{ps}) \cdot \mathbf{r}'_{mn}}{d} - \frac{|\mathbf{r}_o - \mathbf{r}_{ps}|^2}{2d}.
\tag{5.36}
$$

Substituting (5.36) into (5.30) leads to

$$
\Phi(\mathbf{r}_o) = \frac{1}{N^2} e^{-jk_0 d} e^{-ja_0} e^{-jk_0 \frac{|\mathbf{r}_o - \mathbf{r}_{ps}|^2}{2d}} \sum_{m=1}^{N} \sum_{n=1}^{N} e^{jk_0 \frac{(\mathbf{r}_o - \mathbf{r}_{ps}) \cdot \mathbf{r}'_{mn}}{d}}.
\tag{5.37}
$$

By making use of (5.5), the dot product $\frac{k_0}{d}(\mathbf{r}_o - \mathbf{r}_{ps}) \cdot \mathbf{r}'_{mn}$ in (5.37) is

$$
\begin{aligned}
\frac{k_0}{d}(\mathbf{r}_o - \mathbf{r}_{ps}) \cdot \mathbf{r}'_{mn} &= \frac{k_0}{d}[(\mathbf{r}_o - \mathbf{r}_{ps}) \cdot \hat{\mathbf{x}}_s]\left(m - \frac{N+1}{2}\right)s_0 + \frac{k_0}{d}[(\mathbf{r}_o - \mathbf{r}_{ps}) \cdot \hat{\mathbf{z}}_s]\left(n - \frac{N+1}{2}\right)s_0 \\
&= \left(m - \frac{N+1}{2}\right)\xi_x + \left(n - \frac{N+1}{2}\right)\xi_z
\end{aligned}
\tag{5.38}
$$

where

$$
\xi_x = \frac{k_0 s_0}{d}[(\mathbf{r}_o - \mathbf{r}_{ps}) \cdot \hat{\mathbf{x}}_s],
\tag{5.39}
$$

$$
\xi_z = \frac{k_0 s_0}{d}[(\mathbf{r}_o - \mathbf{r}_{ps}) \cdot \hat{\mathbf{z}}_s].
\tag{5.40}
$$

Thanks to (5.13), the constructive coefficient in (5.37) can be evaluated analytically:

$$
\begin{aligned}
\Phi(\mathbf{r}_o) &= \frac{1}{N^2} e^{-jk_0 d} e^{-ja_0} e^{-jk_0 \frac{|\mathbf{r}_o - \mathbf{r}_{ps}|^2}{2d}} \sum_{m=1}^{N} e^{j\left(m - \frac{N+1}{2}\right)\xi_x} \times \sum_{n=1}^{N} e^{j\left(n - \frac{N+1}{2}\right)\xi_z} \\
&= \frac{1}{N^2} e^{-jk_0 d} e^{-ja_0} e^{-jk_0 \frac{|\mathbf{r}_o - \mathbf{r}_{ps}|^2}{2d}} \frac{\sin\left\{\frac{N\xi_x}{2}\right\}}{\sin\left\{\frac{\xi_x}{2}\right\}} \times \frac{\sin\left\{\frac{N\xi_z}{2}\right\}}{\sin\left\{\frac{\xi_z}{2}\right\}}
\end{aligned}
\tag{5.41}
$$

The analytical formulation in (5.41) is very accurate in the context of SSPS applications. In other words, if the constructive coefficient in (5.30) is denoted as Φ^D (with superscript "D" standing for "definition") and the constructive coefficient in (5.41) is denoted as Φ^A (with superscript "A" standing for "analytical"), the difference between Φ^D and Φ^A is very small. In fact, all the derivations from (5.30) to (5.41) are rigorous except that the following phase angle is neglected in (5.33).

$$\frac{k_0 d}{8}\left[\frac{2\hat{\mathbf{y}}\cdot\left[(\mathbf{r}_o-\mathbf{r}_{ps})-\mathbf{r}'_{mn}\right]}{d}+\frac{|(\mathbf{r}_o-\mathbf{r}_{ps})-\mathbf{r}'_{mn}|^2}{d^2}\right]^2 - \frac{k_0 d}{8}\left[-\frac{2\hat{\mathbf{y}}\cdot\mathbf{r}'_{mn}}{d}+\frac{|\mathbf{r}'_{mn}|^2}{d^2}\right]^2$$

$$-\frac{k_0 d}{16}\left[\frac{2\hat{\mathbf{y}}\cdot\left[(\mathbf{r}_o-\mathbf{r}_{ps})-\mathbf{r}'_{mn}\right]}{d}+\frac{|(\mathbf{r}_o-\mathbf{r}_{ps})-\mathbf{r}'_{mn}|^2}{d^2}\right]^3 + \frac{k_0 d}{16}\left[-\frac{2\hat{\mathbf{y}}\cdot\mathbf{r}'_{mn}}{d}+\frac{|\mathbf{r}'_{mn}|^2}{d^2}\right]^3$$

$$+\cdots$$

$$(5.42)$$

The terms in (5.42) are organized with respect to $1/d$ (as $1/d$ is a small value in SSPS applications). The two leading terms in (5.42), $\frac{k_0 d}{8}\left[\frac{2\hat{\mathbf{y}}\cdot\left[(\mathbf{r}_o-\mathbf{r}_{ps})-\mathbf{r}'_{mn}\right]}{d}\right]^2$ and

$\frac{k_0 d}{8}\left[\frac{2\hat{\mathbf{y}}\cdot\mathbf{r}'_{mn}}{d}\right]^2$, cancel each other under the condition of $\hat{\mathbf{y}}\cdot(\mathbf{r}_o-\mathbf{r}_{ps})=0$. As a result, the largest values among the terms in (5.42) have the form of

$$k_0\frac{D\times D\times D}{d^2}=\frac{2\pi}{\lambda_0}\frac{D\times D\times D}{d^2},\qquad(5.43)$$

where D stands for either D_s or D_o. As analyzed in Section 5.3, the value of $(D_s D_o)/(\lambda_0 d)$ is about 2 in the SSPS applications. In addition, D is on the order of 1 km, and d is as large as 36,000 km. It is therefore not difficult to observe that the phase angle in (5.43) is much smaller than 1°, which is negligibly small. Consequently, the difference between Φ^D and Φ^A must be very small. Some numerical results of the difference between Φ^D and Φ^A are presented in Table 5.2.

With the setup of Figure 5.4, Φ^D evaluated via (5.30) and Φ^A evaluated via (5.41) are compared with each other in Table 5.2, with the following parameters: $f=5.8$ GHz, $\lambda_0=5.17$ cm, $s_0=0.7\lambda_0$, $N=27{,}633$, $\Delta\theta=10°$, and $\Delta\phi=20°$. The observation point $\mathbf{r}_o$ has coordinates of $(x=x_o,\ y=d,\ z=z_o)$. When x_o varies in the range of $[-2$ km, 2 km$]$ and z_o varies in the range of $[-2$ km, 2 km$]$ as well, the difference between Φ^D and Φ^A is very small consistently. Because Φ^D and Φ^A are

Table 5.2 Numerical results of the error between Φ^D and Φ^A

	$x_o=-2$ km	$x_o=-1$ km	$x_o=0$	$x_o=1$ km	$x_o=2$ km
$z_o=2$ km	$\Phi=1\times10^{-3}\angle-54°$ Error $=4\times10^{-7}$	$\Phi=5.1\times10^{-2}\angle56°$ Error $=1.5\times10^{-8}$	$\Phi=5.5\times10^{-2}\angle153°$ Error $=2.3\times10^{-8}$	$\Phi=1.6\times10^{-2}\angle56°$ Error $=6\times10^{-8}$	$\Phi=5.4\times10^{-5}\angle126°$ Error $=4.3\times10^{-6}$
$z_o=1$ km	$\Phi=5\times10^{-3}\angle56°$ Error $=2.2\times10^{-7}$	$\Phi=0.35\angle167°$ Error $=1.1\times10^{-9}$	$\Phi=0.6\angle-97°$ Error $=1.2\times10^{-9}$	$\Phi=0.4\angle167°$ Error $=9.5\times10^{-10}$	$\Phi=6.6\times10^{-3}\angle56°$ Error $=2.1\times10^{-7}$
$z_o=0$	$\Phi=9.6\times10^{-3}\angle153°$ Error $=1.7\times10^{-7}$	$\Phi=0.63\angle-96°$ Error $=6.2\times10^{-10}$	$\Phi=1\angle0°$ Error $=0$	$\Phi=0.63\angle-96°$ Error $=6.2\times10^{-10}$	$\Phi=9.6\times10^{-3}\angle153°$ Error $=1.7\times10^{-7}$
$z_o=-1$ km	$\Phi=6.6\times10^{-3}\angle56°$ Error $=2.1\times10^{-7}$	$\Phi=0.4\angle167°$ Error $=9.5\times10^{-10}$	$\Phi=0.6\angle-97°$ Error $=1.2\times10^{-9}$	$\Phi=0.35\angle167°$ Error $=1.1\times10^{-9}$	$\Phi=5\times10^{-3}\angle56°$ Error $=2.2\times10^{-7}$
$z_o=-2$ km	$\Phi=5.4\times10^{-5}\angle126°$ Error $=4.3\times10^{-6}$	$\Phi=1.6\times10^{-2}\angle56°$ Error $=6\times10^{-8}$	$\Phi=5.5\times10^{-2}\angle153°$ Error $=2.3\times10^{-8}$	$\Phi=5.1\times10^{-2}\angle56°$ Error $=1.5\times10^{-8}$	$\Phi=1\times10^{-3}\angle-54°$ Error $=4\times10^{-7}$

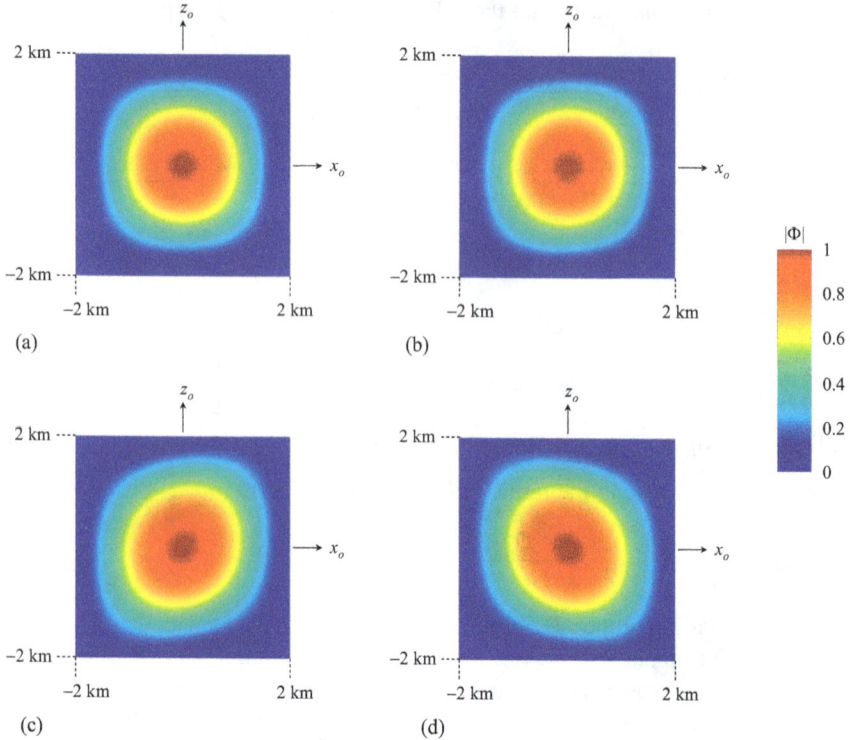

Figure 5.21 *Constructive coefficient $|\Phi|$ calculated via (5.41) when x_o varies in [−2 km, 2 km] and z_o varies in [−2 km, 2 km]: (a) $\Delta\theta = 0$ and $\Delta\phi = 0$; (b) $\Delta\theta = 10°$ and $\Delta\phi = 10°$; (c) $\Delta\theta = 20°$ and $\Delta\phi = 20°$; and (d) $\Delta\theta = 20°$ & $\Delta\phi = −20°$*

so close to each other that they can barely be differentiated from each other, only one of them is shown in Table 5.2 as Φ. The error between Φ^D and Φ^A is calculated as $\frac{|\Phi^D - \Phi^A|}{|\Phi^D|}$. All the error values in Table 5.2 are very small. When Φ has a weak amplitude with $x_o = 2$ km and $z_o = 2$ km, the error is as small as 4.3×10^{-6}. Therefore, it is safe to conclude that the closed-form formulation in (5.41) is very accurate.

Some numerical results evaluated via (5.41) are plotted in Figure 5.21. In Figure 5.21, $f = 5.8$ GHz, $\lambda_0 = 5.17$ cm, $s_0 = 0.7\lambda_0$, and $N = 27,633$, which are the same as the parameters associated with Table 5.2. The amplitude of constructive coefficient $|\Phi|$ is plotted at $x = x_o$, $y = d$, and $z = z_o$, with $x_o \in [−2 \text{ km}, 2 \text{ km}]$ and $z_o \in [−2 \text{ km}, 2 \text{ km}]$. In the four subplots of Figure 5.21, $(\Delta\theta, \Delta\phi)$ have the values of $(\Delta\theta = 0, \Delta\phi = 0)$, $(\Delta\theta = 10°, \Delta\phi = 10°)$, $(\Delta\theta = 20°, \Delta\phi = 20°)$, and $(\Delta\theta = 20°, \Delta\phi = −20°)$, respectively. In Figure 5.21(a), a beam centered at $x_o = 0$ and $z_o = 0$ is clearly visible. Figure 5.21(a) appears almost identical to Figure 5.6 because the difference between the formulations of this section and those in the previous section is as little as 3% in terms of power (as demonstrated in Figure 5.10). When $x_o = 0$ and $z_o = 0$, $\zeta_x = \zeta_z = 0$ and $|\Phi| = 1$ according to (5.39), (5.40), and (5.41). It fulfills the objective of retro-reflective beamforming: The contributions from all the antenna elements are constructive at the location from which the pilot signal is broadcasted. The spatial size of the beam can be determined by letting

$$\frac{Nk_0s_0}{2d}x_{null} = \pi. \tag{5.44}$$

Equation (5.44) results in $x_{null} = \pm 1.86$ km. When $x_o = x_{null}$, Φ is 0 because $\sin(N\xi_x/2) = 0$ in the last line of (5.41). Similarly, Φ is 0 when $z_o = \pm 1.86$ km. Thus, the power beam in Figure 5.21(a) could be fully captured if the ground station covers the region of $(-1.86 \text{ km} < x < 1.86 \text{ km})$ and $(-1.86 \text{ km} < z < 1.86 \text{ km})$ in the $y = d$ plane. This conclusion is consistent with the theoretical analysis in [12,13]. When the size of the ground station's aperture D_o has the value of $|2x_{null}|$, (5.44) becomes

$$\frac{Nk_0s_0}{2d}\frac{D_o}{2} = \pi \quad \rightarrow \quad \frac{2\pi}{\lambda_0}\frac{(Ns_0)}{2d}\frac{D_o}{2} = \pi \quad \rightarrow \quad \frac{D_sD_o}{d\lambda_0} = 2.$$

As found in the theoretical analysis of [12,13], $\tau = (D_sD_o)/(\lambda_0d) = 2$ would result in a power transmission efficiency close to 100%. The power beams in Figures 5.21(b), 5.21(c), and 5.21(d) do not appear much different from the power beam in Figure 5.21 (a). Specifically in Figures 5.21(b), 5.21(c), and 5.21(d), $|\Phi|$ always has the value of 1 at $x_o = 0$ and $z_o = 0$, and the size of the power beam remains to be approximately 3.7 km by 3.7 km. The change in the satellite's attitude probably has a magnitude far below $\Delta\theta = \pm 20°$ or $\Delta\phi = \pm 20°$ in practice. Therefore in principle, the retro-reflective beamforming technique is capable of maintaining high power transmission efficiency from a geostationary satellite to the Earth when the satellite's attitude is in change.

At an observation point $\mathbf{r}_o$ $(x = x_o, y = d, z = z_o)$ over the ground station, the Poynting vector is $\hat{y}|\mathbf{E}|^2/(2\eta_0)$. Therefore, the following integral represents the total amount of power incident upon the ground station's aperture.

$$P_{avl} = \int_{-\frac{D_o}{2}}^{\frac{D_o}{2}}\int_{-\frac{D_o}{2}}^{\frac{D_o}{2}} \hat{y}\frac{|\mathbf{E}(x_o,d,z_o)|^2}{2\eta_0} \cdot \hat{y}dx_odz_o$$

$$= \int_{-\frac{D_o}{2}}^{\frac{D_o}{2}}\int_{-\frac{D_o}{2}}^{\frac{D_o}{2}} \frac{|X_0U_0(x_o,d,z_o)|^2}{2\eta_0} \times \frac{N^4}{(x_o)^2 + d^2 + (z_o)^2} \times |\Phi(x_o,d,z_o)|^2 dx_odz_o$$

$$\tag{5.45}$$

The integral in (5.45) is defined as P_{avl}, standing for "power available to the ground station." Since d is much greater than D_o, $(x_o)^2 + d^2 + (z_o)^2$ can be replaced by d^2 and $U_0(x_o,d,z_o)$ can be replaced by $U_0(0,d,0)$ with negligible inaccuracy. As a result,

$$P_{avl} = \frac{|X_0U_0(0,d,0)|^2}{2\eta_0} \times \frac{N^4}{d^2} \int_{-\frac{D_o}{2}}^{\frac{D_o}{2}}\int_{-\frac{D_o}{2}}^{\frac{D_o}{2}} |\Phi(x_o,d,z_o)|^2 dx_odz_o. \tag{5.46}$$

The two-dimensional integration with respect to x_o and z_o in (5.46) can be evaluated by integrating the data in Figure 5.21 numerically.

The ratio between P_{avl} and P_{rad} is defined as *beam collection efficiency*. The total power radiated by the retro-reflective array, P_{rad}, is found in (5.26). In (5.26), U_0 is measured in the far zone toward the $+y$ direction; in (5.46), U_0 is measured at a point with Cartesian coordinates of $(x = 0, y = d, z = 0)$. U_0 is due to the radiation of one antenna element of the retro-reflective array. As far as one antenna element is concerned, $(x = 0, y = d, z = 0)$ is located in the far zone toward the $+y$ direction. Thus, U_0 in (5.26) and U_0 in (5.46) are identical to each other. It is then not difficult to calculate the beam collection efficiency:

$$\frac{P_{avl}}{P_{rad}} = \frac{1}{d^2 \Omega_{rad}} \int_{-\frac{D_o}{2}}^{\frac{D_o}{2}} \int_{-\frac{D_o}{2}}^{\frac{D_o}{2}} |\Phi(x_o, d, z_o)|^2 dx_o dz_o. \tag{5.47}$$

With the data of Ω_{rad} in Section 5.3 and the data of Φ in Figure 5.21, some numerical results of beam collection efficiency evaluated via (5.47) are tabulated in Table 5.3. As observed from Table 5.3, the beam collection efficiency is insensitive to $\Delta\theta$ or $\Delta\phi$ given a certain value of D_o (which is the aperture size of the ground station), demonstrating that the retro-reflective beamforming technique is capable of aiming the power beam toward the ground station when the satellite's attitude is under constant change. When D_o is 3 km, the beam collection efficiency is about 80%. However, when D_o increases to 3.5 km or 4 km, the beam collection efficiency has little improvement. This means that a ground station with an aperture size of 3 km has roughly covered the entire main power beam. Obviously, if the beam collection efficiency is about 80% when the entire main power beam is captured, the beam collection efficiency would exceed 80% considerably only when some side lobes are captured as well. It does not appear cost-effective to build additional ground stations at the side lobes' locations for the sake of increasing the beam collection efficiency by as little as 10%. It is therefore necessary to seek techniques to suppress the side lobes if a beam collection efficiency greater than 80% is desired. Some research efforts indicate that adjusting the amplitude profile of wireless power excitation is an effective approach to suppress the side lobes of

Table 5.3   *Numerical results of beam collection efficiency (ratio between P_{avl} and P_{rad})*

	$\Delta\theta = 0$ and $\Delta\phi = 0$	$\Delta\theta = 10°$ and $\Delta\phi = 10°$	$\Delta\theta = 20°$ and $\Delta\phi = 20°$
$D_o = 1$ km	25.2%	24.5%	22.6%
$D_o = 1.5$ km	46.6%	45.5%	42.8%
$D_o = 2$ km	64.6%	63.6%	60.8%
$D_o = 2.5$ km	75.8%	75%	73.2%
$D_o = 3$ km	80.5%	80.1%	79.4%
$D_o = 3.5$ km	81.6%	81.3%	81.5%
$D_o = 4$ km	81.6%	81.4%	81.7%

space solar power satellites [14,15]. In Section 5.2, Section 5.3, and this section, the antenna elements of the retro-reflective array are assumed to be excited by a uniform amplitude of wireless power. Undoubtedly, applying a nonuniform amplitude profile would create enormous complications; for instance, the power dispatching network and microwave power amplifiers over the satellite would become a lot more complicated compared with the scenario of a uniform amplitude profile. The pros and cons of adopting a nonuniform amplitude profile are subject to further research efforts and are not elaborated in this book.

It should be noted that "beam collection efficiency" and "power transmission efficiency" are not identical to each other. The beam collection efficiency is defined as the ratio between P_{avl} and P_{rad}, while the power transmission efficiency is defined as the ratio between P_r and P_t in this book. Since $P_{avl} \neq P_r$ and $P_{rad} \neq P_t$ in practice, "beam collection efficiency" is not the same as "power transmission efficiency." The concepts of P_{avl}, P_r, P_{rad}, P_t, beam collection efficiency, and power transmission efficiency are discussed in the rest of this section.

The relationship between P_{rad} and P_t is illustrated in Figure 5.22. Throughout this book, the transmitted power P_t is defined as the microwave power supplied from transmitter circuits to transmitting antenna(s). As there are multiple antenna elements over the satellite, P_t is the sum of microwave power at the circuit port of all the antenna elements. The functionality of the transmitting antenna array is to convert P_t to electromagnetic radiation. When the electromagnetic power radiated by the transmitting antenna array is integrated over a large fictitious spherical surface, the integral is the radiated power P_{rad}, as formulated in (5.23). The transmitted power P_t and radiated power P_{rad} are equal to each other when everything inside the fictitious spherical surface is lossless. In this book, the medium inside the fictitious surface is assumed to be free space, which is lossless. Under the condition

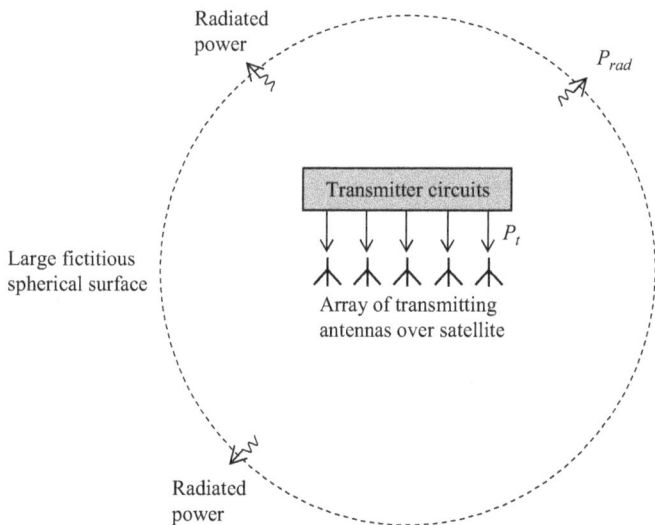

Figure 5.22   Illustration of the relationship between P_t and P_{rad}

of a lossless medium, the ratio between P_{rad} and P_t is defined as the antenna efficiency [10]. Antenna efficiency embodies the thermal loss of transmitting antennas. It is well known that the antenna loss is typically not severe when the operating frequency is below 6 GHz. Thus, the antenna efficiency was assumed to be 100% in the previous few chapters (i.e., Chapters 1 to 4). Nevertheless, probably, the antenna efficiency cannot be assumed to be close to 100% in SSPS applications (as shown by one example in Section 5.5).

The relationship between P_{avl} and P_r in a typical wireless system is illustrated in Figure 5.23(a). In a typical wireless system (such as a cell phone communication system), a wireless transmitter and a wireless receiver reside in each other's far zone. Under the far zone condition, the radiation from the transmitter behaves as a plane wave when it reaches the receiver. Denote the power density of the incident plane wave as W^{inc}. Following the definition of (5.45), the available power P_{avl} is the product between W^{inc} and A_p, where A_p is the surface area of the receiving antenna's physical aperture. When the receiving antenna is terminated by a matched load, the power at the receiving antenna's circuit port, P_r, can be expressed as the product between W^{inc} and A_e, where A_e is the surface area of the receiving antenna's effective aperture. In a typical wireless system, it is possible that P_r is greater than P_{avl}, in other words, it is possible that A_e is greater than A_p. Nevertheless, "$P_r > P_{avl}$" is not probable in SSPS applications. As depicted in Figure 5.23(b), the size of the ground station intends to cover the main power beam radiated by a space solar power satellite. Meanwhile as shown in Figure 5.23(b), the power carried by side lobes is not available to the ground station. If the aperture of the ground station is built with perfect electromagnetic absorber material, the total amount of power absorbed by the ground station is P_{avl},

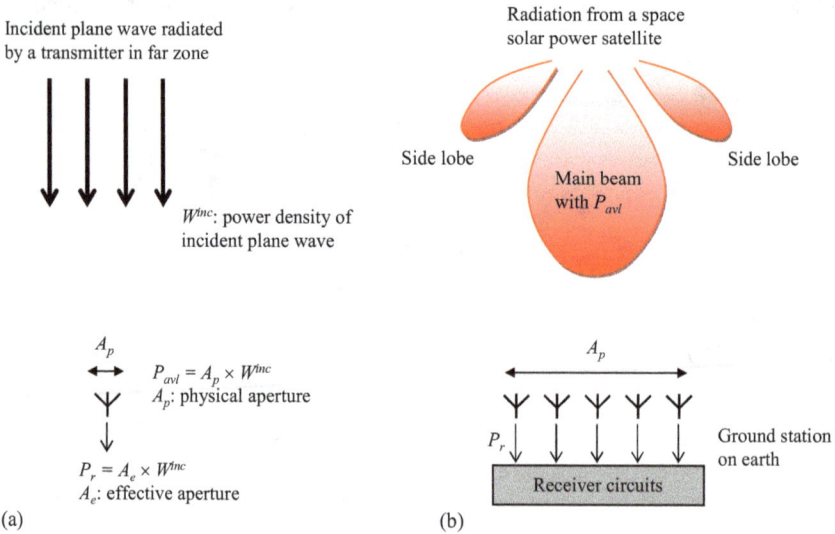

Figure 5.23   *Illustration of the relationship between P_r and P_{avl}: (a) a typical receiving antenna, upon the incidence of electromagnetic plane wave and (b) receiving antenna array in SSPS applications, with array's physical size covering the main beam*

which is approximately the amount of power carried by the main beam. Of course, if the ground station is made of perfect electromagnetic absorber, P_{avl} would be obtained in the form of thermal power, and it is difficult to convert thermal power to electrical power with high efficiency and low cost. Therefore, the practical implementation of a ground station must utilize an array of receiving antennas. The functionality of the receiving antenna array is to convert the incident electromagnetic power to electrical power at the receiving antennas' circuit port. The received power P_r is defined as the sum of microwave power supplied from the receiving antenna array to receiver circuits. The ceiling level of P_r is dictated by P_{avl}, since P_{avl} specifies the pool of power available to the ground station (which is the power carried by the main power beam). Moreover, the interaction between the incident electromagnetic wave and receiving antennas unavoidably causes scattered power and thermal loss. As a result, "$P_r \geq P_{avl}$" is very unlikely in Figure 5.23(b). How to design the receiving antenna array to let P_r approach P_{avl} in SSPS applications is an unprecedented engineering problem; it is addressed briefly (that is, not comprehensively) in Section 5.6.

Because $P_t > P_{rad}$ and $P_{avl} > P_r$ in SSPS applications, the power transmission efficiency (which is P_r/P_t) must be lower than the beam collection efficiency values in Table 5.3. In this book, the power transmission efficiency is defined between the transmitting antenna's circuit port and the receiving antenna's circuit port. It is actually not sufficient to characterize the efficiency of the entire SSPS system. For instance, DC power is generated by the solar panel of a space solar power satellite, and converting it to microwave power inevitably incurs certain conversion losses. Similarly, after microwave power is collected by the receiving antennas over the ground station, converting it to DC power cannot be achieved with 100% efficiency. The technical issues pertinent to the conversion between DC power and microwave power are beyond the scope of this book, as declared in Chapter 1.

5.5 A bench-scale experimental demonstration of wireless power transmission from satellite to Earth

"Wireless power transmission from a geostationary satellite to the Earth" constitutes an extremely complex engineering problem. The theoretical model and analyses presented in Sections 5.2, 5.3, and 5.4 are far from characterizing this problem completely. A large number of technical issues that are not incorporated into the model of Section 5.2 call for systematic research endeavors. Two of them are described below briefly.

First, the retro-reflective antenna array in the numerical examples of Sections 5.3 and 5.4 includes $27,633 \times 27,633 = 763,582,689$ antenna elements, and it is cost-prohibitive to attach one pilot signal receiving circuit and one wireless power transmitting circuit to each antenna element. In practice, the number of circuits must be minimized by grouping antenna elements into sub-arrays. Specifically, each sub-array includes a number of antenna elements, and one set of circuits (for pilot signal reception and wireless power transmission) is attached to each sub-array. If the number of antenna elements in each sub-array is large, the cost of a retro-reflective antenna array drops tremendously. The radiation pattern of each sub-array is fixed and ought to demonstrate a narrow beam toward the Earth. With the increase in the number of antenna elements in each sub-array, the radiation beam associated with each sub-array gets narrower [10]. As a result, the power transmission efficiency may drop if each sub-array's beam is too narrow and if the satellite's attitude has a large change.

Second, the pilot signal propagation and wireless power propagation must be separated from each other by a certain means in practice. Over the satellite, the power level of wireless power transmission is higher than the power level of pilot signal reception by many orders. It is therefore very possible for the circuits of pilot signal reception to get jammed if the pilot signal reception and wireless power transmission are not well separated from each other. In the Internet of Things applications (discussed in Chapter 4), time-division duplexing is a highly effective means to prevent wireless power transmission from jamming pilot signal reception, in which the wireless power transmitting circuits are shut down when pilot signal reception is active. In the SSPS applications, however, turning a power of Giga-Watt level on and off frequently would create tremendous technical complications. Therefore, frequency-division seems to be an excellent candidate for SSPS applications. Specifically, when the pilot signal and wireless power are carried by different frequencies, it is possible to isolate them from each other by taking advantage of filters.

In this section, a bench-scale experimental demonstration of retro-reflective beamforming for wireless power transmission is presented with the two practical issues discussed above incorporated. The retro-reflective beamforming antenna array in the experimental setup has physical dimensions of 0.6 m by 0.6 m, much smaller than the dimension of 1 km^2 studied in Sections 5.3 and 5.4. The bench-scale experiments of this section intend to verify the fundamental scheme of retro-reflective beamforming for wireless power transmission from a satellite to the Earth.

The experimental setup is shown in Figure 5.24. A wireless power transmitter and a wireless power receiver are separated from each other by 4 m in space. The wireless power receiver emulates a ground station. It consists of four parts: a pilot signal generator, a pilot signal transmitting antenna, a wireless power receiving antenna, and a power meter to detect the wireless power. The wireless power transmitter emulates a space solar power satellite. It consists of three parts: four pilot signal receiving antennas, four wireless power transmitting antennas, and four retro-reflective beamforming circuits. The interaction between the wireless power transmitter and the wireless power receiver involves two steps. In the first step, a time-harmonic pilot signal at 2.9 GHz is broadcasted from the wireless power receiver to the wireless power transmitter. In the second step, the wireless power transmitter delivers a power beam at 5.8 GHz to the wireless power receiver, in reaction to the pilot signal. As discussed in the second issue at the beginning of this section, a frequency-division scheme is employed to isolate pilot signal and wireless power from each other (that is, the pilot signal propagation and wireless power propagation are carried by 2.9 GHz and 5.8 GHz, respectively [20]).

All the antenna elements in Figure 5.24 are implemented by rectangular microstrip patches [10]. The four wireless power transmitting antennas are identical to each other. Each of them comprises ($8 \times 8 - 1 = 63$) microstrip patches. In total,

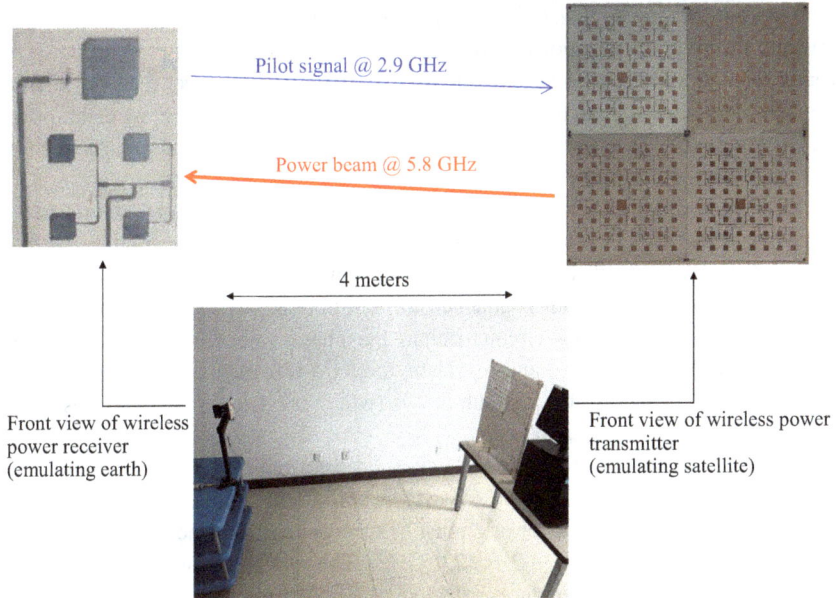

Figure 5.24   A bench-scale experimental setup of retro-reflective beamforming for wireless power transmission in SSPS applications

there are $4 \times 63 = 252$ microstrip patch elements for transmitting wireless power. The physical dimensions of one patch element are about 27 mm by 27 mm. The distance between two adjacent patch elements s_0 is 35.8 mm, which is approximately $0.7\lambda_0$ with λ_0 being the wavelength in free space corresponding to 5.8 GHz. As discussed in the first issue at the beginning of this section, it would be cost-prohibitive to assign circuitry to the 252 patch elements individually. Thus, the 252 patch elements are grouped into four sub-arrays, each including 63 patches. Each sub-array has one port to a wireless power transmitting circuit. The power delivered to the sub-array from the power transmitting circuit is distributed to the 63 patches using a power distribution network made of microstrip lines. The microstrip power distribution network is designed such that the 63 patches are excited with the same phase. Consequently, the radiation pattern of one sub-array has a narrow beam toward its broadside direction. With the power loss over the power distribution network taken into account, the peak gain value of one sub-array is 20.6 dBi at 5.8 GHz. The overall dimension of one sub-array is approximately 0.3 m by 0.3 m. With the aid of a pilot signal, the four sub-arrays jointly construct and steer a beam narrower than an individual sub-array's beam. It must be noted that the optimal number of elements in one sub-array is determined by numerous practical factors; 63 is chosen in the experimental setup of Figure 5.24 merely for the sake of proof of concept. At the center of each sub-array, there is one pilot signal receiving patch. The resonant frequency of the four pilot signal receiving patch antennas is 2.9 GHz, and as a result, they are physically larger than the 5.8-GHz patches. The wireless power transmitting patches have right-hand circular polarization at 5.8 GHz, and the pilot signal receiving patches have left-hand circular polarization at 2.9 GHz. The orthogonality between the two polarizations provides additional isolation between pilot signal propagation and wireless power propagation. At the wireless power receiver, there is one 2.9-GHz pilot signal transmitting antenna and one 5.8-GHz wireless power receiving antenna. The wireless power receiving antenna, including an array of four microstrip patches, has a broadside gain value of 9 dBi at 5.8 GHz.

In the experimental demonstration of Figure 5.24, a time-harmonic pilot signal at 2.9 GHz is generated by an oscillator connected to the pilot signal transmitting antenna. The pilot signal is received by the four pilot signal receiving antennas. One retro-reflective beamforming circuit is attached to each pilot signal receiving antenna. The retro-reflective beamforming circuits obtain the phase values of the received pilot signals, denoted as α_1, α_2, α_3, and α_4. Then, the retro-reflective beamforming circuits supply 5.8-GHz power to the four sub-arrays (which are wireless power transmitting antennas) with phase values of $-2\alpha_1$, $-2\alpha_2$, $-2\alpha_3$, and $-2\alpha_4$, respectively. As the fundamental mechanism of retro-reflective beamforming, the phase conjugation operation guarantees that a 5.8-GHz power beam would be constructed toward the wireless power receiver. A factor of 2 must be incorporated into the phase conjugation operation because (5.8 GHz) = 2 × (2.9 GHz). The wireless power received by the wireless power receiving antenna is detected by a power meter.

Two types of retro-reflective beamforming circuits are designed and implemented using commercial off-the-shelf components. The basic architecture and principles of these two retro-reflective beamforming circuits are presented in Section 3.6. In the context of the bench-scale experimental demonstration of Figure 5.24, their performances are not differentiable from each other. However, the differences between them are anticipated to become obvious with the increase of the scale (for instance, when the physical dimension of the retro-reflective antenna array reaches 10 m or larger). These two types of retro-reflective beamforming circuits are described next one by one.

The block diagram of the first type of retro-reflective beamforming circuit is shown in Figure 5.25. Since the four retro-reflective beamforming circuits in the experimental setup are identical to one another, only one of them is plotted in

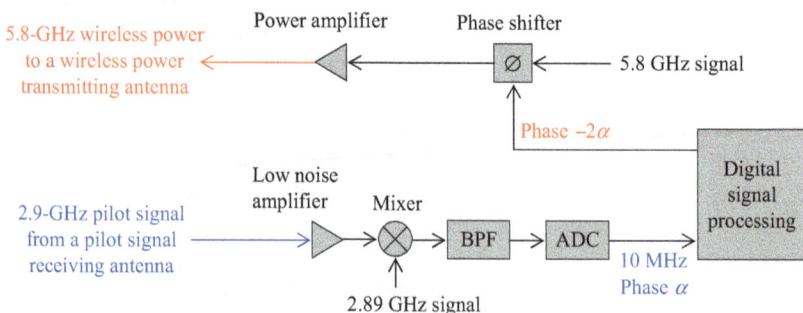

Figure 5.25   Block diagram of retro-reflective beamforming circuit with DSP in the bench-scale experimental demonstration

Figure 5.25. When the pilot signal of 2.9 GHz is received by a pilot signal receiving antenna, it is mixed with a signal at 2.89 GHz generated by a local oscillator. Since the mixer behaves as a multiplier, the output of the mixer includes two frequency components: 10 MHz and 5.79 GHz. The 10-MHz frequency component is selected by a band-pass-filter (BPF) and then converted to the digital format by an analog-to-digital converter (ADC). The digital output of the ADC is analyzed by a digital signal processing (DSP) unit to obtain the phase angle α of the pilot signal. The phase angle of wireless power at 5.8 GHz to the corresponding wireless power transmitting antenna is adjusted to be -2α by the DSP through a phase shifter. The retro-reflective beamforming circuit in Figure 5.25 follows the architecture of [21], with a DSP unit serving as the buffer between the pilot signal receiving circuit and the wireless power transmitting circuit. While the DSP unit offers great flexibility, it would become more complex, expensive, and power-consuming with more antenna elements included in the retro-reflective array. Digital circuits are avoided in the second type of retro-reflective beamforming circuit, which is described in the next paragraph. As a consequence of avoiding DSP, the second type of retro-reflective beamforming circuit has lower complexity, cost, and power consumption.

The second type of retro-reflective beamforming circuit, depicted in Figure 5.26, follows the heterodyne architecture proposed by Pon [22]. When the pilot signal of 2.9 GHz is received by a pilot signal receiving antenna, it is mixed with a signal at 5.8 GHz generated by a local oscillator. When the phase of the received pilot signal is α, the output of the mixer includes a 2.9-GHz signal with phase value of $-\alpha$. After the 2.9-GHz signal with phase value of $-\alpha$ is selected by a BPF, it passes through a frequency doubler. The output of the frequency doubler has a frequency of 5.8 GHz and phase of -2α, which is amplified and supplied to the corresponding wireless power transmitting antenna. Compared with Figure 5.25, the retro-reflective beamforming circuit in Figure 5.26 is simpler as it does not involve any digital circuits. Nevertheless, various technical complications must be managed carefully to minimize the error in phase conjugation. Some of them are discussed in [23] and are not elaborated in this section.

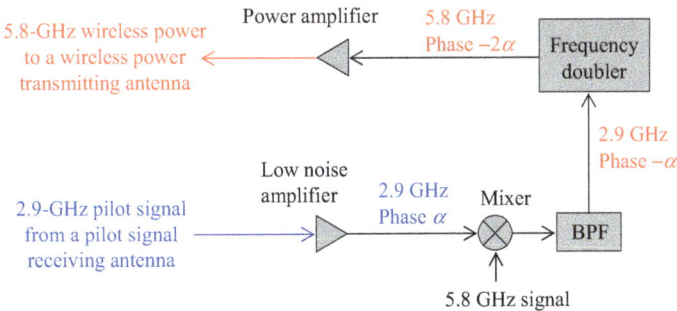

Figure 5.26   Block diagram of retro-reflective beamforming circuit without DSP in the bench-scale experimental demonstration

Some baseline data are measured with the setup shown in Figure 5.27. In Figure 5.27, a time-harmonic power of 15 dBm at 5.8 GHz is supplied to one sub-array residing at the spatial origin. The wireless power receiving antenna changes its location in the region of $x_o \in [-1\text{ m}, 1\text{ m}]$ and $z_o \in [-1\text{ m}, 1\text{ m}]$, and the power value detected by the power meter attached to the wireless power receiving antenna is denoted as P_r. The measured P_r data are plotted in Figure 5.27. Clearly, a beam toward $+y$ direction is radiated by the sub-array. At the beam center (that is, $x_o = 0$ and $z_o = 0$), the measured P_r value is -15.7 dBm. To verify the measured data, the P_r value at $x_o = 0$ and $z_o = 0$ is calculated from the Friis transmission equation:

$$\frac{P_r}{P_t} = \left(\frac{\lambda_0}{4\pi d}\right)^2 G_t G_r.$$

With $\lambda_0 = 5.17$ cm, $d = 4$ m, the gain value of one sub-array $G_t = 20.6$ dBi, the gain value of the wireless power receiving antenna $G_r = 9$ dBi, and transmitted power $P_t = 15$ dBm, the received power P_r is calculated to be -15.2 dBm via the Friis transmission equation, which agrees with the measured data very well.

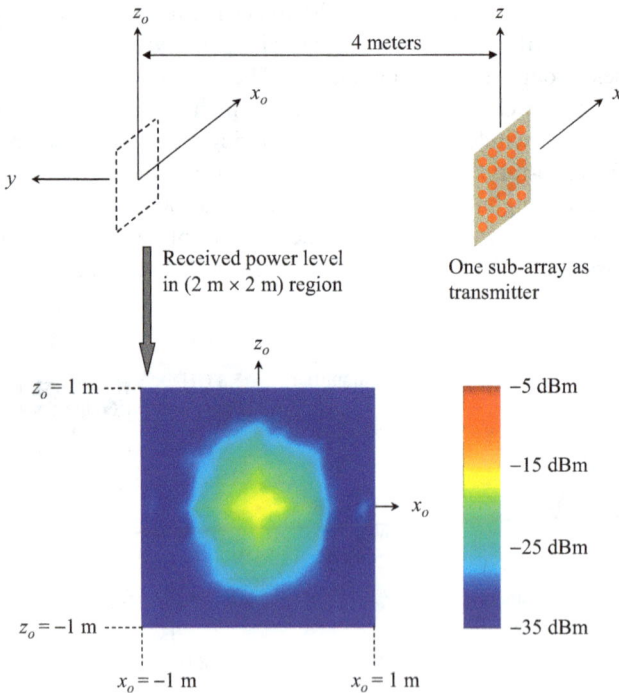

Figure 5.27   Experimental results of radiation from one sub-array

Figures 5.28 and 5.29 present some experimental data on retro-reflective beamforming. The experimental data are obtained by the following procedure. First, a pilot signal at 2.9 GHz is broadcasted by the pilot signal transmitting antenna at a certain location in the $y = 4$ m plane. Then, the pilot signal is received by the four pilot signal receiving antennas located in the $y = 0$ plane; the phase values of the four received pilot signals are obtained as α_1, α_2, α_3, and α_4. Next, 5.8-GHz wireless power is transmitted by the four sub-arrays collaboratively; the power fed to each sub-array is 15 dBm; and, the four phase values of wireless power fed to the four sub-arrays are $-2\alpha_1$, $-2\alpha_2$, $-2\alpha_3$, and $-2\alpha_4$, respectively. Finally, the wireless power is detected by the wireless power receiving antenna with its location changing in the $y = 4$ m plane; the wireless power value measured by the power meter attached to the wireless power receiving antenna is recorded and plotted.

In Figure 5.28, the pilot signal is broadcasted from $\mathbf{r}_{ps}$ with coordinates of ($x_o = 0$, $z_o = 0$) in the $y = 4$ m plane. Due to the geometrical symmetry, the phase values of the four pilot signals received by the pilot signal receiving antennas are almost identical to each other. In reaction, the four sub-arrays are excited by a uniform amplitude and uniform phase. As a result, a power beam toward $+y$ direction is generated, which can be clearly observed in Figure 5.28. Suppose the electric field produced by each sub-array at $\mathbf{r}_{ps}$ ($x_o = 0$, $y = 4$ m, $z_o = 0$) is $\mathbf{E}_0$, the total electric

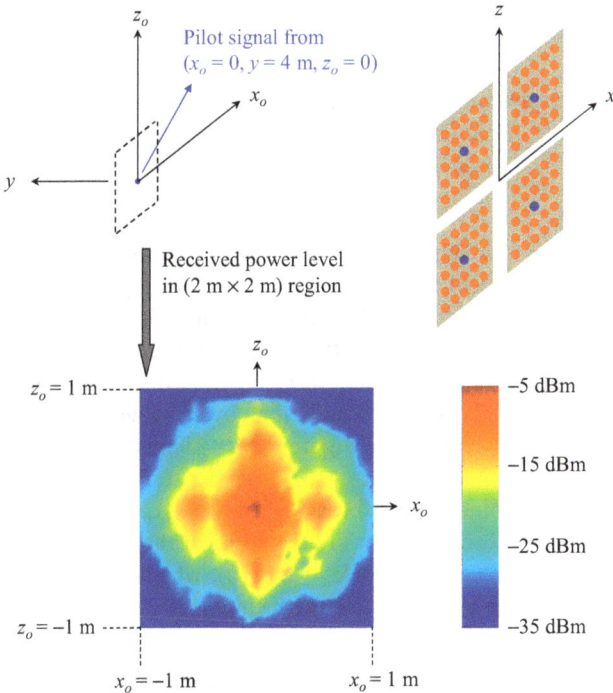

Figure 5.28   Experimental results of retro-reflective beamforming when the pilot signal is broadcasted from ($x_o = 0$, $z_o = 0$)

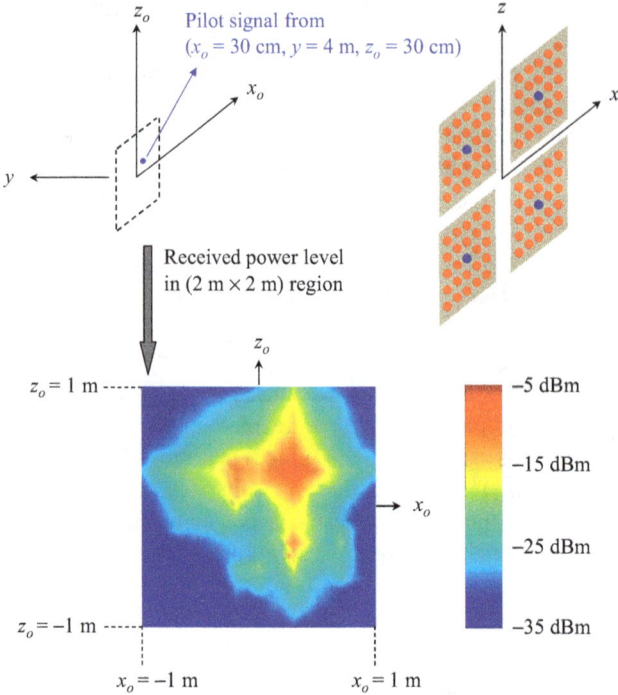

Figure 5.29   Experimental results of retro-reflective beamforming when the pilot signal is broadcasted from (x_o = 30 cm, z_o = 30 cm)

field produced by the four sub-arrays at $\mathbf{r}_{ps}$ is $4\mathbf{E}_0$. It is noticed that $\mathbf{E}_0$ is slightly weaker than the electric field measured at (x_o = 0, y = 4 m, z_o = 0) in Figure 5.27 because $\mathbf{r}_{ps}$ deviates slightly from the broadside direction of each sub-array in Figure 5.28 whereas it is located along the broadside direction of one sub-array in Figure 5.27. The power value measured at $\mathbf{r}_{ps}$ is −5.4 dBm in Figure 5.28 and is −15.7 dBm in Figure 5.27. In other words, the power value at $\mathbf{r}_{ps}$ is enhanced by 10.3 dB in Figure 5.28 with respect to Figure 5.27, which is slightly smaller than 4^2 = 16 = 12 dB.

In Figure 5.29, the pilot signal is broadcasted from (x_o = 30 cm, z_o = 30 cm) in the y = 4 m plane. It is observed from Figure 5.29 that the wireless power beam is steered to aim at (x_o = 30 cm, y = 4 m, z_o = 30 cm), which explicitly demonstrates the fundamental principle of retro-reflective beamforming.

5.6   Wireless power reception on the Earth

As discussed in Section 5.4, the ground station on the Earth includes an array of antennas to harvest the wireless power transmitted from a space solar power satellite. The physical aperture of the ground station is on the order of square kilometers, in order to capture the main power beam radiated by the satellite. The top design criterion of the receiving antenna array over the ground station is to maximize P_r/P_{avl}, where P_r is the amount of microwave power at the circuit ports of the receiving antenna array and P_{avl} is the power "available" to the ground station as illustrated in Figure 5.23(b).

The spatial distribution of the main power beam radiated by a space solar power satellite can be represented by the constructive coefficient Φ defined in (5.29). In Figure 5.30, the amplitude of $\Phi(\mathbf{r}_o)$ and phase of $\Phi(\mathbf{r}_o)$ are plotted at an observation point $\mathbf{r}_o$ ($x = x_o$, $y = d$, $z = z_o$). Φ_0 in Figure 5.30(b) is the constructive coefficient with $x_o = 0$ and $z_o = 0$. The plots in Figure 5.30 follow the coordinate system established in Figure 5.4. The geometrical center of a retro-reflective antenna array over the satellite is designated as the spatial origin, and the center of the ground station is ($x = 0$, $y = d$, $z = 0$) with $d = 36{,}000$ km. The data in Figure 5.30 are calculated using (5.41) with the set of parameters of Section 5.4: The frequency of wireless power propagation is $f = 5.8$ GHz, the wavelength in free space is $\lambda_0 = 5.17$ cm, the spatial spacing between two adjacent antenna elements over the satellite is $s_0 = 0.7\lambda_0$, the number of antenna elements included in the retro-reflective array is $N \times N = 27{,}633 \times 27{,}633$, and a pilot signal is broadcasted from $\mathbf{r}_{ps}$ ($x = 0$, $y = d$, $z = 0$). In Figure 5.30, $\Delta\theta = 0$ and $\Delta\phi = 0$ are assumed, and therefore, Figure 5.30(a) is identical to Figure 5.21(a). The phase of Φ does not depend on $\Delta\theta$ or $\Delta\phi$, which is obvious from observing (5.41). When the values of $\Delta\theta$ and $\Delta\phi$ deviate from 0, the amplitude distribution of Φ is only slightly different from that in Figure 5.30(a), as shown in Figure 5.21.

As illustrated in Figure 5.31, suppose there is only one receiving antenna element located at ($x = 0$, $y = d$, $z = 0$) under the wireless power illumination displayed in Figure 5.30. Assume the receiving antenna element is a regular planar antenna with a physical size close to $0.5\lambda_0$. As far as the receiving antenna element's small

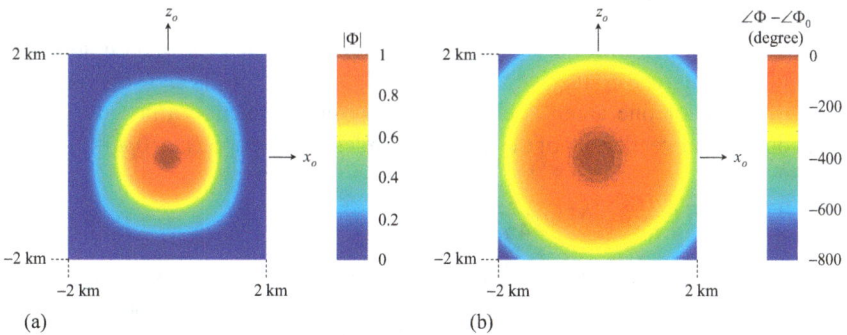

(a)   (b)

Figure 5.30   *Amplitude distribution and phase distribution of constructive coefficient Φ over the aperture of the ground station with $\Delta\theta = 0$ and $\Delta\phi = 0$: (a) amplitude of constructive coefficient and (b) phase of constructive coefficient*

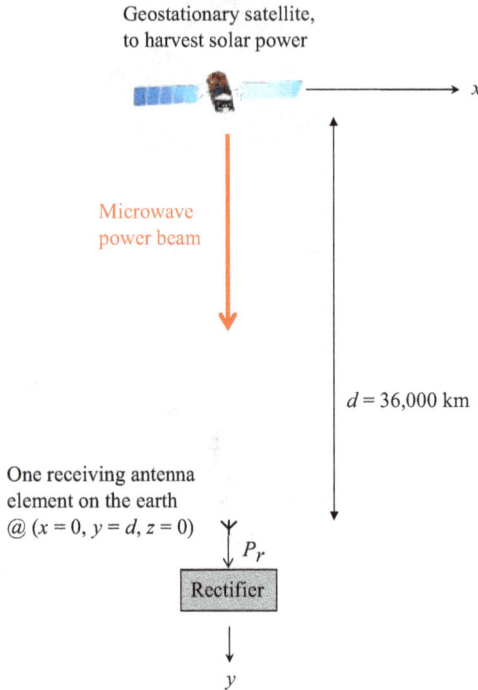

Figure 5.31   Illustration of wireless power reception with one receiving antenna element located at (x = 0, y = d, z = 0)

physical size is concerned, the incident wireless power is a plane wave with uniform amplitude and uniform phase. The radiations from the $N \times N$ elements of the retro-reflective array are constructive at $(x = 0, y = d, z = 0)$, because $(x = 0, y = d, z = 0)$ is the location from which the pilot signal is broadcasted. If the mutual coupling among the antenna elements of the retro-reflective array is ignored, the gain value of the retro-reflective array at $(x = 0, y = d, z = 0)$ is $G_t = N^2 G_0$, where G_0 is the gain of one antenna element of the retro-reflective array along the $\hat{y}$ direction. The power density of the incident wireless power upon $(x = 0, y = d, z = 0)$ is

$$W^{inc} = G_t \frac{P_t}{4\pi d^2} = N^2 G_0 \frac{P_t}{4\pi d^2}$$

where P_t the total amount of power transmitted by the retro-reflective array over the satellite. When P_t is assumed to be 1 GW = 10^9 Watts and G_0 is assumed to be 4, W^{inc} is calculated to be 19 mW per $(cm)^2$. The power density of 19 mW per $(cm)^2$ is not intimidating, actually. When a cell phone is communicating with a remote cell tower actively, it is not uncommon that the power radiated by the cell phone's antenna is above 2 Watts. Meanwhile, a cell phone's screen size can be reasonably estimated to be 100 $(cm)^2$. If half of the 2-Watt power (which is 1 Watt) is radiated outward through the cell phone's screen, the power density at some spots of the screen may exceed 10 mW per $(cm)^2$. Therefore, as an interesting fact, being under the direct illumination of a space solar power satellite does not seem to be much more risky than holding a cell phone to the head. If the receiving antenna element's location deviates from $(x = 0, y = d, z = 0)$, the incident power density attenuates as prescribed by Figure 5.30(a). When the receiving antenna element moves out of the 4 km × 4 km region of Figure 5.30(a) along either the x direction or z direction, it approaches a side lobe. The position and magnitude of the side lobes are plotted in Figures 5.13, 5.16, and 5.19. The amplitude of Φ at the four strongest side lobes is about 0.2. This means that the incident power density associated with the strongest side lobes is approximately $(0.2)^2 \times 19$ mW per $(cm)^2 = 0.76$ mW per $(cm)^2$. In the frequency band around 5.8 GHz, a power density lower than 1 mW per $(cm)^2$ is commonly believed to be safe for the general public [24]. Thus, the side lobes generated by a space solar power satellite seem to have no biological hazards. If the receiving antenna element's gain value toward $-\hat{y}$ direction is $G_r = 4$, the corresponding effective aperture is

$$A_e = \frac{(\lambda_0)^2}{4\pi} G_r = \frac{\lambda_0}{\sqrt{\pi}} \times \frac{\lambda_0}{\sqrt{\pi}} = (0.56\lambda_0) \times (0.56\lambda_0) = 8.4 \ (cm)^2,$$

with

$$\lambda_0 = \frac{c}{f} = 5.17 \ cm$$

If the load of the receiving antenna element (which is a rectifier circuit) is matched to the receiving antenna element, the received power P_r at the receiving antenna element's circuit port is calculated to be $A_e \times W^{inc} = 0.16$ W, consistent with the outcome of Equation (1.10). To be specific, with $P_t = 1$ GW, $d = 36,000$ km, $L = 1$ km, $W = 1$ km, $G_0 = 4$, and $G_r = 4$, Equation (1.10) yields the same value of P_r:

$$P_r = P_t \left(\frac{1}{4\pi d}\right)^2 (2LW) G_0 G_r = 0.16 \ W.$$

In Figure 5.31, the received microwave power P_r is converted to DC power by a rectifier circuit. Typically, a rectifier's performance depends on P_r heavily, and the rectifier behaves as a nonlinear load of the receiving antenna element (in other words, the input impedance of the rectifier depends on P_r). At the same time, the load impedance influences how much P_r the receiving antenna element harvests from the incident wave. Consequently, it is a general practice for a receiving antenna element and a rectifier circuit to be co-designed, resulting in a rectenna, which stands for "rectifier + antenna." Enormous research efforts have been published on the design and development of rectennas [25]. One example of a rectifier circuit is presented in Figure 4.10; the data of its microwave-to-DC conversion efficiency versus the input microwave power are plotted in Figure 4.11.

As it is impossible for one rectenna to collect power on the order of Giga-Watts, the ground station on the Earth must include an array of rectennas. As illustrated in Figure 5.32, a two-dimensional array of rectennas is deployed along

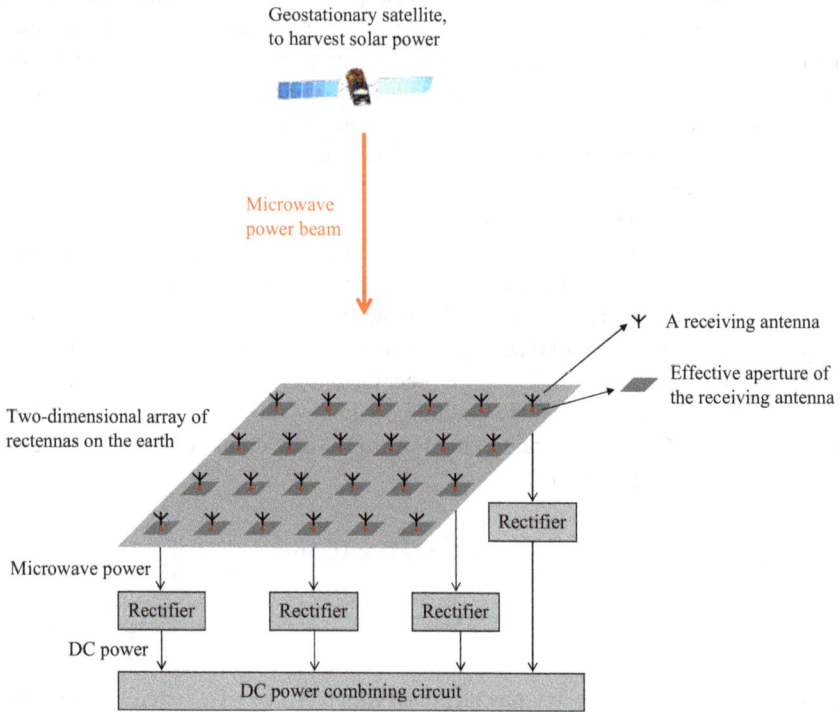

Figure 5.32   Illustration of a two-dimensional rectenna array over the ground station

the x and z directions over the ground station's aperture. The DC power outputs of the rectennas are combined by a DC power combining circuit. Spatial spacing among the rectenna elements is perhaps the most critical parameter in the design of a rectenna array [26]. In Figure 5.32, each receiving antenna's effective aperture is represented by a fictitious shaded square region; if the broadside gain value of a receiving antenna element is 4, the side length of the square region is $0.56\lambda_0$ as calculated above. When the spacing among the rectenna elements is large with respect to their effective aperture (as plotted in Figure 5.32), there is little coupling among the receiving antenna elements. As far as each receiving antenna element is concerned, the incident wave is a plane wave. Thus, each rectenna harvests power from the incident plane wave individually, with the microwave power output as $A_e \times W^{inc}$ where A_e is the effective aperture and W^{inc} is the power density of the incident plane wave (under the condition that the rectifier circuit is matched to the receiving antenna element). It should be noted that Equation (1.10) is only valid for the rectenna elements located in the neighborhood of $(x = 0, y = d, z = 0)$. The large spacing among rectenna elements in Figure 5.32 is unable to achieve high power transmission efficiency, as the incident power not covered by the shaded regions (which represent the effective apertures) is not harvested by the rectenna array.

When the spacing gets smaller, the effective apertures of receiving antenna elements approach one another, and the power transmission efficiency is expected to improve. Nevertheless, the coupling among the rectenna elements becomes strong with small spacing, which renders the design of the rectenna array more involved.

Very probably, the rectenna array architecture of Figure 5.32 is not the best candidate for the ground station in SSPS applications. This is because the incident power density is not uniform over the aperture of the ground station. As shown in Figure 5.30(a), $|\Phi| = 1$ at ($x = 0, y = d, z = 0$) and $|\Phi|$ drops to 0.7 roughly at ($x = 1$ km, $y = d, z = 0$). As a result, the rectenna at ($x = 0, y = d, z = 0$) and the rectenna at ($x = 1$ km, $y = d, z = 0$) are illuminated by different incident power densities, and the difference is roughly 3 dB. If the two rectennas are identical to each other, they would demonstrate different microwave-to-DC conversion efficiencies due to the nonlinearity of rectifier circuits. It is particularly difficult for a rectenna near the edge of the ground station, which is under a weak incident wave, to achieve high

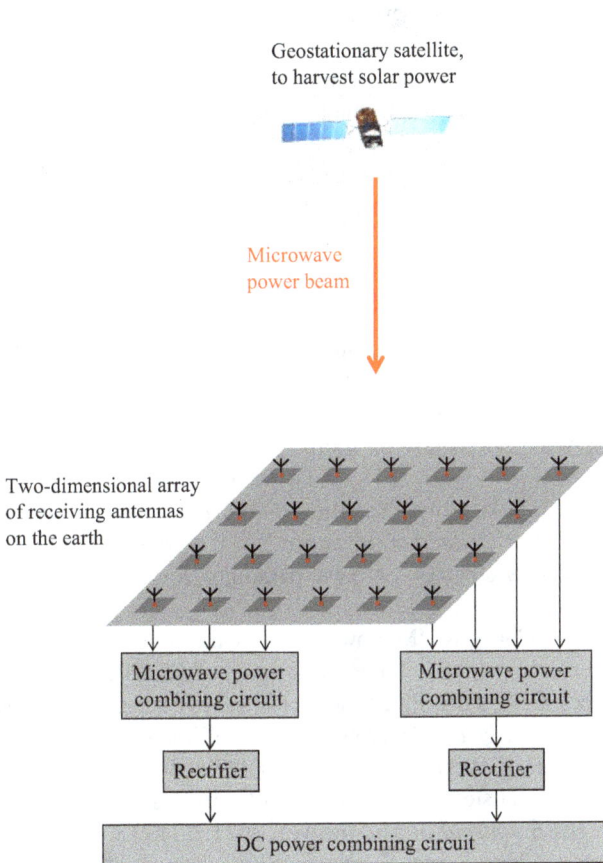

Figure 5.33   Illustration of two-dimensional receiving antenna array followed by microwave power combining and DC power combining

microwave-to-DC conversion efficiency. Therefore, it is necessary to combine the microwave power outputs of multiple receiving antenna elements near the edge of the ground station before rectification, as depicted in Figure 5.33. The number of antenna elements involved by one microwave power combining circuit should be determined with the criterion that the combined microwave power level facilitates a rectifier behind it to achieve the optimal microwave-to-DC conversion efficiency [27]. Actually, microwave power combining may be applied throughout the aperture of the ground station. Employing microwave power combining circuits results in fewer rectifiers in Figure 5.33, compared with Figure 5.32. Because microwave power combining circuits do not include semiconductor devices but rectifiers do, the architecture in Figure 5.33 is anticipated to have a lower cost than Figure 5.32. However, generally speaking, combining power at 5.8 GHz causes more loss than combining power at DC. Identifying the optimal trade-off between microwave power combining and DC power combining in SSPS applications therefore requires delicate and systematic endeavors [28].

This section touches upon several technical issues pertinent to the reception of wireless power in SSPS applications. The discussions of this section are far from comprehensive or profound, as the reception of wireless power is not directly related to the retro-reflective beamforming technique (which is the theme of this book). The readers are referred to Professor Shinohara's book [29] or other references in this section to gain more knowledge of wireless power reception in SSPS applications.

References

[1] Glaser P. 'Power from the sun: its future'. *Science*. 1968;162(3856):857–61.
[2] *URSI White Paper on Solar Power Satellite (SPS) Systems and Report of the URSI Inter-Commission Working Group on SPS*. URSI (Union Radio Scièntifique Internationale); 2007.
[3] Hou X., Wang L. 'Study on multi-rotary joints space power satellite concept'. *Aerospace China*. 2018;19(1):19–26.
[4] Sangster A.J. Solar Power Satellites (SPS). *Electromagnetic Foundations of Solar Radiation Collection*. Springer; 2014. p. 207–40.
[5] Cash I. 'CASSIOPeiA – A new paradigm for space solar power'. *Acta Astronautica*. 2019;159:170–8.
[6] Strassner B., Chang K. 'Microwave power transmission: historical milestones and system components'. *Proceedings of the IEEE*. 2013;101(6):1379–96.
[7] Rodenbeck C.T., Jaffe P.I., Strassner II B.H., Hausgen P.E., McSpadden J. O., Kazemi H., *et al.* 'Microwave and millimeter wave power beaming'. *IEEE Journal of Microwaves*. 2021;1(1):229–59.
[8] Shinohara N. 'History and innovation of wireless power transfer via microwaves'. *IEEE Journal of Microwaves*. 2021;1(1):218–28.
[9] Baraskar A., Yoshimura Y., Nagasaki S., Hanada T. 'Space solar power satellite for the Moon and Mars mission'. *Journal of Space Safety Engineering*. 2022;9(1):96–105.

[10] Balanis C.A. *Antenna Theory: Analysis and Design*. 3rd ed. New York: Wiley-Interscience; 2005.

[11] Wu B., Wang D., Poh E.K. 'High precision satellite attitude tracking control via iterative learning control'. *Journal of Guidance, Control, and Dynamics*. 2015;38(3):528–33.

[12] Goubau G. 'Microwave power transmission from an orbiting solar power station'. *Journal of Microwave Power*. 1970;5(4):224–31.

[13] Goubau G., Schwering F. Free space beam transmission. In: Okress EC (ed.). *Microwave Power Engineering: Generation, Transmission, Rectification*. New York: Academic Press (Electrical Science); 1968.

[14] Sugawara A., Kami Y., Takano T., Hanayama E. 'An investigation of the radiation characteristics of an ultra-large antenna for space use'. *Presented at 21st International Communications Satellite Systems Conference and Exhibit*; Yokohama, Japan. 2003.

[15] Li X., Duan B., Song L., Zhang Y., Xu W. 'Study of stepped amplitude distribution taper for microwave power transmission for SSPS'. *IEEE Transactions on Antennas and Propagation*. 2017;65(10):5396–405.

[16] Dong S.W., Dong Y., Li C., Wang Y., Lv P., Li X. 'Design concept of microwave power transmitting antenna array and its SCS promotion'. *Advances in Astronautics Science and Technology*. 2022;5:3–10.

[17] Wang Y., Dong Y., Dong S., Fu W. 'Analysis of antenna beam efficiency in solar power satellite system'. *Presented at 2016 IEEE International Conference on Electronic Information and Communication Technology*; Harbin, China. 2016.

[18] Alogla A., Eleiwa M.A.H., Alshortan H. 'Design and evaluation of transmitting antennas for solar power satellite systems'. *Engineering, Technology and Applied Science Research*. 2021;11(6):7950–6.

[19] Raza M., Tanaka K. 'Demonstration of digital retrodirective method for solar power satellite'. *Electronics*. 2021;10(4):498.

[20] Hsieh L.H., Strassner B.H., Kokel S.J., *et al*. 'Development of a retrodirective wireless microwave power transmission system'. *Presented at IEEE International Symposium on Antennas and Propagation*; Columbus, OH. 2003.

[21] Wang X., Sha S., He J., Guo L., Lu M. 'Wireless power delivery to low-power mobile devices based on retro-reflective beamforming'. *IEEE Antennas and Wireless Propagation Letters*. 2014;13:919–22.

[22] Pon C. 'Retrodirective array using the heterodyne technique'. *IEEE Transactions on Antennas and Propagation*. 1964;12:176–80.

[23] Cao X., Cheng H., Wang X., Lu M. 'Retro-reflective beamforming for wireless power transmission based on harmonic mixing scheme'. Presented at *The 11th International Conference on Microwave and Millimeter Wave Technology*; Guangzhou, China. 2019.

[24] Cleveland R.F. 'Human exposure to RF electromagnetic fields: policies of the U.S. federal communications commission'. *Presented at Workshop on Mobile Telephony and Health*; Tokyo, Japan. 2006.

[25]   Surender D., Khan T., Talukdar F.A., Antar Y.M.M. 'Rectenna design and development strategies for wireless applications: a review'. *IEEE Antennas and Propagation Magazine*. 2022;64(5):16–29.

[26]   Otsuka M., Omuro N., Kakizaki K. *et al.* 'Relation between spacing and receiving efficiency of finite rectenna array'. *Electronics and Communications in Japan (Part I: Communications)*. 1991;74(2):88–96.

[27]   Erkmen F., Ramahi O.M. 'A scalable, dual-polarized absorber surface for electromagnetic energy harvesting and wireless power transfer'. *IEEE Transactions on Microwave Theory and Techniques*. 2021;69(9):4021–8.

[28]   Shen S., Clerckx B. 'Beamforming optimization for MIMO wireless power transfer with nonlinear energy harvesting: RF combining versus DC combining'. *IEEE Transactions on Wireless Communications*. 2021;20(1):199–213.

[29]   Shinohara N. *Wireless Power Transfer via Radiowaves*. London: ISTE Ltd and John Wiley & Sons, Inc.; 2014.

Chapter 6

Retro-reflective beamforming technique for microwave power transmission in fully-enclosed space

In this chapter, "fully-enclosed space" stands for a spatial region fully enclosed by conducting walls. When electrical sources reside in a fully-enclosed region, the conducting walls prevent the electrical power from radiating out of the region. As a result, the power transmission efficiency between a wireless power transmitter and a wireless power receiver can approach 100% inside a fully-enclosed space. Moreover, any negative impacts associated with wireless power transmission (including possible biological hazards and electromagnetic interferences, as discussed in Section 1.1 of this book) are confined in the fully-enclosed space; in other words, wireless power transmission in a fully-enclosed space does not influence the exterior region. This chapter investigates the retro-reflective beamforming technique for potential wireless power transmission applications in fully-enclosed space. In Section 6.1, the basic concepts of wireless power transmission in fully-enclosed space are proposed. Feasibility studies of efficient wireless power transmission in fully-enclosed space are conducted by the theoretical means in Section 6.2 and by the experimental means in Section 6.3, respectively. Two retro-reflective beamforming schemes for wireless power transmission in fully-enclosed space are analyzed in Section 6.4; one is based on phased arrays and the other is based on parasitic arrays. Section 6.5 studies a special case of Section 6.4, in which a wireless power transmitter includes two antenna elements; specifically, closed-form formulations are derived for the two retro-reflective beamforming schemes and the two retro-reflective beamforming schemes are compared with each other in terms of power transmission efficiency. Some preliminary experimental results of wireless power transmission based on parasitic arrays are presented in Section 6.6.

6.1 Technical concept of wireless power transmission in fully-enclosed space

A microwave oven is fully enclosed by conducting walls. The conducting walls prevent microwave power from leaking out of the oven. The electrical blockage due to conducting walls not only increases the efficiency of heating inside the oven but also avoids possible hazardous impacts on the exterior region. The practical application of this technical concept can be extended from heating to wireless charging [1]. As proposed in Figure 6.1, an oven-like box fully enclosed by conducting walls is fabricated for the purpose of wireless charging: When multiple electronic devices are placed inside the box, their rechargeable batteries get charged without any wiring. The wireless charging box in Figure 6.1 offers higher charging capacity than the commercially available wireless charging pads (which are shown in Section 1.2 of this book). Since the wireless charging pads rely on nonradiative magnetic fields [2], electronic devices must be physically placed on a pad to be charged efficiently. In other words, a wireless charging pad only takes advantage of a two-dimensional surface. In contrast, the wireless charging box in Figure 6.1 employs a three-dimensional region and thus would enable more electronic devices to be charged simultaneously.

As a drawback of the scheme in Figure 6.1, the electronic devices are not mobile or portable when they are charged in the wireless charging box. Rather, the electronic devices must be collected and placed in the wireless charging box manually before they can be charged wirelessly. Moreover, all the functionalities of the electronic devices must be paused while the wireless charging is in process. Although not being able to preserve the mobility or portability of electronic devices, the wireless charging box in Figure 6.1 offers three distinctive benefits. First, high wireless power transmission efficiency might be accomplished in the wireless charging box. Second, wireless power transmission in the wireless charging box generates no hazards in the exterior region. Third, a large number of electronic devices can be charged simultaneously without any wiring.

Below is one example of potential applications of the wireless charging box proposed in Figure 6.1. Many tourism resorts utilize audio tour guides to assist visitors. A typical audio tour guide device is a wireless device with a rechargeable battery. If the audio tour guide devices are collected by a museum staff and placed in the wireless charging box after the museum is closed, they would be charged at night and be ready for the next day. Indeed, the wireless charging box depicted in

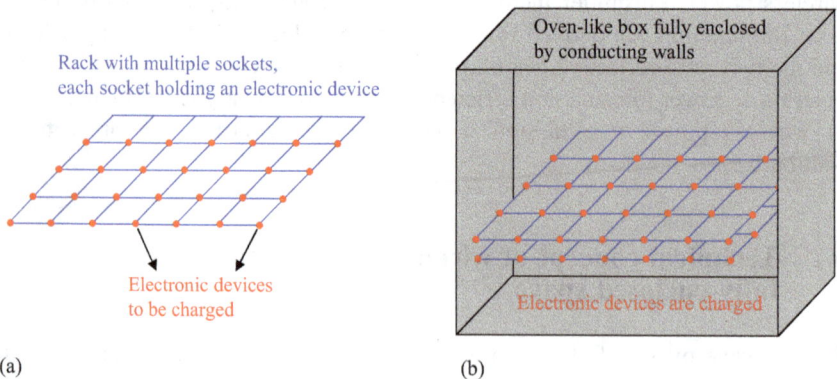

(a)   (b)

Figure 6.1   Depiction of wireless charging box: (a) electronic devices are placed over a rack and (b) racks are placed inside an oven-like box

Figure 6.1 would be valuable for practical applications where a large number of electronic devices are necessities but are not in need constantly.

When a set of permanent positions is prescribed for wireless charging (by the racks and by the electronic device holders in Figure 6.1, for instance), it seems unnecessary to reconfigure the wireless power distribution in a wireless charging box. Nevertheless, it is always desirable to adjust the pattern of wireless power in real-time. For instance, if Device A's rechargeable battery is almost full while Device B's rechargeable battery is nearly empty, illuminating Device B with stronger wireless power than Device A would improve the system's efficiency obviously. Ideally, the wireless charging box ought to be "smart": The wireless charging box ought to communicate with the electronic devices periodically and then reconfigure the wireless power distribution in the box according to the devices' needs.

In addition to microwave ovens, plenty of practical environments are fully enclosed or almost fully enclosed by conducting walls, such as spacecraft [3], submarines, engine compartments [4,5], and some greenhouses [6]. Wireless power transmission has potential applications in these fully-enclosed environments. For instance, if a wireless sensor attached to a rotary shaft in an engine compartment has access to wireless power, it would be unnecessary to stop the engine for the sake of charging the sensor. As another example, wireless power transmission in fully-enclosed space has enabled powering devices implanted in animal bodies wirelessly in medical experiments [7–9]. The technology proposed in Figure 6.1 may be applicable to equip the above-mentioned environments with a wireless charging module. Because electronic devices' mobility or portability should not be sacrificed when they acquire wireless power in practical environments like spacecraft or submarines, intuitively the wireless power transmission must be reconfigured to maintain the optimal power transmission efficiency when the devices do not have fixed locations. Even if the electronic devices are stationary with fixed locations, the spatial distribution of wireless power should be adjusted in real-time ideally, as argued in the previous paragraph.

In summary, it seems that wireless power transmission would not reach the optimal performance without reconfigurability in a fully-enclosed space (such as a wireless charging box, a space station, etc.).

Though the technical concept of Figure 6.1 resembles wireless charging pads, the underlying technology of wireless charging pads cannot be extended to Figure 6.1 straightforwardly. Because wireless charging pads intend to minimize radiative fields, their operating frequency is typically below 1 MHz [10]. When the frequency is as low as 1 MHz, the fully-enclosed box in Figure 6.1 behaves as a large capacitive load to the wireless power transmitter. To achieve conjugate matching, the wireless power transmitter must incorporate a large inductance to neutralize the capacitive load [11,12], which leads to a range of technical complications. Thus, it seems more optimal to upgrade the operating frequency to be close to the natural resonant frequencies of the fully-enclosed box. When the physical dimension of a fully-enclosed box is on the order of 1 m, the resonant frequencies are higher than 100 MHz [13]. Therefore, this chapter adopts the microwave frequency band for wireless power transmission in fully-enclosed space. As the retro-reflective beamforming technique has been demonstrated to offer close-to-optimal efficiency and reconfigurability in the previous five chapters, this chapter investigates applying the retro-reflective beamforming technique to achieve efficient and reconfigurable wireless power transmission in fully-enclosed space.

As a relatively new research topic, wireless power transmission in fully-enclosed space has not been explored comprehensively or thoroughly. Therefore, most of the analyses and experiments in this chapter are feasibility studies by nature; in other words, the preliminary conclusions drawn in this chapter are subject to further investigations.

6.2   Feasibility study of efficient microwave power transmission in fully-enclosed space: theoretical analysis

As hypothesized in the previous section, the efficiency of wireless power transmission can approach 100% in a fully-enclosed space since the conducting walls prevent electrical power from radiating to the exterior region. This hypothesis is verified theoretically in this section.

The physical configuration studied in this section is illustrated in Figure 6.2. A rectangular cavity has its six walls made of perfect electric conductors. There is only air inside the cavity except one transmitting probe and one receiving probe. Both the transmitting and receiving probes are thin conducting wires oriented along the z direction, behaving as monopole antennas. The two probes are assumed to have the same length l_p and same radius r_p. The transmitting and receiving probes are terminated at "$z = 0$" plane by co-axial connectors, which are defined as Port 1 and Port 2, respectively. The transmitting probe and receiving probe are located at $(x = x_t, y = y_t)$ and $(x = x_r, y = y_r)$, respectively. It is assumed that Port 1 is connected to a time-harmonic source and Port 2 is connected to a passive load.

The physical configuration in Figure 6.2 is selected for feasibility studies because it enables a semi-analytical modal analysis that yields explicit physical insights toward efficient wireless power transmission in fully-enclosed space. To conduct the modal analysis, an equivalent configuration corresponding to the setup of Figure 6.2 is constructed in Figure 6.3. In Figure 6.3, the cavity is fully enclosed by perfect conducting walls. Electric surface currents $\mathbf{J}_t$ and $\mathbf{J}_r$ reside at $(x = x_t, y = y_t)$ and $(x = x_r, y = y_r)$, respectively, extended along the z direction. If the probes in Figure 6.2 are thin with respect to the wavelength, $\mathbf{J}_t$ and $\mathbf{J}_r$ are assumed to have z components only. Magnetic surface currents $\mathbf{M}_t$ and $\mathbf{M}_r$ reside around $(x = x_t, y = y_t)$ and $(x = x_r, y = y_r)$, respectively, corresponding to the co-axial openings in Figure 6.2 [14].

An electric field integral equation can be formulated for the equivalent configuration in Figure 6.3.

$$\widehat{\mathbf{z}} \cdot \int \bar{\mathbf{G}}_{EM}(\mathbf{r}, \mathbf{r}') \cdot [\mathbf{M}_t(\mathbf{r}') + \mathbf{M}_r(\mathbf{r}')]d\mathbf{r}' + \widehat{\mathbf{z}} \cdot \int \bar{\mathbf{G}}_{EJ}(\mathbf{r}, \mathbf{r}') \cdot [\mathbf{J}_t(\mathbf{r}') + \mathbf{J}_r(\mathbf{r}')]d\mathbf{r}' = 0,$$

when $\mathbf{r}$ is on probe surface.

$$(6.1)$$

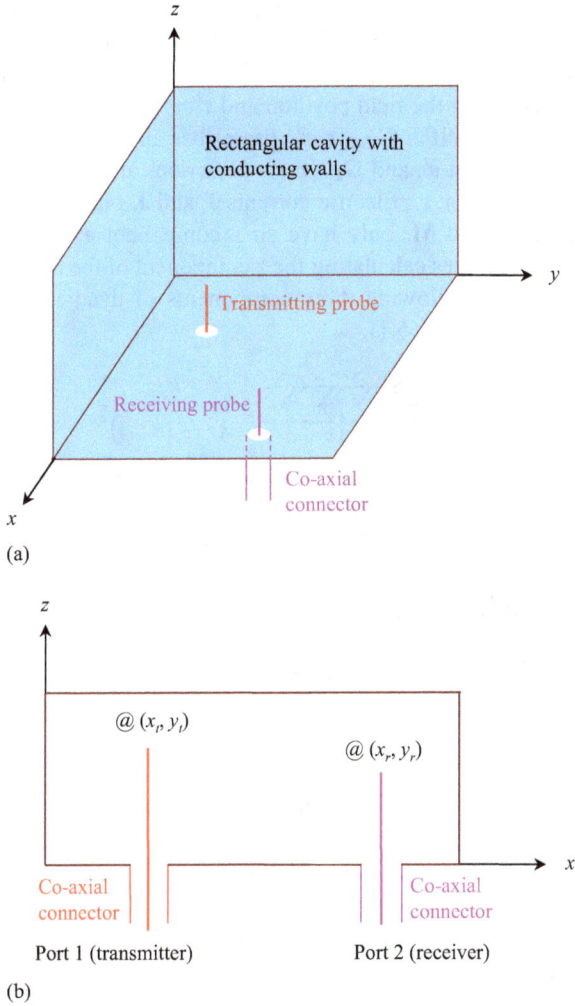

Figure 6.2   *Physical configuration of theoretical feasibility studies of Section 6.2.*
(a) Three-dimensional view and (b) View in the x-z plane.

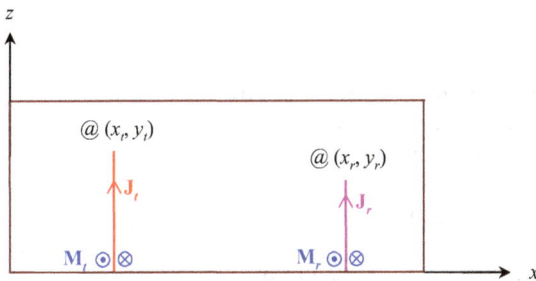

Figure 6.3   *An equivalent configuration to Figure 6.2*

Equation (6.1) enforces that the tangential electric field over the probe surface is zero. The electric field in Figure 6.3 is generated by four sources: $\mathbf{J}_t$, $\mathbf{J}_r$, $\mathbf{M}_t$, and $\mathbf{M}_r$. In (6.1), $\bar{\mathbf{G}}_{EM}$ and $\bar{\mathbf{G}}_{EJ}$ are dyadic Green's functions of the rectangular cavity, with $\mathbf{r} = (x, y, z)$ denoting the field position and $\mathbf{r}' = (x', y', z')$ denoting the source position. To be more specific, $\bar{\mathbf{G}}_{EM}(\mathbf{r}, \mathbf{r}')$ prescribes an electric field at $\mathbf{r}$ due to a magnetic current dipole at $\mathbf{r}'$, and $\bar{\mathbf{G}}_{EJ}(\mathbf{r}, \mathbf{r}')$ prescribes an electric field at $\mathbf{r}$ due to an electric current dipole at $\mathbf{r}'$. Electric currents $\mathbf{J}_t$ and $\mathbf{J}_r$ only have a z component; magnetic currents $\mathbf{M}_t$ and $\mathbf{M}_r$ only have an x component and y component, and Equation (6.1) only requires calculating the z component of the electric field. Based on the above facts, the following three components of dyadic Green's functions $\bar{\mathbf{G}}_{EM}$ and $\bar{\mathbf{G}}_{EJ}$ are needed in (6.1).

$$
\hat{\mathbf{z}} \cdot \bar{\mathbf{G}}_{EM} \cdot \hat{\mathbf{x}} = \frac{8}{abd} \sum_{m=1}^{\infty} \sum_{n=1}^{\infty} \sum_{p=0}^{\infty} \frac{1}{\xi_p} \frac{k_{y,n}}{k^2 - (k_{mnp})^2}
$$

$$
\cdot \sin(k_{x,m}x) \sin(k_{y,n}y) \cos(k_{z,p}z) \sin(k_{x,m}x') \cos(k_{y,n}y') \cos(k_{z,p}z')
$$
(6.2)

$$
\hat{\mathbf{z}} \cdot \bar{\mathbf{G}}_{EM} \cdot \hat{\mathbf{y}} = \frac{8}{abd} \sum_{m=1}^{\infty} \sum_{n=1}^{\infty} \sum_{p=0}^{\infty} \frac{1}{\xi_p} \frac{-k_{x,m}}{k^2 - (k_{mnp})^2}
$$

$$
\cdot \sin(k_{x,m}x) \sin(k_{y,n}y) \cos(k_{z,p}z) \cos(k_{x,m}x') \sin(k_{y,n}y') \cos(k_{z,p}z')
$$
(6.3)

$$
\hat{\mathbf{z}} \cdot \bar{\mathbf{G}}_{EJ}(\mathbf{r}, \mathbf{r}') \cdot \hat{\mathbf{z}} = \frac{1}{j\omega\varepsilon_0} \frac{8}{abd} \sum_{m=1}^{\infty} \sum_{n=1}^{\infty} \sum_{p=0}^{\infty} \frac{1}{\xi_p} \frac{(k_{z,p})^2 - k^2}{k^2 - (k_{mnp})^2}
$$

$$
\cdot \sin(k_{x,m}x) \sin(k_{y,n}y) \cos(k_{z,p}z) \sin(k_{x,m}x') \sin(k_{y,n}y') \cos(k_{z,p}z')
$$
(6.4)

The expressions in (6.2) to (6.4) follow the classic eigenfunction expansion derivations in [15]. Time dependence $e^{j\omega t}$ is suppressed with $j = \sqrt{-1}$; $\omega = 2\pi f$ is the angular frequency; f is the frequency of the source connected to Port 1; a, b, and d are the cavity dimensions along the x, y, and z axes, respectively (in the theoretical examples of this section, $a = b = d = 1$ m); $\xi_0 = 2$, and $\xi_p = 1$ when $p \neq 0$; $k = \omega\sqrt{\varepsilon_0\mu_0}$ is the wavenumber in free space; ε_0 and μ_0 are the permittivity and permeability of free space, respectively; and $k_{mnp} = \sqrt{(k_{x,m})^2 + (k_{y,n})^2 + (k_{z,p})^2}$, with $k_{x,m} = m\pi/a$, $k_{y,n} = n\pi/b$, and $k_{z,p} = p\pi/d$.

With k_{mnp} denoting the wavenumber associated with the mode (m, n, p) of the rectangular cavity, $f_{mnp} = k_{mnp}/(2\pi\sqrt{\varepsilon_0\mu_0})$ is used to denote the resonant frequency associated with the mode (m, n, p). It is possible to identify excellent wireless channels when the operating frequency f is around $f_{m^*n^*p^*}$, the resonant frequency of a specific mode (m^*, n^*, p^*) (in this section, the superscript "*" does not stand for complex conjugation). The TM$_{z220}$ mode is studied as a specific example in this section. In other words, in this section, it is assumed that $f \rightarrow f_{m^*n^*p^*}$ (that is, f approaches $f_{m^*n^*p^*}$) with $m^* = 2$, $n^* = 2$, and $p^* = 0$. With $m^* = 2$, $n^* = 2$, and $p^* = 0$, the resonant frequency $f_{m^*n^*p^*}$ is approximately 424 MHz. When f approaches $f_{m^*n^*p^*}$ (which is equivalent to k approaching $k_{m^*n^*p^*}$), mode (m^*, n^*, p^*) dominates the fields inside the cavity. Mathematically speaking, the terms associated with indices (m^*, n^*, p^*) are outstanding in (6.2) to (6.4). By defining $u^* = k^2 - (k_{m^*n^*p^*})^2$ and multiplying u^* onto (6.1), it is not difficult to find that, when $f \to f_{m^*n^*p^*}$

$$Q^* = k_{y,n^*}\left[(\tilde{M}_t)_{x,m^*n^*p^*} + (\tilde{M}_r)_{x,m^*n^*p^*}\right]$$
$$- k_{x,m^*}\left[(\tilde{M}_t)_{y,m^*n^*p^*} + (\tilde{M}_r)_{y,m^*n^*p^*}\right] \tag{6.5}$$
$$+ \frac{(k_{z,p^*})^2 - k^2}{j\omega\varepsilon_0}\left[(\tilde{J}_t)_{z,m^*n^*p^*} + (\tilde{J}_r)_{z,m^*n^*p^*}\right] \to 0$$

where

$$(\tilde{M}_t)_{x,m^*n^*p^*} = \int \hat{x}\cdot M_t(\mathbf{r}')\sin\left(k_{x,m^*}x'\right)\cos\left(k_{y,n^*}y'\right)\cos\left(k_{z,p^*}z'\right)d\mathbf{r}' \tag{6.6}$$

$$(\tilde{M}_r)_{x,m^*n^*p^*} = \int \hat{x}\cdot M_r(\mathbf{r}')\sin\left(k_{x,m^*}x'\right)\cos\left(k_{y,n^*}y'\right)\cos\left(k_{z,p^*}z'\right)d\mathbf{r}' \tag{6.7}$$

$$(\tilde{M}_t)_{y,m^*n^*p^*} = \int \hat{y}\cdot M_t(\mathbf{r}')\cos\left(k_{x,m^*}x'\right)\sin\left(k_{y,n^*}y'\right)\cos\left(k_{z,p^*}z'\right)d\mathbf{r}' \tag{6.8}$$

$$(\tilde{M}_r)_{y,m^*n^*p^*} = \int \hat{y}\cdot M_r(\mathbf{r}')\cos\left(k_{x,m^*}x'\right)\sin\left(k_{y,n^*}y'\right)\cos\left(k_{z,p^*}z'\right)d\mathbf{r}' \tag{6.9}$$

$$(\tilde{J}_t)_{z,m^*n^*p^*} = \int \hat{z}\cdot J_t(\mathbf{r}')\sin\left(k_{x,m^*}x'\right)\sin\left(k_{y,n^*}y'\right)\cos\left(k_{z,p^*}z'\right)d\mathbf{r}' \tag{6.10}$$

$$(\tilde{J}_r)_{z,m^*n^*p^*} = \int \hat{z}\cdot J_r(\mathbf{r}')\sin\left(k_{x,m^*}x'\right)\sin\left(k_{y,n^*}y'\right)\cos\left(k_{z,p^*}z'\right)d\mathbf{r}' \tag{6.11}$$

As the physical meaning of (6.5), the sum of all the sources' projections onto mode (m^*, n^*, p^*) must approach zero when f approaches $f_{m^*n^*p^*}$. This phenomenon has been investigated in [16,17]. Moreover, a robust method of moments scheme is developed in [16] to avoid the numerical singularities when f is around $f_{m^*n^*p^*}$, which can be readily employed to obtain the numerical solutions to the integral equation in (6.1).

The method of moments solutions indicates that the contributions from J_t and J_r dominate in (6.5); to be more specific, the contributions from M_t and M_r are typically weaker than those from J_t and J_r by two orders. In other words, the contributions from M_t and M_r can be neglected in (6.5), leading to

$$(\tilde{J}_t)_{z,m^*n^*p^*} \cong -(\tilde{J}_r)_{z,m^*n^*p^*}, \quad \text{when } f \to f_{m^*n^*p^*}. \tag{6.12}$$

The relationship in (6.12) is verified by a numerical example below. The transmitting probe is located at $(x_t = 20 \text{ cm}, y_t = 20 \text{ cm})$, the receiving probe is located at $(x_r = 80 \text{ cm}, y_r = 80 \text{ cm})$, both probes have a radius of $r_p = 0.6$ mm, and both probes have a length of $l_p = 17$ cm. The operating frequency is $f = 424$ MHz, which is very close to the resonant frequency of mode $(m^* = 2, n^* = 2, p^* = 0)$. A voltage source with $V_1 = 1$ V is applied at Port 1 as the transmitter, and Port 2 is terminated by a 50-Ω load as the receiver. Simulation results of current distribution over the two probes, $I_t(z) = 2\pi r_p \mathbf{J}_t(z) \cdot \hat{\mathbf{z}}$ and $I_r(z) = 2\pi r_p \mathbf{J}_r(z) \cdot \hat{\mathbf{z}}$, are plotted in Figure 6.4. It is observed from Figure 6.4 that $\mathbf{J}_t \cong -\mathbf{J}_r$. The transmitting probe and receiving probe are symmetric to each other as far as mode $(m^* = 2, n^* = 2, p^* = 0)$ is concerned. Specifically, $\sin(k_{x,m^*} x_t)\sin(k_{y,n^*} y_t)$ and $\sin(k_{x,m^*} x_r)\sin(k_{y,n^*} y_r)$ have identical values. Moreover, the two probes share the same length and same radius. Straightforwardly, the facts above lead to $(\tilde{J}_t)_{z,m^* n^* p^*} \cong -(\tilde{J}_r)_{z,m^* n^* p^*}$, that is, Equation (6.12) holds true. Because the current at Port 1, I_1, is $I_t(z = 0)$ and the current at Port 2, I_2, is $I_r(z = 0)$, Figure 6.4 further reveals that $I_1 = -I_2$ approximately.

If I_1 and I_2 are both real-valued, "$I_1 = -I_2$" results in 100% power transmission efficiency from Port 1 to Port 2. The proof is as follows. Because Port 2 is terminated by a 50-Ω load, the voltage at Port 2 is $V_2 = -(50 \, \Omega) \times I_2$. If there is no power loss over the cavity walls or over the cavity interior, $\text{Re}\{V_1 \cdot \text{conj}(I_1)\} = -\text{Re}\{V_2 \cdot \text{conj}(I_2)\}$ (here, the operator "conj" stands for "complex conjugate"). Then because of "$I_1 = -I_2$," $V_1/I_1 = 50 \, \Omega$, meaning that the load impedance is mapped to the transmitter without any distortion. The reflection coefficient at Port 1 with respect to the standard 50-Ω is zero. Since there is no return loss at Port 1 and the cavity is lossless, all the power is delivered from the transmitter to the receiver, in other words, the power transmission efficiency is 100%.

In Figure 6.4, the imaginary parts of I_1 and I_2 are on the same order as the real parts. As a result, the power transmission efficiency is considerably close to 100% though it is not exactly 100%. Specifically, $I_1 = (11.27 - j8.33)$ mA. The input

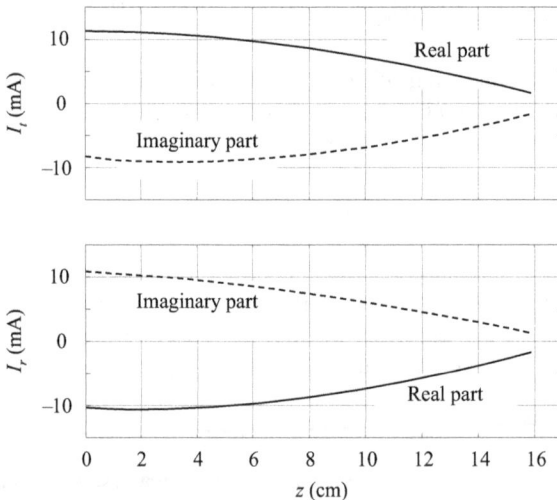

Figure 6.4   Simulation results of current distribution over probes with ($x_t = 20$ cm, $y_t = 20$ cm) and ($x_r = 80$ cm, $y_r = 80$ cm) at 424 MHz, the resonant frequency of mode TM_{z220}. Reproduced from [18], courtesy of the Electromagnetics Academy

impedance at Port 1 is $V_1/I_1 = (57.4 + j42.4)$ Ω. The corresponding reflection coefficient at Port 1 is $S_{11} = 0.19 + j0.32$, the return loss is $|S_{11}|^2 = 0.14$, and the power transmission efficiency is $|S_{21}|^2 = 1 - |S_{11}|^2 = 0.86 = 86\%$. The imaginary parts of I_1 and I_2 can be effectively controlled by l_p, and unsurprisingly, the imaginary parts of I_1 and I_2 are close to zero when $l_p \cong \lambda_0/4$, with $\lambda_0 = 2\pi/k$ denoting the wavelength in free space.

Figure 6.5 displays some numerical results when the frequency f varies around 424 MHz. When $f = 424$ MHz, $u^* \cong 0$, $Q^* \cong 0$, and $|S_{21}| = 0.93$, as articulated in

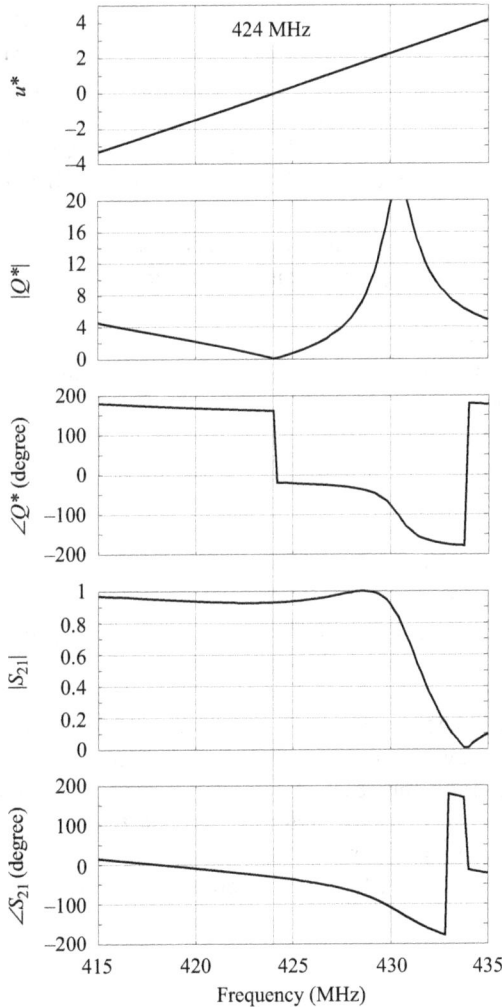

Figure 6.5   *Numerical results with ($x_t = 20$ cm, $y_t = 20$ cm) and ($x_r = 80$ cm, $y_r = 80$ cm) around 424 MHz, the resonant frequency of mode TM$_{z220}$. Reproduced from [18], courtesy of the Electromagnetics Academy*

the several paragraphs above. When f drops from 424 MHz to 415 MHz, $|Q^*|$ (amplitude of Q^*) changes so slightly that $|S_{21}|$ remains greater than 0.9 (corresponding to power transmission efficiency greater than 80%). When f drops from 424 MHz to 415 MHz, $\angle Q^*$ (phase of Q^*) increases slowly. As indicated by (6.5), Q^* embodies the "difference" between I_t and I_r. As $\angle Q^*$ changes slowly, $\angle S_{21}$ (phase of S_{21}) changes slowly as well in the frequency band of [415 MHz, 424 MHz]. When f increases from 424 MHz, $|Q^*|$ changes slowly initially and $|S_{21}|$ remains greater than 0.9 till the frequency 430 MHz. Beyond 430 MHz, $|Q^*|$ changes drastically and $|S_{21}|$ becomes poor. The drastic change is because of the interference from mode ($m = 2$, $n = 2$, $p = 1$), whose resonant frequency resides at 450 MHz. In other words, the frequency band [430 MHz, 435 MHz] in Figure 6.5 is impacted by two modes heavily: mode ($m = 2$, $n = 2$, $p = 0$) and mode ($m = 2$, $n = 2$, $p = 1$). Whereas in the frequency band of [415 MHz, 430 MHz], mode ($m = 2$, $n = 2$, $p = 0$) dominates, as 367 MHz is the nearest resonant frequency (associated with mode ($m = 2$, $n = 1$, $p = 1$)) to the left of 424 MHz. The curve of $\angle Q^*$ has a 180-degree phase jump at frequency 424 MHz, because u^* changes its sign across 424 MHz. When f increases from 424 MHz to 430 MHz, $\angle Q^*$ and $\angle S_{21}$ both decrease smoothly. Overall within the frequency band of [415 MHz, 430 MHz], $|S_{21}|$ is close to 1 and $\angle S_{21}$ appears linear with respect to the frequency, revealing an excellent wireless channel for not only wireless power transmission but also wireless communication.

It is worth noting that wireless communication is an inevitable element in wireless power transmission applications, in other words, certain information exchange or handshaking is very necessary to achieve the optimal performance of wireless power transmission in practice. For instance, when multiple devices call for wireless charging simultaneously, the coordination among the multiple devices must rely on wireless communication. It is well known that a distortion-less wireless communication requires a frequency band with little dispersion; in contrast, the bandwidth associated with wireless power transmission could be zero in principle. The wireless channels in fully-enclosed spaces are highly dispersive typically, which presents challenges to wireless communication [19,20]. The wireless channel in Figure 6.5 demonstrates little dispersion in the frequency band of [415 MHz, 430 MHz], and thus has the potential to enable reliable wireless communication between a wireless power transmitter and a wireless power receiver [18].

Before the end of this section, another numerical example is presented. Compared with the previous example, the only difference is that the receiving probe is located at ($x_r = 90$ cm, $y_r = 60$ cm). As the transmitting probe stays at ($x_t = 20$ cm, $y_t = 20$ cm), the transmitting probe and receiving probe are no longer symmetric to each other with respect to mode ($m^* = 2, n^* = 2, p^* = 0$). Specifically,

$$\frac{\sin(k_{x,m^*}x_t)\sin(k_{y,n^*}y_t)}{\sin(k_{x,m^*}x_r)\sin(k_{y,n^*}y_r)} = \chi = 2.62 \tag{6.13}$$

Simulation results of current distribution at 424 MHz over the two probes, $I_t(z)$ and $I_r(z)$, are plotted in Figure 6.6. It is observed that $\mathbf{J}_t \cong -\mathbf{J}_r/\chi$, which makes

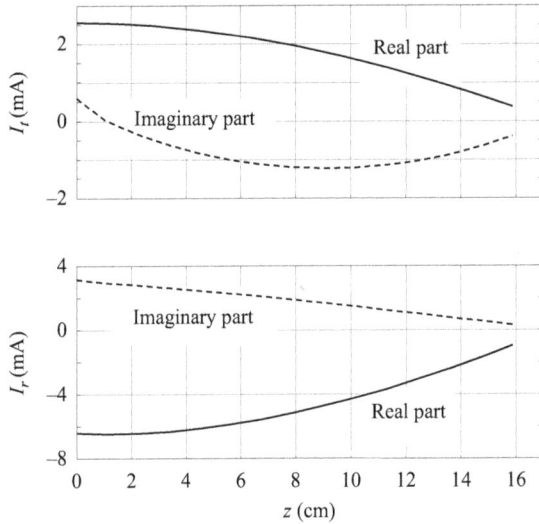

Figure 6.6   Simulation results of current distribution over probes with ($x_t = 20$ cm, $y_t = 20$ cm) and ($x_r = 90$ cm, $y_r = 60$ cm) at 424 MHz, the resonant frequency of mode TM_{z220}. Reproduced from [18], courtesy of the Electromagnetics Academy

Equation (6.12) valid. In the previous example, the value of χ is 1; as a result, the load of Port 2 is mapped to Port 1 with little distortion and the return loss at Port 1 is minimal. In this example, the load of Port 2 is mapped to Port 1 with a factor of χ; as a result, there is an impedance mismatching at Port 1, the return loss at Port 1 is $|S_{11}|^2 = 0.6$, and the power transmission efficiency is $|S_{21}|^2 = 1 - |S_{11}|^2 = 0.4 = 40\%$.

When the frequency f varies around 424 MHz, some numerical results are displayed in Figure 6.7. The phenomena in Figure 6.7 are similar to those of Figure 6.5, except that $|S_{21}|$ in Figure 6.7 has smaller values. Though the $|S_{21}|$ curve of Figure 6.7 is lower than the $|S_{21}|$ curve of Figure 6.5, it is flat between 415 MHz and 430 MHz. Meanwhile, the $\angle S_{21}$ curve of Figure 6.7 is linear with respect to the frequency between 415 MHz and 430 MHz. Therefore, the S_{21}'s behavior within the frequency band of [415 MHz, 430 MHz] in Figure 6.7 appears to be an excellent channel for wireless communication. Obviously, this channel is not optimal for wireless power transmission, since $|S_{21}|$ is not close to 1. As analyzed above, the small $|S_{21}|$ values of Figure 6.7 are due to the impedance mismatching at Port 1, which is further due to χ.

Two wireless channels are demonstrated by numerical results in this section. The channel in Figures 6.4 and 6.5 is when the receiving probe is at ($x_r = 80$ cm, $y_r = 80$ cm), and the channel in Figures 6.6 and 6.7 is when the receiving probe is at ($x_r = 90$ cm, $y_r = 60$ cm). Relatively, the coupling between the transmitting probe and receiving probe is weaker when ($x_r = 90$ cm, $y_r = 60$ cm). To be specific, the

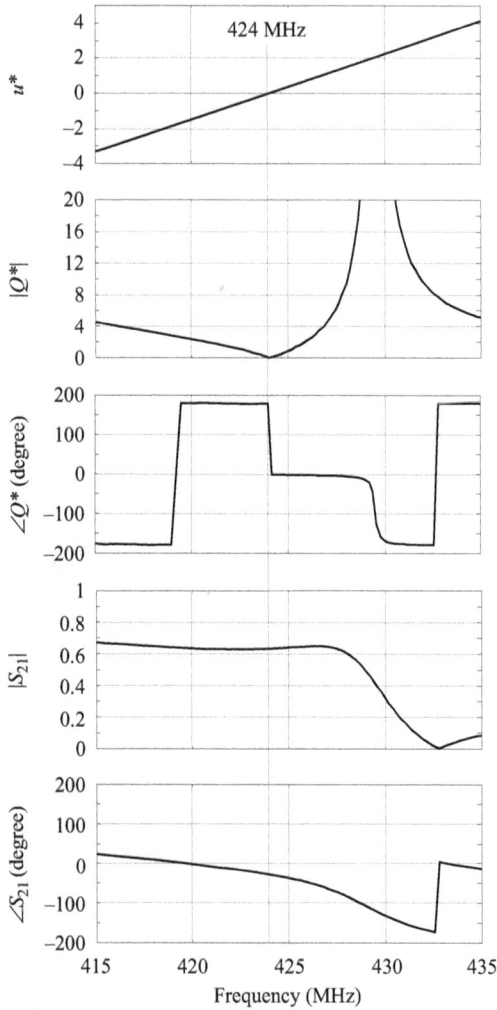

Figure 6.7   Numerical results with (x_t = 20 cm, y_t = 20 cm) and (x_r = 90 cm, y_r = 60 cm) around 424 MHz, the resonant frequency of mode TM$_{z220}$. Reproduced from [18], courtesy of the Electromagnetics Academy

magnitude of S_{21} associated with (x_r = 90 cm, y_r = 60 cm) drops by about 3 dB compared with the magnitude of S_{21} associated with (x_r = 80 cm, y_r = 80 cm). As analyzed above, it is because $\sin(k_{x,m^*}x_r)\sin(k_{y,n^*}y_r)$ has a smaller value with (x_r = 90 cm, y_r = 60 cm). In the two numerical examples above, the TM$_{z220}$ mode dominates the electromagnetic fields in the cavity. Thus unsurprisingly, the coupling between the transmitting probe and receiving probe heavily depends on the standing wave pattern of the TM$_{z220}$ mode. If the receiving probe is located at a "dark spot" of the TM$_{z220}$ mode, (x_r = 50 cm, y_r = 50 cm) for instance, $\sin(k_{x,m^*}x_r)\sin(k_{y,n^*}y_r) = 0$

and $|S_{21}|$ would be close to zero. As a matter of fact, similar phenomena appear in the microwave oven as well. Since the electromagnetic field distribution exhibits certain standing wave patterns inside a microwave oven, the magnitude of heating heavily depends on the spatial location. If the food were not rotated in a microwave oven, some parts would remain frozen while others burned.

The wireless channels in Figures 6.4 to 6.7 are around 424 MHz, the resonant frequency of mode ($m = 2, n = 2, p = 0$). The channel properties become poor when the frequency is above 430 MHz, because of the interference from mode ($m = 2$, $n = 2$, $p = 1$). With the increase of the mode order, the resonant frequencies of cavity modes are more and more densely populated (the asymptotic relationship between mode density and frequency is given in [21]), and the interference among cavity modes would become stronger. Consequently, it would be difficult to identify excellent wireless channels associated with high-order modes. Therefore, wireless channels associated with low-order modes (such as $m = 2$, $n = 2$, and $p = 0$) are taken advantage of for efficient wireless power transmission in this chapter. As a benefit of low-order modes, their resonant frequencies are low such that ohmic loss can be ignored in the theoretical and experimental studies of this chapter. To be specific, the modes investigated in this chapter have resonant frequencies lower than 1 GHz. The ohmic loss over cavity walls is not significant below 1 GHz, as the experimental results in Section 6.3 demonstrate power transmission efficiencies close to 100%. However, the ohmic loss may become not negligible beyond 1 GHz.

6.3 Feasibility study of efficient microwave power transmission in fully-enclosed space: measurement results

Based on the theoretical analysis in Section 6.2, some experiments are carried out, and the measurement results presented in this section reveal that the efficiency of wireless power transmission can approach 100% in a fully-enclosed space.

The experimental setup is illustrated by a photo in Figure 6.8. It follows the theoretical setup in Figure 6.2. The cubic cavity in Figure 6.8 has physical dimensions of 1 m × 1 m × 1 m, with all of its six walls made of aluminum. In Figure 6.8, z-axis is downward, to facilitate routing the cables. One transmitting probe and one receiving probe are connected to measurement equipment through co-axial cables.

In Figure 6.9, the scattering parameter $|S_{21}|$ (with respect to 50 Ω) is plotted, with the following geometrical parameters.

- Port 1: ($x_t = 20$ cm, $y_t = 20$ cm), probe length 17 cm
- Port 2: ($x_r = 80$ cm, $y_r = 80$ cm), probe length 17 cm

The simulation curve in Figure 6.9 is obtained by the method of moments described in Section 6.2. The measurement data in Figure 6.9 are measured by a vector network analyzer. The data from simulation and measurement exhibit excellent agreement. It is observed from Figure 6.9 that $|S_{21}|$ is above -1.5 dB roughly

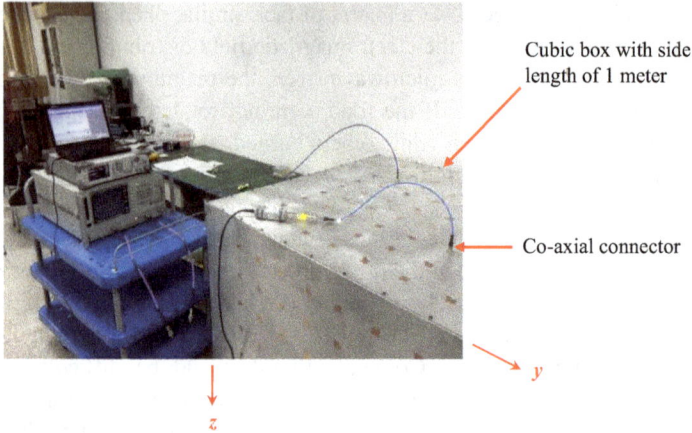

Figure 6.8   A photo of the experimental setup in Section 6.3. Reproduced from [3], courtesy of John Wiley and Sons

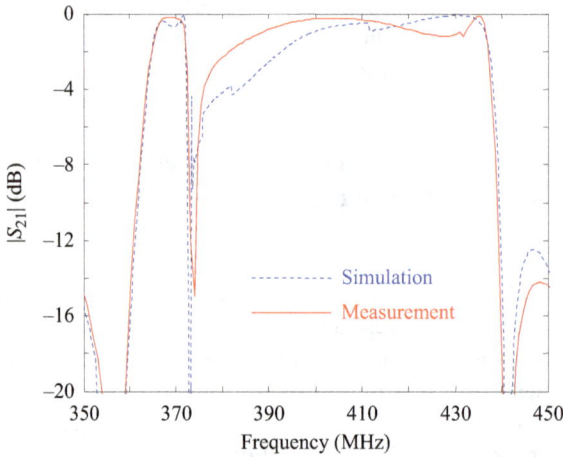

Figure 6.9   Simulated and measured $|S_{21}|$ results with Port 1 located at ($x_t =$ 20 cm, $y_t = 20$ cm) and Port 2 located at ($x_r = 80$ cm, $y_r = 80$ cm). Reproduced from [3], courtesy of John Wiley and Sons

between 410 MHz and 430 MHz, corresponding to power transmission efficiency greater than 70%.

When the receiver's location is moved to ($x_r = 70$ cm, $y_r = 70$ cm) and with all the other conditions unchanged, simulation results and measurement results of $|S_{21}|$ are plotted in Figure 6.10. Because ($x_r = 80$ cm, $y_r = 80$ cm) and ($x_r = 70$ cm, $y_r = 70$ cm) are symmetric to each other with respect to the E_z pattern of TM$_{z220}$ mode, Figure 6.10 is fairly similar to Figure 6.9.

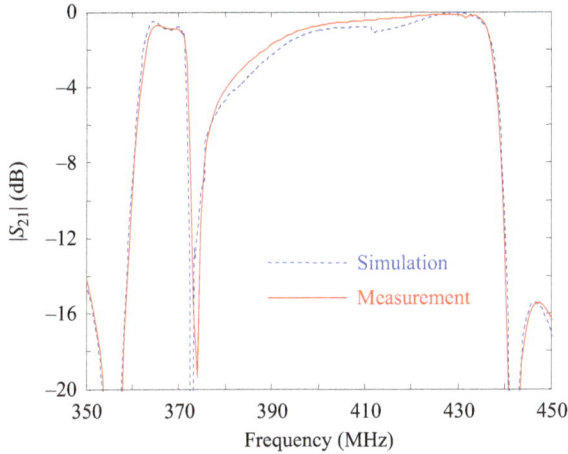

Figure 6.10   Simulated and measured |S$_{21}$| results with Port 1 located at (x$_t$ = 20 cm, y$_t$ = 20 cm) and Port 2 located at (x$_r$ = 70 cm, y$_r$ = 70 cm). Reproduced from [3], courtesy of John Wiley and Sons

Since network analyzers do not offer high precision in terms of power measurement, a signal generator and a power meter are used for measurement when the transmitting probe is at (x$_t$ = 20 cm, y$_t$ = 20 cm) and the receiving probe is at (x$_r$ = 70 cm, y$_r$ = 70 cm). First, the signal generator (which generates a 20-dBm continuous-wave signal) is connected to Port 1, the power meter is connected to Port 2, and the power meter's reading is recorded as P$_r$, received power. Second, the power meter is directly connected to the signal generator with the cavity bypassed, and the power meter's reading is recorded as P$_t$, transmitted power. Finally, the power transmission efficiency is calculated as P$_r$/P$_t$. Simulated and measured data of power transmission efficiency are plotted in Figure 6.11. Overall, Figure 6.11 is consistent with Figure 6.10. It is observed from Figure 6.11 that the power transmission efficiency is greater than 80% over a 20-MHz bandwidth around 425 MHz. At several frequencies, the measured efficiency values reach 100%.

It must be noted that the research efforts reported in Section 6.2 and this section are classified as feasibility studies due to two reasons. First, the theoretical and experimental results are collected below 1 GHz with negligible ohmic loss in Section 6.2 and this section. Second, electric probes (i.e., long straight wires) are employed in Section 6.2 and this section, whereas magnetic probes (i.e., wire loops) with smaller size and higher loss are preferred in practice obviously [13,22,23].

Next, two scenarios with multiple receivers are studied experimentally. The measured data in Figure 6.12 are collected with the following parameters.

- Transmitter (Port 1): (x = 20 cm, y = 20 cm), probe length 17 cm
- The first receiver (Port 2): (x = 70 cm, y = 70 cm), probe length 17 cm
- The second receiver (Port 3): (x = 70 cm, y = 80 cm), probe length 17 cm

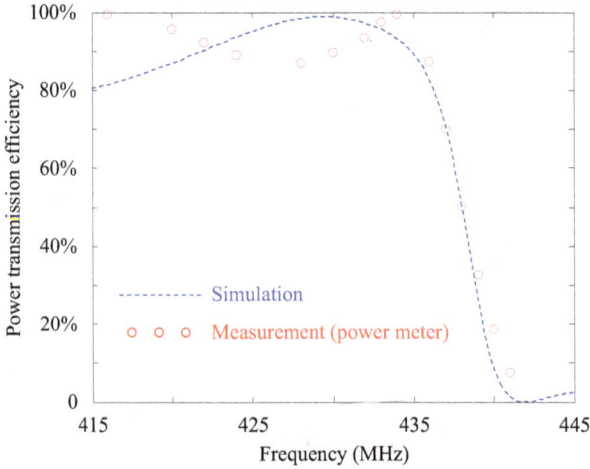

Figure 6.11   Simulated and measured power transmission efficiency results with transmitter located at ($x_t = 20$ cm, $y_t = 20$ cm) and receiver located at ($x_r = 70$ cm, $y_r = 70$ cm). Reproduced from [3], courtesy of John Wiley and Sons.

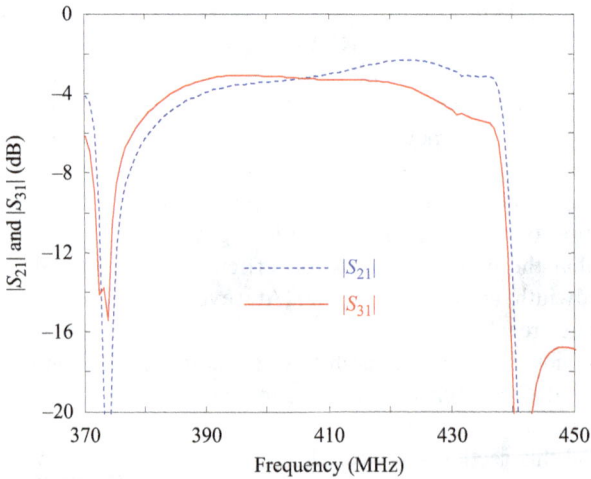

Figure 6.12   Measured $|S_{21}|$ and $|S_{31}|$ results with Port 1 at ($x = 20$ cm, $y = 20$ cm), Port 2 at ($x = 70$ cm, $y = 70$ cm), and Port 3 at ($x = 70$ cm, $y = 80$ cm). Reproduced from [3], courtesy of John Wiley and Sons

It is observed from Figure 6.12 that both $|S_{21}|$ and $|S_{31}|$ are close to -3.5 dB around 410 MHz, corresponding to power transmission efficiency of 44%. The results plotted in Figure 6.13 are measured with the following parameters.

- Transmitter (Port 1): ($x = 20$ cm, $y = 20$ cm), probe length 17 cm
- The first receiver (Port 2): ($x = 70$ cm, $y = 70$ cm), probe length 17 cm
- The second receiver (Port 3): ($x = 70$ cm, $y = 80$ cm), probe length 17 cm
- The third receiver (Port 4): ($x = 80$ cm, $y = 70$ cm), probe length 17 cm

Around 410 MHz, measured $|S_{21}|$, $|S_{31}|$, and $|S_{41}|$ values are all close to -5.5 dB. Specifically at 410 MHz, $|S_{21}| = -5.39$ dB (corresponding to power transmission efficiency of 29%), $|S_{31}| = -5.9$ dB (corresponding to power transmission efficiency of 26%), and $|S_{41}| = -5.94$ dB (corresponding to power transmission efficiency of 25%). The combined power transmission efficiency is 29% + 26% + 25% = 80%.

The feasibility studies in Section 6.2 and this section demonstrate that the wireless power transmission efficiency can approach 100% in fully-enclosed space. Nevertheless, the performance of wireless power transmission heavily depends on the spatial location of the wireless power transmitter and wireless power receiver, as the electromagnetic fields inside a fully-enclosed space are in the form of standing waves rather than traveling waves. In other words, there may exist "dark spots" in a fully-enclosed space in which a wireless power receiver receives little power. If a wireless power receiver is not stationary or if a wireless power receiver's location is unknown to the wireless power transmitter, the standing wave pattern must be reconfigured in real-time to avoid dark spots from the wireless power receiver's location. Mechanical stirring is applied in [24] to reconfigure the

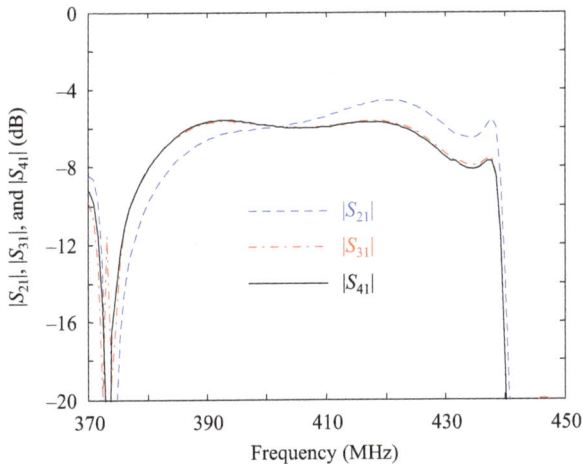

Figure 6.13   *Measured $|S_{21}|$, $|S_{31}|$, and $|S_{41}|$ results with Port 1 at (x = 20 cm, y = 20 cm), Port 2 at (x = 70 cm, y = 70 cm), Port 3 at (x = 70 cm, y = 80 cm), and Port 4 at (x = 80 cm, y = 70 cm). Reproduced from [3], courtesy of John Wiley and Sons*

standing wave patterns, and in turn, to improve the uniformity of field distribution in the statistical sense. The research work in [22,25] takes advantage of multiple frequencies/modes to relieve the dark spots' impact. The retro-reflective beamforming technique has been proven to be an effective approach to reconfiguring electromagnetic fields in the previous few chapters of this book. The retro-reflective beamforming technique could operate at a single frequency and does not have to resort to wideband antennas or circuits. Moreover in retro-reflective beamforming, electromagnetic fields are reconfigured by electronic means rather than mechanical means. Therefore, retro-reflective beamforming may be an appealing candidate to accomplish high-performance wireless power transmission in fully-enclosed space. The next few sections, that is, Sections 6.4, 6.5, and 6.6, are devoted to the retro-reflective beamforming technique for wireless power transmission in fully-enclosed space.

6.4 Retro-reflective beamforming based on phased arrays versus retro-reflective beamforming based on parasitic arrays: general theoretical analysis

The retro-reflective beamforming technique discussed in Chapters 1 to 5 of this book is based on phased arrays. Whereas the retro-reflective beamforming technique based on phased arrays is applicable in fully-enclosed space, it may yield poor performance as analyzed in this section and the next section (i.e., Section 6.5). In this section, another retro-reflective beamforming scheme that is based on parasitic arrays is proposed for wireless power transmission in fully-enclosed space. Moreover, the two retro-reflective beamforming schemes (based on phased arrays and parasitic arrays respectively) are compared with each other theoretically.

Figure 6.14 illustrates the retro-reflective beamforming technique based on a phased array for wireless power transmission in fully-enclosed space. A wireless

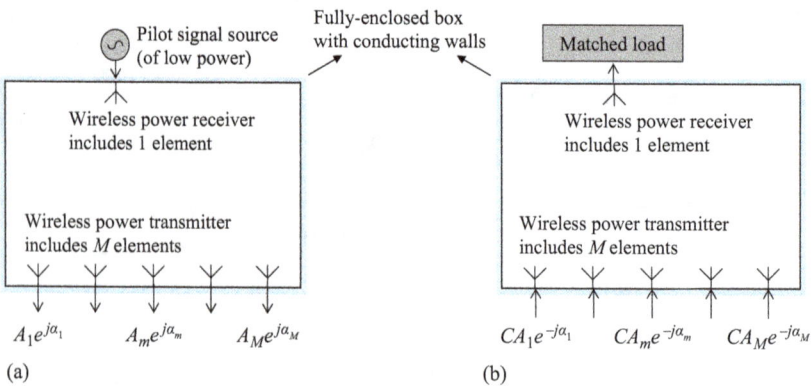

Figure 6.14   Illustration of the retro-reflective beamforming technique based on a phased array for wireless power transmission in a fully-enclosed space: (a) Step (i): pilot signal propagation and (b) Step (ii): wireless power propagation

power transmitter includes M antenna elements. Each wireless power receiver includes one antenna element. The retro-reflective beamforming scheme based on a phased array includes two steps. In the first step, a low-power time-harmonic pilot signal is broadcasted by a wireless power receiver. The pilot signal is detected by the M antenna elements of the wireless power transmitter. The pilot signals detected by the M antenna elements are denoted as $A_m e^{j\alpha_m}$, $m = 1, 2, \ldots, M$. In the second step of retro-reflective beamforming, wireless power is transmitted from the wireless power transmitter to the wireless power receiver. The M antenna elements of the wireless power transmitter are excited by $CA_m e^{-j\alpha_m}$, $m = 1, 2, \ldots, M$, at the same frequency as the pilot signal. In other words, the profile of wireless power excitation is conjugate to the profile of pilot signal reception. The real constant C is far greater than 1 in practice, and it is determined by the power budget of the wireless power transmitter.

Although the retro-reflective beamforming scheme in Figure 6.14 appears the same as those described in the previous few chapters of this book, there is actually one distinctive difference: The wireless channels in Figure 6.14 are a lot more complicated. In Chapters 2 to 5 of this book, the wireless power transmitter and wireless power receiver are assumed to reside in free space. In free space, the wireless channel between the wireless power receiver and the m-th antenna element of the wireless power transmitter is primarily determined by the distance between the wireless power receiver and the m-th antenna element of the wireless power transmitter. In Figure 6.14, the line-of-sight distance is far from enough to characterize the wireless channel between the wireless power receiver and the m-th antenna element of the wireless power transmitter because the cavity walls create multiple paths for them to interact with each other. For instance, it is possible that the signals through multiple paths cancel each other, resulting in weak A_m. As another consequence of complicated wireless channels in Figure 6.14, there might be strong mutual coupling among the antenna elements of wireless power transmitter, whereas mutual coupling was assumed to be negligible in the previous few chapters. Because of the complicated channels, the performance of retro-reflective beamforming based on a phased array in Figure 6.14 is not as predictable as those in Chapters 2 to 5 of this book.

As an alternative, a retro-reflective beamforming technique based on parasitic arrays is proposed in Figure 6.15 for wireless power transmission in fully-enclosed space. A wireless power transmitter includes M antenna elements. One of the M elements is the *driver element*. The other $(M - 1)$ elements, termed as "parasitic elements," are terminated by tunable purely reactive loads. Each wireless power receiver includes one antenna element. The wireless power receivers acquire wireless power from the wireless power transmitter via two steps. In the first step (Figure 6.15(a)), a low-power time-harmonic pilot signal is broadcasted by a wireless power receiver. A power detector is attached to the driver element of the wireless power transmitter. The loads of parasitic elements are adjusted until the power detector's output is maximal. The reactance values of the $(M - 1)$ loads corresponding to the maximal power detector's output are recorded as $jX_1, jX_2, \ldots, jX_{M-1}$. In the second step (Figure 6.15(b)), the driver element is

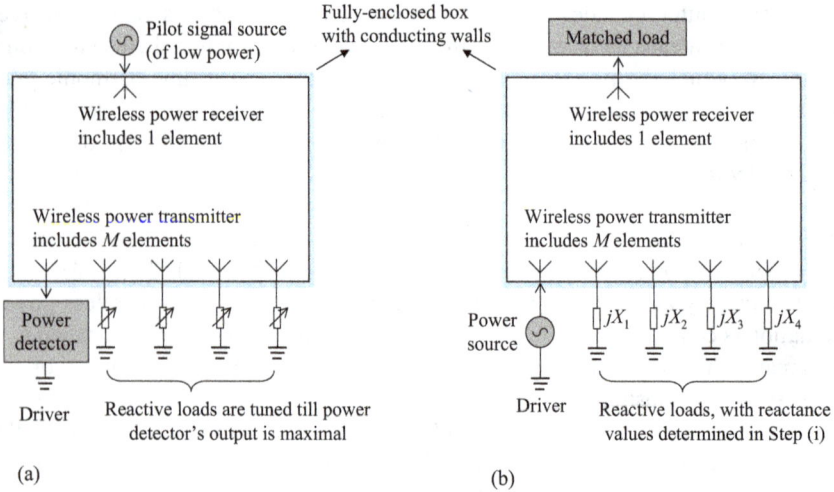

Figure 6.15   Illustration of the retro-reflective beamforming technique based on a parasitic array for wireless power transmission in a fully-enclosed space: (a) Step (i): pilot signal propagation and (b) Step (ii): wireless power propagation

excited by a time-harmonic power source with the same frequency as the pilot signal and the wireless power receiver is terminated by a matched load. The $(M-1)$ loads' values are fixed as $jX_1, jX_2, \ldots, jX_{M-1}$. (The symbol "$X$" stands for "reactance" in Chapter 6, while it stands for "excitation amplitude" in the previous few chapters.)

The M-element parasitic array in Figure 6.15 resembles the classic Yagi-Uda antenna. In a Yagi-Uda antenna, one element of the array is the driver element, and the other elements are parasitic dipoles terminated by short [26]. When a parasitic array is employed as a transmitting antenna, the driver element is the only active element, that is, the driver element is excited by a power source. After the source's power is radiated by the driver element, a certain portion of the power is coupled to the parasitic elements. The coupled power is re-radiated, as the parasitic elements are terminated by purely reactive loads. The total electromagnetic field radiated by the parasitic array is the sum of direct radiation from the driver element and re-radiation from the parasitic elements. After the reactive loads are tuned, the re-radiation would be altered and the total field would be reconfigured [27,28].

Though a parasitic array has significantly lower cost and complexity than a phased array, in practice parasitic arrays are not as popular as phased arrays. In a parasitic array, the contribution of a certain parasitic element to the total electro-magnetic field depends on how much power it couples from the driver element. In other words, if the coupling between the driver element and a certain parasitic element is weak, the parasitic element's role is minimal. On the other hand, each element's contribution in a phased array is controlled individually, and as a result, a phased array is more flexible and scalable. Nevertheless, as far as wireless power transmission in fully-enclosed space is concerned, a parasitic array appears to be an excellent candidate. Since radiation is blocked by conducting walls in fully-enclosed space, it is possible for the coupling between the driver element and parasitic elements to be very strong. Consequently, a parasitic element could play an important role even when it is far away from the driver element in space. On the other hand, strong coupling among array elements is harmful to phased arrays, as it prevents each element from being adjusted independently. An intuitive comparison below between Figures 6.14(b) and 6.15(b) reveals that it seems easier to achieve higher power transmission efficiency with the configuration of Figure 6.15(b). Assume that the fully-enclosed space is lossless and its conducting walls are lossless too. At the same time, assume that the $(M - 1)$ parasitic ports' terminations in Figure 6.15(b) are lossless as well. The power delivered to the wireless power receiver would be maximized when the outgoing/reflected power at the driver port is minimized. In Figure 6.14(b), however, the power delivered to the wireless power receiver would be maximized when the total outgoing power at the wireless power transmitter's M ports is minimized. As a result, it appears easier for a parasitic array to achieve high power transmission efficiency than a phased array in fully-enclosed space, especially when M is large.

The theoretical analysis of the retro-reflective beamforming technique based on phased arrays in Section 3.4 can be applied to Figure 6.14 without any alteration. In the following, the retro-reflective beamforming technique based on parasitic arrays in Figure 6.15 is modeled using scattering parameters, which is analogous to the theoretical analysis in Section 3.4.

The retro-reflective beamforming scheme based on a parasitic array in Figure 6.15 can be modeled as an $(M + 1)$-port network:

$$\begin{bmatrix} b_i \\ \mathbf{b}_r \\ b_o \end{bmatrix} = \begin{bmatrix} S_{ii} & \mathbf{S}_{ir} & S_{io} \\ \mathbf{S}_{ri} & \bar{\mathbf{S}}_{rr} & \mathbf{S}_{ro} \\ S_{oi} & \mathbf{S}_{or} & S_{oo} \end{bmatrix} \begin{bmatrix} a_i \\ \mathbf{a}_r \\ a_o \end{bmatrix} \qquad (6.14)$$

Figure 6.16 provides a pictorial explanation of (6.14). In (6.14), scalar a_i represents the incident wave at the driver port of the wireless power transmitter, and scalar b_i represents the outgoing wave at the driver port of the wireless power transmitter. The subscript "i" in a_i and b_i stands for "input," as the driver is connected to a power source in the second step of retro-reflective beamforming. a_o in (6.14) represents the incident wave at the port of the wireless power receiver, and b_o in (6.14) represents the outgoing wave at the port of the wireless power receiver, with subscript "o" standing for "output." $\mathbf{a}_r = [a_1, a_2, \ldots, a_{M-1}]^T$ and $\mathbf{b}_r = [b_1, b_2, \ldots, b_{M-1}]^T$ are the incident and outgoing wave vectors at the $M - 1$ parasitic ports respectively, where the superscript "T" is the transpose operator. Since the parasitic ports are terminated by reactive loads and their job is "to reflect the power," subscript "r" is adopted to stand for "reflection." In (6.14), $\bar{\mathbf{S}}_{rr}$ is an $(M - 1) \times (M - 1)$ matrix, $\mathbf{S}_{ri}$ and $\mathbf{S}_{ro}$ are column vectors with $(M - 1)$ elements, and $\mathbf{S}_{ir}$ and $\mathbf{S}_{or}$ are row vectors with $(M - 1)$ elements.

In the first step of retro-reflective beamforming (Figure 6.15(a)), a pilot signal is broadcasted by the wireless power receiver and the driver element is terminated

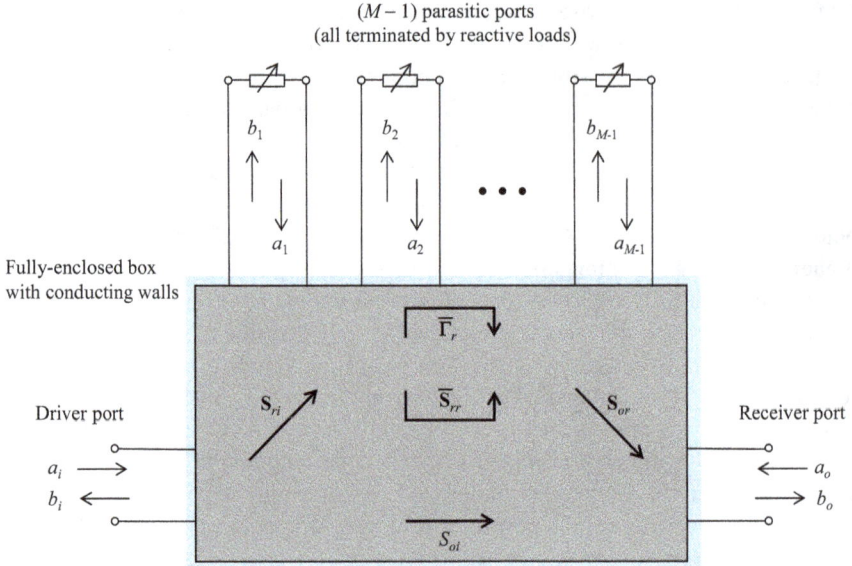

Figure 6.16 A circuit model of retro-reflective beamforming based on a parasitic array using scattering parameters

by a power detector. Assume the power detector is a matched load at the driver port such that $a_i = 0$. The excitation of the pilot signal is assumed to be $a_o = 1$. From (6.14),

$$b_i = S_{ir}\mathbf{a}_r + S_{io}a_o = S_{ir}\mathbf{a}_r + S_{io} \tag{6.15}$$

$$\mathbf{b}_r = \bar{\mathbf{S}}_{rr}\mathbf{a}_r + \mathbf{S}_{ro}a_o = \bar{\mathbf{S}}_{rr}\mathbf{a}_r + \mathbf{S}_{ro} \tag{6.16}$$

Since the $(M - 1)$ parasitic ports are terminated by purely reactive loads, vectors $\mathbf{a}_r$ and $\mathbf{b}_r$ are related to each other by

$$\mathbf{a}_r = \bar{\mathbf{\Gamma}}_r \mathbf{b}_r = \begin{bmatrix} \Gamma_1 & 0 & \cdots & & 0 \\ 0 & \Gamma_2 & 0 & & \vdots \\ \vdots & 0 & \ddots & & 0 \\ 0 & \cdots & & 0 & \Gamma_{M-1} \end{bmatrix} \mathbf{b}_r \tag{6.17}$$

where

$$\Gamma_m = e^{j\phi_m}, \quad m = 1, 2, \ldots, M - 1 \tag{6.18}$$

and ϕ_m is the phase of the reflection coefficient at the m-th parasitic port. By substituting (6.16) into (6.17), it is easy to arrive at

$$\mathbf{a}_r = \left((\bar{\mathbf{\Gamma}}_r)^{-1} - \bar{\mathbf{S}}_{rr} \right)^{-1} \mathbf{S}_{ro} \tag{6.19}$$

Next, by substituting (6.19) into (6.15), b_i can be obtained as

$$b_i = \mathbf{S}_{ir}\left((\bar{\Gamma}_r)^{-1} - \bar{\mathbf{S}}_{rr}\right)^{-1}\mathbf{S}_{ro} + S_{io} \tag{6.20}$$

In the first step of retro-reflective beamforming (Figure 6.15(a)), the reactive termination of the $(M - 1)$ parasitic elements is adjusted such that the output of the power detector is maximized. Obviously, the output of the power detector is

$$|b_i|^2 = \left|\mathbf{S}_{ir}\left((\bar{\Gamma}_r)^{-1} - \bar{\mathbf{S}}_{rr}\right)^{-1}\mathbf{S}_{ro} + S_{io}\right|^2 \tag{6.21}$$

In the second step of retro-reflective beamforming (Figure 6.15(b)), the driver element is excited by an incoming power a_i. The wireless power receiver is terminated by a matched load, in other words, $a_o = 0$. From (6.14)

$$b_o = S_{oi}a_i + \mathbf{S}_{or}\mathbf{a}_r \tag{6.22}$$

$$\mathbf{b}_r = \mathbf{S}_{ri}a_i + \bar{\mathbf{S}}_{rr}\mathbf{a}_r \tag{6.23}$$

Following a set of equations similar to (6.17), (6.18), and (6.19), it is not difficult to conclude that

$$b_o = \left[S_{oi} + \mathbf{S}_{or}\left((\bar{\Gamma}_r)^{-1} - \bar{\mathbf{S}}_{rr}\right)^{-1}\mathbf{S}_{ri}\right]a_i \tag{6.24}$$

In the second step of retro-reflective beamforming (Figure 6.15(b)), the power transmission efficiency is

$$\frac{|b_o|^2}{|a_i|^2} = \frac{\left|\left[S_{oi} + \mathbf{S}_{or}\left((\bar{\Gamma}_r)^{-1} - \bar{\mathbf{S}}_{rr}\right)^{-1}\mathbf{S}_{ri}\right]a_i\right|^2}{|a_i|^2} = \left|S_{oi} + \mathbf{S}_{or}\left((\bar{\Gamma}_r)^{-1} - \bar{\mathbf{S}}_{rr}\right)^{-1}\mathbf{S}_{ri}\right|^2 \tag{6.25}$$

The term $S_{oi} + \mathbf{S}_{or}\left((\bar{\Gamma}_r)^{-1} - \bar{\mathbf{S}}_{rr}\right)^{-1}\mathbf{S}_{ri}$ in (6.25) is a scalar. Thus,

$$S_{oi} + \mathbf{S}_{or}\left((\bar{\Gamma}_r)^{-1} - \bar{\mathbf{S}}_{rr}\right)^{-1}\mathbf{S}_{ri} = S_{oi} + (\mathbf{S}_{ri})^T\left[\left((\bar{\Gamma}_r)^{-1} - \bar{\mathbf{S}}_{rr}\right)^{-1}\right]^T(\mathbf{S}_{or})^T \tag{6.26}$$

The terms in (6.26) and the terms in (6.20) are identical to each other under the following conditions.

$$S_{oi} = S_{io} \tag{6.27}$$

$$(\mathbf{S}_{ri})^T = \mathbf{S}_{ir} \tag{6.28}$$

$$\bar{\mathbf{S}}_{rr} = (\bar{\mathbf{S}}_{rr})^T \tag{6.29}$$

$$(\mathbf{S}_{or})^T = \mathbf{S}_{ro} \tag{6.30}$$

Apparently, the four conditions in (6.27) to (6.30) are equivalent to the condition of "the wireless channels in the fully-enclosed space are reciprocal." Thus, under the condition of reciprocal channels, maximizing (6.21) in the first step of retro-reflective beamforming ensures that the power transmission efficiency in (6.25) is maximized in the second step of retro-reflective beamforming.

6.5   Retro-reflective beamforming based on phased arrays versus retro-reflective beamforming based on parasitic arrays: theoretical analysis of a special case with the wireless power transmitter including two antenna elements

In the previous section (i.e., Section 6.4), two retro-reflective beamforming schemes, which are based on phased arrays and parasitic arrays respectively, are compared with each other theoretically. This section studies a special case of the previous section. Specifically, it is assumed that the wireless power transmitter includes two antenna elements in this section. The maximum values of power transmission efficiency associated with a two-element phased array and a two-element parasitic array are available in closed form. Some numerical results obtained from the closed-form formulations reveal that the magnitude of mutual coupling between the two elements determines which scheme (that is, retro-reflective beamforming based on a phased array or retro-reflective beamforming based on a parasitic array) is more favorable.

As depicted in Figure 6.17, a wireless power transmitter includes a phased array with two antenna elements. Port 1 and Port 2 correspond to the two antenna elements of the phased array, respectively. Port 3 corresponds to the wireless power receiver. According to the analysis in Section 3.4, the maximum power

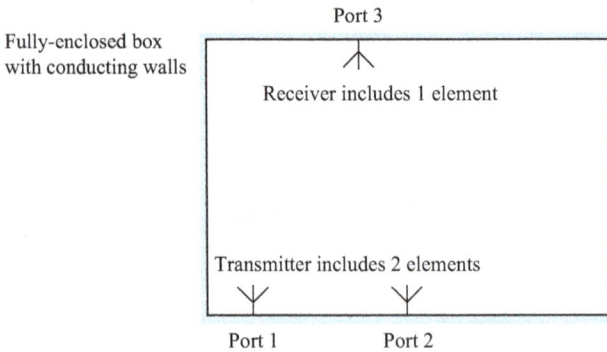

Figure 6.17   Depiction of a wireless power transmitter including a 2-element phased array

Port 3

Fully-enclosed box
with conducting walls

Receiver includes 1 element

Transmitter includes 2 elements

Port 1
(driver)

Port 2
(parasitic element, terminated
by tunable reactive load)

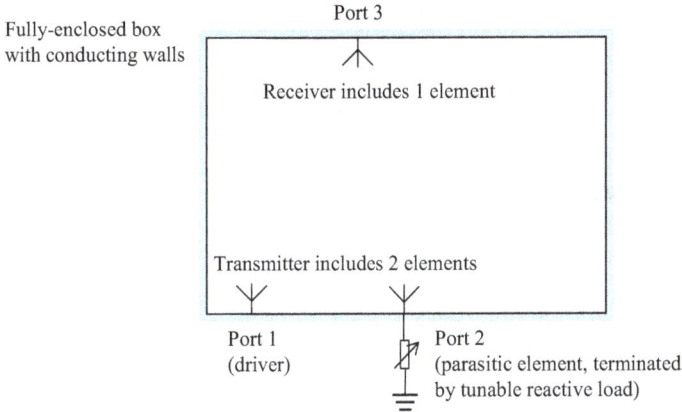

Figure 6.18   Depiction of a wireless power transmitter including a 2-element parasitic array

transmission efficiency (PTE) $_{\text{phased, max}}$ between the phased array and wireless power receiver is

$$(\text{PTE})_{\text{phased, max}} = |S_{13}|^2 + |S_{23}|^2. \tag{6.31}$$

In Figure 6.18, a wireless power transmitter includes a parasitic array with two antenna elements. Port 1 corresponds to the driver element, Port 2 corresponds to the parasitic element, and Port 3 corresponds to the wireless power receiver. With $M = 2$, the power transmission efficiency (PTE)$_{\text{parasitic}}$ between the parasitic array and wireless power receiver in (6.25) reduces to

$$(\text{PTE})_{\text{parasitic}} = \left| S_{31} + \frac{S_{32}S_{21}}{e^{-j\phi_r} - S_{22}} \right|^2 \tag{6.32}$$

where ϕ_r is the phase of the reflection coefficient (Γ_r) of the reactive load at Port 2. Equation (6.32) indicates that the power launched from the driver element reaches the wireless power receiver through two paths: One is directly from the driver to the receiver (i.e., Port 1 → Port 3), and the other is a detour via the parasitic element (i.e., Port 1 → Port 2 → Port 3). The closed-form expression of (PTE)$_{\text{parasitic, max}}$, the maximum power transmission efficiency between the parasitic array and wireless power receiver, is derived next.

Generally, all the scattering parameters in (6.32) are complex numbers. To facilitate the derivation below, it is assumed that the phases of S_{21}, S_{22}, and S_{32} are zero, which can always be realized by shifting the reference planes at the three ports. In other words, it is assumed that S_{21}, S_{22}, and S_{32} are non-negative real numbers. It is also assumed that the circuit network in Figure 6.18 is reciprocal as well as lossless. Then, the 3×3 scattering matrix of the network in Figure 6.18 is a symmetric unitary matrix that satisfies

$$
\begin{bmatrix} S_{11} & S_{21} & S_{31} \\ S_{21} & S_{22} & S_{32} \\ S_{31} & S_{32} & S_{33} \end{bmatrix} \begin{bmatrix} (S_{11})^* & S_{21} & (S_{31})^* \\ S_{21} & S_{22} & S_{32} \\ (S_{31})^* & S_{32} & (S_{33})^* \end{bmatrix} = \bar{I},
\tag{6.33}
$$

where $\bar{I}$ is the identity matrix and the fact that S_{21}, S_{22}, and S_{32} are real numbers is utilized. In this section, the superscript "*" stands for complex conjugation. Specifically from (6.33),

$$
(S_{21})^2 + (S_{22})^2 + (S_{32})^2 = 1
\tag{6.34}
$$

$$
|S_{31}|^2 + (S_{32})^2 + |S_{33}|^2 = 1
\tag{6.35}
$$

$$
S_{31}S_{21} + S_{32}S_{22} + S_{33}S_{32} = 0
\tag{6.36}
$$

Equation (6.36) can be re-arranged to be

$$
S_{33} = -S_{31}\left(\frac{S_{21}}{S_{32}}\right) - S_{22}
\tag{6.37}
$$

It is noticed that S_{33} and S_{31} are complex numbers while S_{21} and S_{22} are assumed to be real-valued in (6.37). After squaring the amplitude of both sides of (6.37),

$$
\begin{aligned}
|S_{33}|^2 &= \left[S_{31}\left(\frac{S_{21}}{S_{32}}\right) + S_{22}\right]\left[(S_{31})^*\left(\frac{S_{21}}{S_{32}}\right) + S_{22}\right] \\
&= |S_{31}|^2\left(\frac{S_{21}}{S_{32}}\right)^2 + 2\left(\frac{S_{21}}{S_{32}}\right)S_{22}\mathrm{Re}[S_{31}] + (S_{22})^2
\end{aligned}
\tag{6.38}
$$

where the operator "Re[·]" selects the real part of the argument. By substituting (6.38) into (6.35) and utilizing (6.34),

$$
|S_{31}|^2 + 2\left(\frac{S_{21}S_{32}\sqrt{1 - (S_{21})^2 - (S_{32})^2}}{(S_{21})^2 + (S_{32})^2}\right)\mathrm{Re}[S_{31}] = \frac{(S_{21})^2(S_{32})^2}{(S_{21})^2 + (S_{32})^2}
\tag{6.39}
$$

or equivalently

$$
\left|S_{31} + \frac{\sqrt{1 - (S_{21})^2 - (S_{32})^2}}{\frac{S_{21}}{S_{32}} + \frac{S_{32}}{S_{21}}}\right|^2 = \frac{1}{\left(\frac{S_{21}}{S_{32}} + \frac{S_{32}}{S_{21}}\right)^2}
\tag{6.40}
$$

After letting

$$
\gamma = \frac{S_{21}}{S_{32}} + \frac{S_{32}}{S_{21}}
\tag{6.41}
$$

and

$$\beta = (S_{21})^2 + (S_{32})^2, \tag{6.42}$$

Equation (6.40) becomes

$$\left| S_{31} + \frac{\sqrt{1 - \beta}}{\gamma} \right|^2 = \frac{1}{\gamma^2}. \tag{6.43}$$

Equation (6.43) implies that S_{31} must be located on a circle with a radius of $1/\gamma$ and centered at $(-\sqrt{1 - \beta}/\gamma, \ 0)$ over the complex plane. Consequently, S_{31} can be expressed as

$$S_{31} = \frac{1}{\gamma} e^{j\xi} - \frac{\sqrt{1 - \beta}}{\gamma}, \tag{6.44}$$

where ξ is an angle on the circle with a radius of $1/\gamma$ and centered at $(-\sqrt{1 - \beta}/\gamma, \ 0)$. Based on the definitions of γ in (6.41) and β in (6.42), the following relationship can be found:

$$\frac{\beta}{\gamma} = \frac{(S_{21})^2 + (S_{32})^2}{\frac{S_{21}}{S_{32}} + \frac{S_{32}}{S_{21}}} = S_{32} S_{21} \tag{6.45}$$

Also by making use of (6.34) and (6.42), it is easy to obtain

$$S_{22} = \sqrt{1 - (S_{21})^2 - (S_{32})^2} = \sqrt{1 - \beta} \tag{6.46}$$

Substituting (6.44), (6.45), and (6.46) into (6.32) leads to

$$\begin{aligned}
(\text{PTE})_{\text{parasitic}} &= \left| S_{31} + \frac{S_{32} S_{21}}{e^{-j\phi_r} - S_{22}} \right|^2 \\
&= \left| \frac{1}{\gamma} e^{j\xi} - \frac{\sqrt{1 - \beta}}{\gamma} + \frac{\beta/\gamma}{e^{-j\phi_r} - \sqrt{1 - \beta}} \right|^2 \\
&= \frac{1}{\gamma^2} \left| e^{j\xi} + \frac{e^{j\phi_r} - \sqrt{1 - \beta}}{e^{-j\phi_r} - \sqrt{1 - \beta}} e^{-j\phi_r} \right|^2
\end{aligned} \tag{6.47}$$

From Figure 6.19, it can be observed that

$$\frac{e^{j\phi_r} - \sqrt{1 - \beta}}{e^{-j\phi_r} - \sqrt{1 - \beta}} = e^{j(2\phi_r + 2\theta)}, \tag{6.48}$$

where θ is the angle between $e^{j\phi_r}$ and $e^{j\phi_r} - \sqrt{1 - \beta}$ over the complex plane. Note that θ is positive when $0 < \phi_r < \pi$ and is negative when $-\pi < \phi_r < 0$.

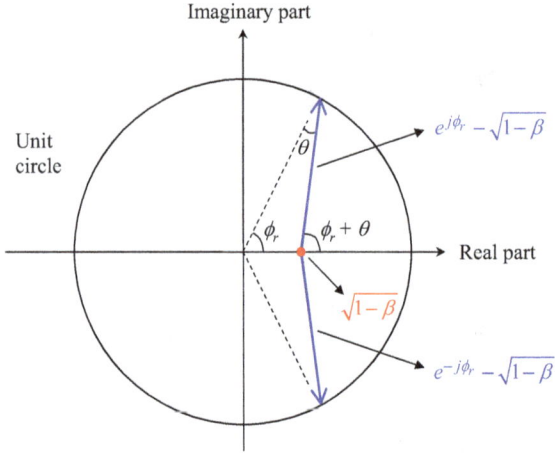

Figure 6.19   Illustration of Equation (6.48) over the complex plane

Substituting (6.48) into (6.47) yields

$$(\text{PTE})_{\text{parasitic}} = \frac{1}{\gamma^2}\left|e^{j\xi} + e^{j(\phi_r + 2\theta)}\right|^2. \tag{6.49}$$

As ϕ_r's value changes from $-\pi$ to π, the value of $\phi_r + 2\theta$ also changes between $-\pi$ and π continuously. Therefore, for any value of ξ in (6.49), it is always possible to find an optimal ϕ_r^{opt} satisfying

$$\phi_r^{opt} + 2\theta = \xi \tag{6.50}$$

With $\phi_r = \phi_r^{opt}$, the maximum power transmission efficiency (PTE)$_{\text{parasitic, max}}$ is

$$(\text{PTE})_{\text{parasitic, max}} = \frac{1}{\gamma^2}\left|e^{j\xi} + e^{j\xi}\right|^2 = \frac{4}{\left(\frac{S_{21}}{S_{32}} + \frac{S_{32}}{S_{21}}\right)^2} \tag{6.51}$$

Equation (6.51) is derived by assuming S_{21}, S_{22}, and S_{32} are all non-negative real numbers. When S_{21}, S_{22}, and S_{32} are complex numbers, Equation (6.51) should be modified to be

$$(\text{PTE})_{\text{parasitic, max}} = \frac{4}{\left(\left|\frac{S_{21}}{S_{32}}\right| + \left|\frac{S_{32}}{S_{21}}\right|\right)^2} \tag{6.52}$$

The closed-form expression derived in (6.52) for the maximum power trans-mission efficiency associated with a 2-element parasitic array reveals a "balanced condition." Specifically, $|S_{21}|$ stands for the coupling between the driver element and parasitic element, and $|S_{32}|$ stands for the coupling between the parasitic element and wireless power receiver. When the two links above (that is, driver-to-parasitic and parasitic-to-receiver) are balanced, the power transmission efficiency can reach 100%.

The expression in (6.52) does not show explicit dependence on S_{31}. In fact, the range of S_{31} is restricted by S_{21} and S_{32}. Specifically from (6.39),

$$\text{Re}[S_{31}] = \frac{S_{21}S_{32} - \left(\frac{S_{32}}{S_{21}} + \frac{S_{21}}{S_{32}}\right)|S_{31}|^2}{2\sqrt{1 - (S_{21})^2 - (S_{32})^2}} \tag{6.53}$$

By using (6.53) and the fact of $|\text{Re}[S_{31}]|^2 < |S_{31}|^2$,

$$\frac{1 - \sqrt{1 - (S_{21})^2 - (S_{32})^2}}{\left(\frac{S_{21}}{S_{32}} + \frac{S_{32}}{S_{21}}\right)} \leq |S_{31}| \leq \frac{1 + \sqrt{1 - (S_{21})^2 - (S_{32})^2}}{\left(\frac{S_{21}}{S_{32}} + \frac{S_{32}}{S_{21}}\right)} \tag{6.54}$$

When S_{21} and S_{32} are complex numbers, (6.54) should be modified to be

$$\frac{1 - \sqrt{1 - |S_{21}|^2 - |S_{32}|^2}}{\left(\left|\frac{S_{21}}{S_{32}}\right| + \left|\frac{S_{32}}{S_{21}}\right|\right)} \leq |S_{31}| \leq \frac{1 + \sqrt{1 - |S_{21}|^2 - |S_{32}|^2}}{\left(\left|\frac{S_{21}}{S_{32}}\right| + \left|\frac{S_{32}}{S_{21}}\right|\right)} \tag{6.55}$$

Similarly, with $|S_{21}|$ and $|S_{31}|$ given, the range of $|S_{32}|$ is

$$\frac{1 - \sqrt{1 - |S_{21}|^2 - |S_{31}|^2}}{\left(\left|\frac{S_{21}}{S_{31}}\right| + \left|\frac{S_{31}}{S_{21}}\right|\right)} \leq |S_{32}| \leq \frac{1 + \sqrt{1 - |S_{21}|^2 - |S_{31}|^2}}{\left(\left|\frac{S_{21}}{S_{31}}\right| + \left|\frac{S_{31}}{S_{21}}\right|\right)} \tag{6.56}$$

In Figure 6.20, some numerical results calculated from (6.31) and (6.52) are displayed, in order to compare the wireless power transmission performance of a 2-element phased array and a 2-element parasitic array in fully-enclosed space. Figure 6.20 includes six subplots, in which the value of $|S_{21}|^2$ is fixed from 0.1 to 0.6, respectively. It is noted that $|S_{21}|^2$ stands for the coupling between the two elements of the 2-element array: The larger $|S_{21}|^2$ is, the stronger the coupling is. When $|S_{21}|^2$ is fixed, the range of $|S_{32}|^2$ is between 0 and $1 - |S_{21}|^2$. With $|S_{21}|^2$ and $|S_{32}|^2$ given, the range of $|S_{31}|^2$ is determined by (6.55). In each subplot, the maximum power transmission efficiency associated with a 2-element phased array (i.e., from (6.31)) and the maximum power transmission efficiency associated with a 2-element parasitic array (i.e., from (6.52)) are plotted.

From Figure 6.20, it is obvious that a 2-element parasitic array appears more and more favorable than a 2-element phased array with the increase of $|S_{21}|^2$. Particularly when $|S_{21}|^2 > 0.4$, a 2-element parasitic array always outperforms a 2-element phased array in terms of maximum power transmission efficiency. This observation agrees with the discussion of the previous section: Strong coupling between the two elements of the wireless power transmitter benefits the parasitic array but penalizes the phased array.

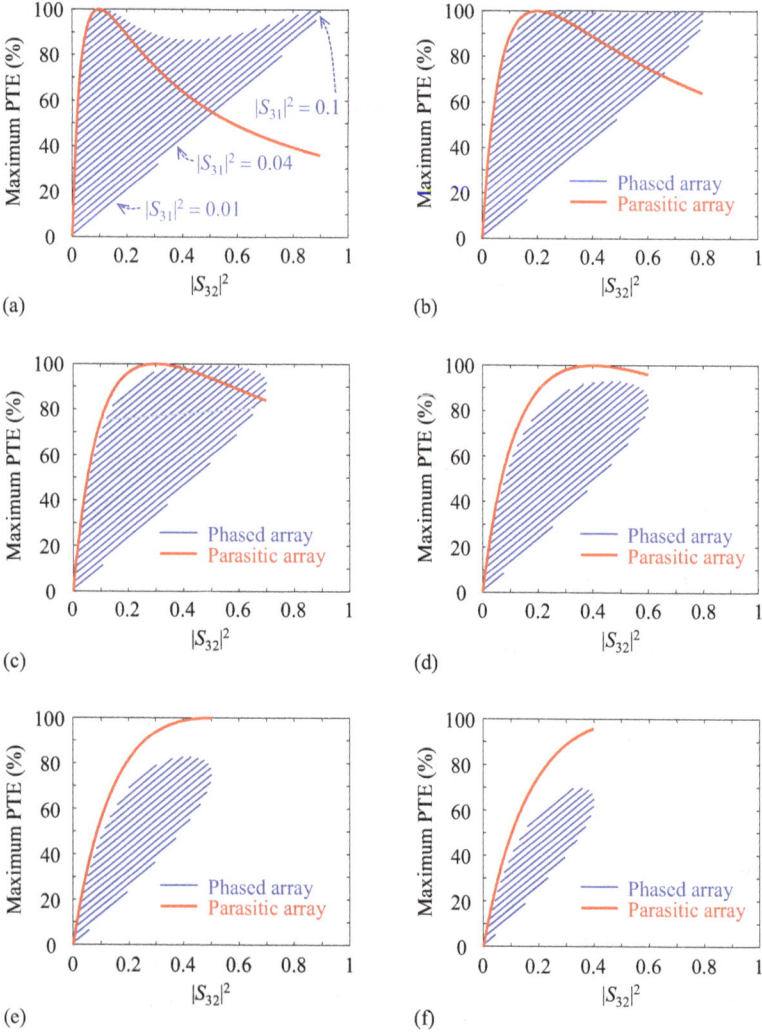

Figure 6.20   Numerical results of maximum power transmission efficiency (PTE) associated with a 2-element phased array and a 2-element parasitic array. (a) When $|S_{21}|^2 = 0.1$; (b) when $|S_{21}|^2 = 0.2$; (c) when $|S_{21}|^2 = 0.3$; (d) when $|S_{21}|^2 = 0.4$; (e) when $|S_{21}|^2 = 0.5$ and (f) when $|S_{21}|^2 = 0.6$.

In Figure 6.20(a), $|S_{21}|^2$ is as small as 0.1. Though the coupling between the driver element and parasitic element is weak, in many cases a 2-element parasitic array demonstrates better power transmission efficiency than a 2-element phased array. Particularly when the balanced condition (that is, $|S_{32}|^2 = |S_{21}|^2 = 0.1$) is satisfied, the maximum power transmission efficiency of a 2-element parasitic array reaches 100%, which can be explained from (6.32). When $|S_{32}|^2 = |S_{21}|^2 = 0.1$, the numerator of the second term on the right-hand side of (6.32) is small. Meanwhile because $|S_{22}|^2 = 1 - |S_{32}|^2 - |S_{21}|^2 = 0.8$ is a large value, the denominator of the second term on the right-hand side of (6.32) could be very small with certain ϕ_r. It is therefore not surprising that the maximum power transmission efficiency of a 2-element parasitic array could reach 100% even when $|S_{21}|^2$ is as small as 0.1. Meanwhile as observed from Figure 6.20(a), the maximum power transmission efficiency of a 2-element parasitic array drops quickly when $|S_{32}|^2$ deviates from 0.1. This means that the probability for a 2-element parasitic array to accomplish large power transmission efficiency is low when $|S_{21}|^2$ is as small as 0.1.

When the value of $|S_{21}|^2$ increases to 0.2 in Figure 6.20(b), the probability for a 2-element parasitic array to accomplish large power transmission efficiency is much greater than in Figure 6.20(a). Specifically, the maximum power transmission efficiency of a 2-element parasitic array is above 80% when $|S_{32}|^2$ resides within a wide range (from 0.1 to 0.5, to be specific), as shown from Figure 6.20(b). When $|S_{21}|^2 = 0.3$ in Figure 6.20(c), when $|S_{21}|^2 = 0.4$ in Figure 6.20(d), and when $|S_{21}|^2 = 0.5$ in Figure 6.20(e), it is very probable for a 2-element parasitic array to accomplish power transmission efficiency above 80% as well. Nevertheless when $|S_{21}|^2 > 0.5$, it is impossible to reach the balanced condition and thus it is impossible for the maximum power transmission efficiency of a 2-element parasitic array to reach 100%.

Based on the numerical results in Figure 6.20, the following two conclusions can be drawn.

(i) A 2-element phased array may offer better power transmission efficiency than a 2-element parasitic array when the coupling between the two elements of the wireless power transmitter is weak. On the other hand, when the coupling between the two elements of the wireless power transmitter is strong, a 2-element parasitic array is likely to outperform a 2-element phased array. With the same number of antenna elements, a parasitic array has lower complexity and lower cost than a phased array. Therefore, the retro-reflective beam-forming technique based on a parasitic array in Figure 6.15 appears to be an excellent candidate to accomplish efficient wireless power transmission in fully-enclosed space.

(ii) It seems that a 2-element parasitic array would offer the optimal performance when $0.2 < |S_{21}|^2 < 0.5$, that is, when the coupling between the driver element and the parasitic element is within the range of $(-7$ dB, -3 dB).

When there are more than two elements in a parasitic array, pursuing closed-form expressions of maximum power transmission efficiency would become unaffordable. However, the insights obtained from analyzing 2-element arrays in this section can be extended to the "$M > 2$" scenarios, as demonstrated by some experimental results in the next section.

6.6   Preliminary numerical and experimental results of microwave power transmission in fully-enclosed space based on parasitic arrays

As analyzed in the previous two sections, the retro-reflective beamforming technique based on a parasitic array portrayed in Figure 6.15 has excellent potential to enable efficient wireless power transmission in fully-enclosed space. Nevertheless, the practical implementation of the retro-reflective beamforming scheme based on a parasitic array relies on the availability of low-cost tunable reactive loads. Ideally, the reflection coefficient associated with the reactive loads ought to have an amplitude value of 1 and a phase value tunable in the range of [0, 360°], which is a fairly difficult task in practice. Fortunately, it seems possible that microwave power transmission based on a parasitic array may demonstrate superb performance without the ideal requirements being fulfilled completely, as shown by some pre-liminary numerical and experimental results in this section.

On the basis of the theoretical model of Figure 6.2, some numerical simula-tions are conducted for microwave power transmission based on a parasitic array [29]. Inside a cubic box with a side length of 1 m and with its six walls made of perfect conductors, there are one driver element and two parasitic elements. Each element is an electric probe extended along the z direction with a length of 17 cm, as described in Figure 6.2. The driver element is excited by a time-harmonic source at 425 MHz. The two parasitic elements are terminated by either short or open. No receiving element is present. Locations of the driver element and parasitic elements are specified in Figure 6.21. Electromagnetic field distribution in the box is simu-lated at the frequency of 425 MHz. The normalized amplitude of E_z in the region of

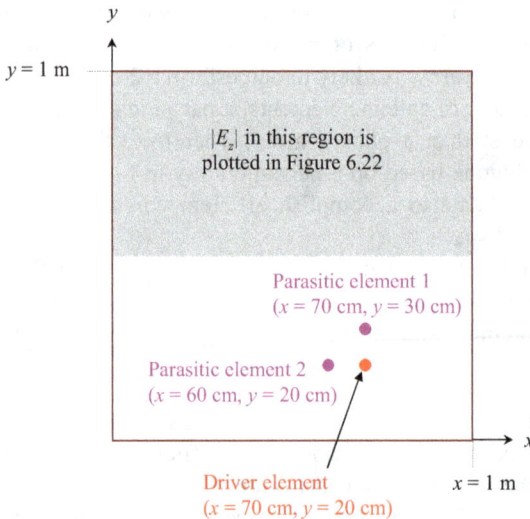

Figure 6.21   Illustration of numerical simulations on the basis of Figure 6.2

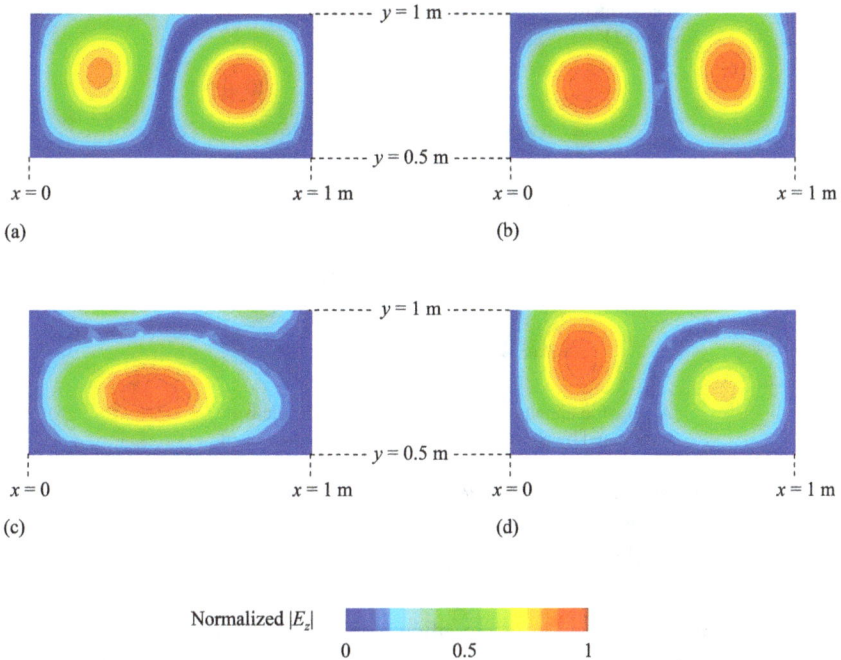

Figure 6.22   Numerical results corresponding to the setup of Figure 6.21. ©
[2017] IEEE. Reprinted, with permission, from [29]. (a) Parasitic
element 1 terminated by open; parasitic element 2 terminated by
open; (b) parasitic element 1 terminated by open; parasitic element 2
terminated by short; (c) parasitic element 1 terminated by short;
parasitic element 2 terminated by open; and (d) parasitic element 1
terminated by short; parasitic element 2 terminated by short.

$(0 < x < 1$ m, 0.5 m $< y < 1$ m) over "$z = 0$ plane" is plotted in Figure 6.22 (normalization is with respect to the largest $|E_z|$ value in the region). The four plots in Figure 6.22 are obtained with the following four scenarios, respectively.

(a) Parasitic element 1 terminated by open; parasitic element 2 terminated by open
(b) Parasitic element 1 terminated by open; parasitic element 2 terminated by short
(c) Parasitic element 1 terminated by short; parasitic element 2 terminated by open
(d) Parasitic element 1 terminated by short; parasitic element 2 terminated by short

It is observed from Figure 6.22 that terminating the parasitic elements by various combinations of open and short results in drastically different field distribution inside

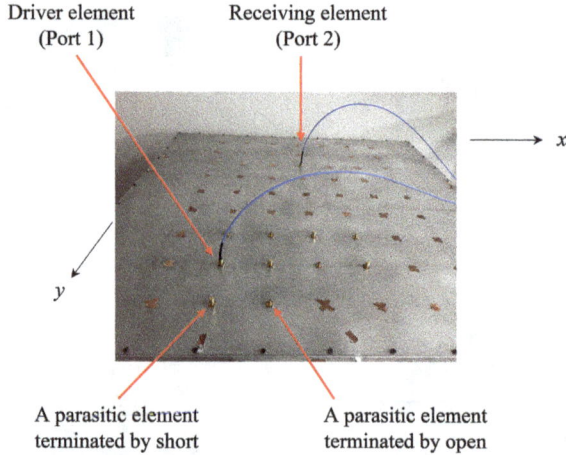

Figure 6.23   A photo of the experimental setup with parasitic elements terminated by either short or open. © [2017] IEEE. Reprinted, with permission, from [29].

the box. This means that enormous reconfigurability could be achieved even when the termination of parasitic elements only has two possible states (that is, open and short).

Encouraged by the numerical results in Figure 6.22, an experiment is conducted as shown by a photo in Figure 6.23 [29]. The cubic box in Figure 6.23 has a side length of 1 m and its six walls are made of aluminum. The z-axis is downward in Figure 6.23, in order to facilitate routing the cables and adjusting the parasitic elements' termination. The driver element, parasitic elements, and receiving element all have a length of 17 cm along the z direction. Scattering parameters with respect to the standard 50-Ω characteristic impedance are measured between the driver element (Port 1) and receiving element (Port 2) at 425 MHz by a network analyzer. When there are one driver element located at ($x = 70$ cm, $y = 80$ cm), one receiving element located at ($x = 50$ cm, $y = 30$ cm), and no parasitic elements in the experimental setup, the measured $|S_{21}|$ is as low as −16.4 dB, corresponding to a power transmission efficiency of 2.3%. Next, nine parasitic elements are incorporated into the experimental setup one by one. Locations of the nine parasitic elements are illustrated in Figure 6.24 and are articulated below.

Parasitic element 1: ($x = 70$ cm, $y = 70$ cm)
Parasitic element 2: ($x = 70$ cm, $y = 90$ cm)
Parasitic element 3: ($x = 60$ cm, $y = 90$ cm)
Parasitic element 4: ($x = 60$ cm, $y = 80$ cm)
Parasitic element 5: ($x = 60$ cm, $y = 70$ cm)
Parasitic element 6: ($x = 50$ cm, $y = 70$ cm)
Parasitic element 7: ($x = 50$ cm, $y = 80$ cm)
Parasitic element 8: ($x = 40$ cm, $y = 80$ cm)
Parasitic element 9: ($x = 40$ cm, $y = 70$ cm)

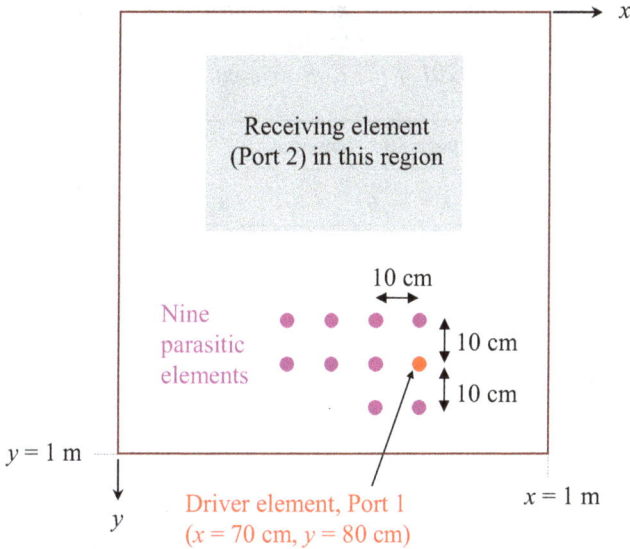

Figure 6.24   Illustration of the experimental setup in Figure 6.23. © [2017] IEEE. Reprinted, with permission, from [29].

With only one parasitic element (which is parasitic element 1) incorporated, two $|S_{21}|$ values are measured when the parasitic element's termination is open and short, respectively; $|S_{21}| = -11.9$ dB (corresponding to a power transmission efficiency of 6.5%) when the parasitic element is terminated by open is the larger value, and this value is recorded as the maximum power transmission efficiency. When parasitic element 1 and parasitic element 2 are incorporated, four $|S_{21}|$ values are measured when the two parasitic elements' terminations are adjusted between open and short; the maximum power transmission efficiency among the four values is recorded. A similar procedure is carried out with more parasitic elements incorporated. The maximum power transmission efficiencies with respect to the number of parasitic elements are tabulated in Table 6.1. When the number of parasitic elements reaches 8 (with $2^8 = 256$ combinations of reconfigurability in total) or 9 (with $2^9 = 512$ combinations of reconfigurability in total), the maximum power transmission efficiency is higher than 80%.

In the results plotted in Figure 6.25, the receiving element's location varies in the shaded region of Figure 6.24. Measurement described in the previous paragraph is conducted at each of the receiving element's locations. Measurement data of three scenarios are presented in Figure 6.25: "0 parasitic," "6 parasitic," and "9 parasitic." In the "0 parasitic" scenario, there are no parasitic elements. In the "6 parasitic" scenario, parasitic elements 1 to 6 are deployed. In the "9 parasitic" scenario, all nine parasitic elements are present. With no parasitic elements, power transmission is very poor at certain "dark spots." In contrast, via tuning the nine parasitic elements it is always possible to achieve a power transmission efficiency

Table 6.1   *Power transmission efficiency with various numbers of parasitic elements when the receiving element is located at (x = 50 cm, y = 30 cm). © [2017] IEEE. Reprinted, with permission, from [29].*

Status of nine parasitic elements									Power transmission efficiency
1	2	3	4	5	6	7	8	9	
									2.3%
o									6.5%
o	o								7.4%
s	o	o							8.9%
s	o	o	o						12.6%
s	o	o	o	o					12.9%
o	o	o	o	o	s				65.9%
o	o	o	o	o	s	o			64.7%
o	s	o	s	o	s	o	s		81.7%
o	s	o	s	o	s	o	s	o	80.4%

Notes: Blank cell stands for "parasitic element not present." "o" stands for "parasitic element terminated by open." "s" stands for "parasitic element terminated by short."

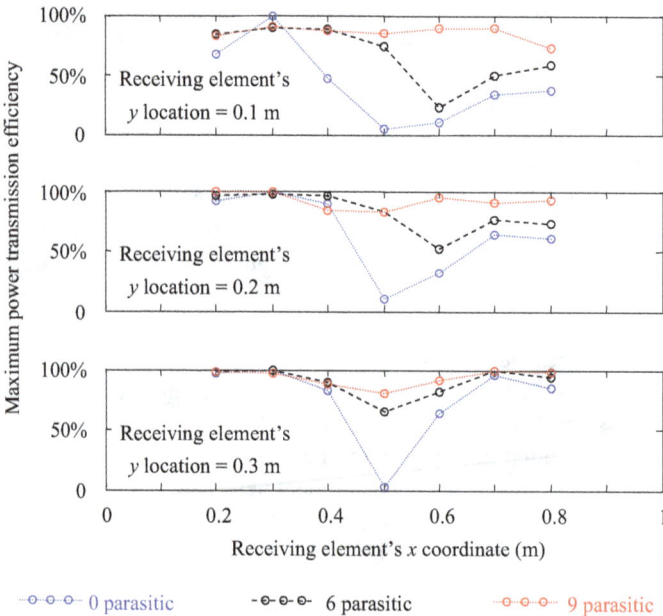

Figure 6.25   *Experimental results corresponding to the setup of Figures 6.23 and 6.24. © [2017] IEEE. Reprinted, with permission, from [29].*

greater than 70%. As an intermediate scenario between "0 parasitic" and "9 parasitic," the power transmission efficiency with six parasitic elements is generally greater than "0 parasitic" but smaller than "9 parasitic."

Another set of experimental results is presented before the end of this section [30]. As the primary improvement with respect to the experiments in Figure 6.23, the reactive load of each parasitic element has four possible states: Open, short, an inductor, and a capacitor. In the experimental setup in Figure 6.26, the cubic cavity is the same cavity as in Figure 6.23 but with a different Cartesian coordinate system. One driver element and multiple parasitic elements are mounted in the "$z = 0$" plane, and a receiving element resides over the "$z = -100$ cm" plane in Figure 6.26. All the elements are electric probes oriented along the z direction with a length of 17 cm. The driver element and receiving element are connected to a vector network analyzer for the measurement of power transmission efficiency. Each parasitic element is terminated by one of the four reactive loads (including open, short, an inductor, and a capacitor). As shown in Figure 6.26, the inductor/capacitor is soldered over a printed circuit board and then connected to the parasitic element through a PCB-to-SMA adapter. Open has a reflection coefficient of 1, and short has a reflection coefficient of -1. The inductive load and capacitive load are selected such that they have reflection coefficients j and $-j$, respectively, at the frequency of 425 MHz. In other words, the four discrete loads correspond to $\phi_r = 0$, $\phi_r = 90°$, $\phi_r = 180°$, and $\phi_r = 270°$, respectively, with ϕ_r representing the phase of reflection coefficient associated with the reactive load. Switching the

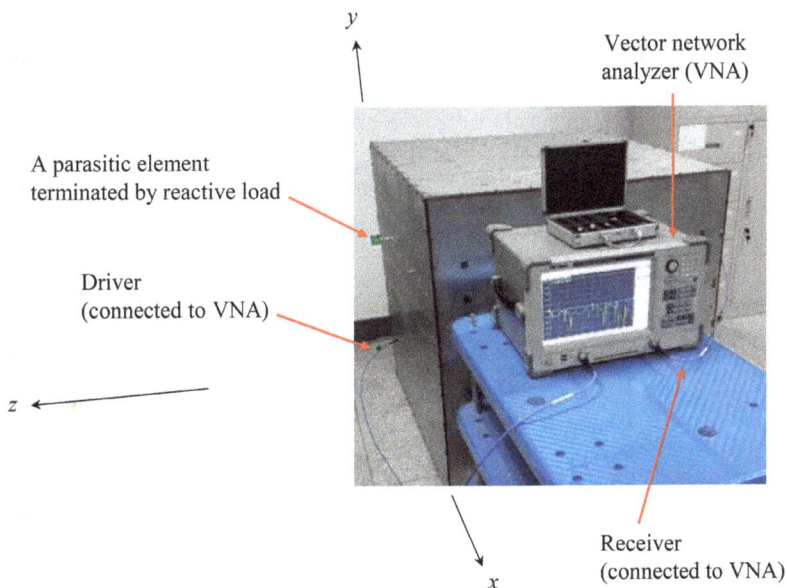

Figure 6.26   A photo of experimental setup with parasitic elements terminated by short, or open, or an inductor, or a capacitor

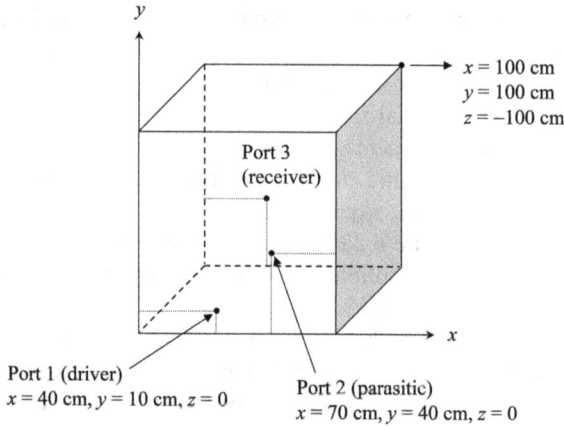

Figure 6.27 *An experimental configuration with one driver element, one parasitic element, and one receiving element*

parasitic elements' termination among the four discrete reactive loads enables verifying the analysis of Sections 6.4 and 6.5 to a large extent, as evidenced by the experimental results below.

As illustrated by an experimental configuration in Figure 6.27, the wireless power transmitter includes two elements located over the "$z = 0$" wall. One of them is the driver element (Port 1), and the other one is a parasitic element (Port 2), with locations specified in Figure 6.27. The wireless power receiver includes one element (Port 3) located over the "$z = -100$ cm" wall. When the receiving element's location is fixed at ($x = 30$ cm, $y = 30$ cm, $z = -100$ cm) and when the parasitic element is absent, the power transmission efficiency between Port 1 and Port 3 is measured to be 31.2%.

When the receiving element's location is fixed at ($x = 30$ cm, $y = 30$ cm, $z = -100$ cm) and with the presence of the parasitic element, the 3×3 scattering parameter matrix is measured as

	Port 1 (driver)	Port2	Port 3 (receiver)
Port 1 (driver)	$S_{11} = -0.38 + j0.59$	$S_{12} = -0.48 - j0.24$	$S_{13} = -0.35 + j0.17$
Port 2	$S_{21} = -0.49 - j0.25$	$S_{22} = -0.23 + j0.5$	$S_{23} = 0.58 - j0.1$
Port 3 (receiver)	$S_{31} = -0.36 + j0.17$	$S_{32} = 0.60 - j0.1$	$S_{33} = 0.47 + j0.48$

If the two elements at Port 1 and Port 2 form a phased array, the maximum power transmission efficiency from the phased array to the receiver is calculated to be 52.4% using (6.31). The 2-element phased array has better power transmission efficiency than the 1-element transmitter, but not significantly.

Suppose the two elements at Port 1 and Port 2 construct a parasitic array, with Port 1 as the driver element and Port 2 as the parasitic element. The maximum power transmission efficiency from the parasitic array to the receiver is calculated

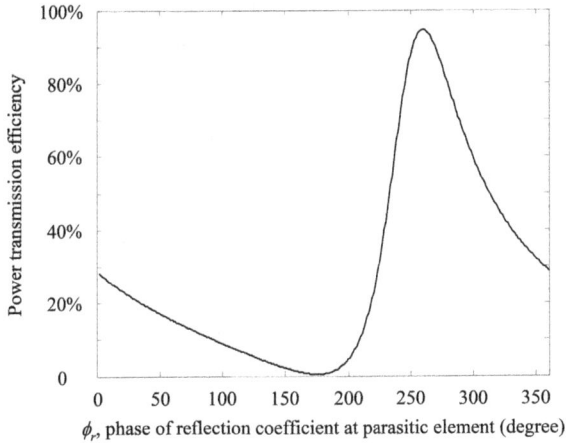

Figure 6.28   *Power transmission efficiency with respect to ϕ_r, for the 2-element parasitic array in Figure 6.27*

to be 99.1% using (6.52). In the configuration of Figure 6.27, $|S_{21}|$ is fairly close to $|S_{32}|$. Thus according to the theoretical derivations of Section 6.5, the parasitic array's power transmission efficiency is able to approach 100%. Figure 6.28 plots the power transmission efficiency calculated from (6.32) when ϕ_r (the phase of reflection coefficient at Port 2) changes its value from 0 to 360°. It is observed from Figure 6.28 that the maximum power transmission efficiency of 94.6% is reached when $\phi_r = 260°$, which verifies the theoretical derivations in Section 6.5. The power transmission efficiency between Port 1 and Port 3 is measured when Port 2 is terminated by the four discrete reactive loads; and, the largest power transmission efficiency among the four measured values is 87.5% when the load is the capacitor. The capacitive load has 270 degrees as the phase of its reflection coefficient, very close to the optimal ϕ_r value observed in Figure 6.28. Unsurprisingly, the capacitive load is able to approach the optimal power transmission efficiency. The capacitive load used in the experiments is not a purely reactive load; specifically, its return loss is measured to be -0.17 dB. As a result, the measured power transmission efficiency of 87.5% is slightly lower than the maximum power transmission efficiency predicted by the theory.

The measurement and data analysis procedure above is repeated when the receiver's y coordinate changes with its z coordinate fixed as -100 cm and its x coordinate fixed as 30 cm. The measured and calculated data are tabulated in Table 6.2. When the transmitter (TX) includes only one element (that is, when the parasitic element is absent), the power transmission efficiency is always poor in Table 6.2. The 2-element phased array occasionally enhances the power transmission efficiency, yet it may also exhibit lower power transmission efficiency when compared to the 1-element transmitter. In all the cases of Table 6.2, the 2-element parasitic array demonstrates higher power transmission efficiency than the 2-element phased array. The maximum power transmission efficiency values from (6.32) and

(6.52) always match each other very well. The measured values of power transmission efficiency are always lower than those predicted by the theory, but not substantially. Out of the seven cases listed in Table 6.2, the 2-element parasitic array
achieves power transmission efficiency greater than 80% in three cases. However, in
the other four cases, the 2-element parasitic array is not capable of accomplishing
highly efficient wireless power transmission. In the case listed at the bottom row of
Table 6.2, the measured $|S_{21}|^2 \approx 0.5$, the measured $|S_{32}|^2 \approx 0.06$, and the power
transmission efficiency data in Table 6.2 agree with Figure 6.20(e) excellently.

On the basis of Figure 6.27, two more parasitic elements are incorporated, as
shown in Figure 6.29. The wireless power transmitter includes four elements located
over the "$z = 0$" wall, and the receiver includes one element located over the "$z = -100$ cm" wall. In the experiments, the locations of the four elements in the wireless

Table 6.2   *Power transmission efficiency (PTE) associated with the 2-element*
array in Figure 6.27

y (cm)	PTE with only 1 element in the transmitter	Maximum PTE of 2-element phased array, calculated from (6.31)	Maximum PTE of 2-element parasitic array, calculated from (6.52)	Maximum PTE of 2-element parasitic array, calculated from (6.32)	Maximum PTE of 2-element parasitic array, measured
20	24.7%	35.2%	86.8%	83.1%	81.4%
30	31.2%	52.4%	99.1%	94.6%	87.5%
40	38.3%	65.6%	97.6%	88.6%	85.9%
50	0.25%	26.2%	77.2%	67.5%	62.1%
60	26.9%	20.8%	28.9%	27.9%	26.2%
70	24.4%	18.3%	41.8%	37.9%	27.9%
80	19.7%	14.9%	39.3%	38.3%	21.9%

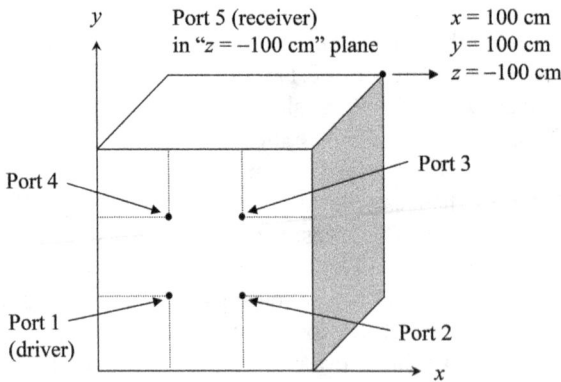

Figure 6.29   *An experimental configuration with one driver element, three*
parasitic elements, and one receiving element

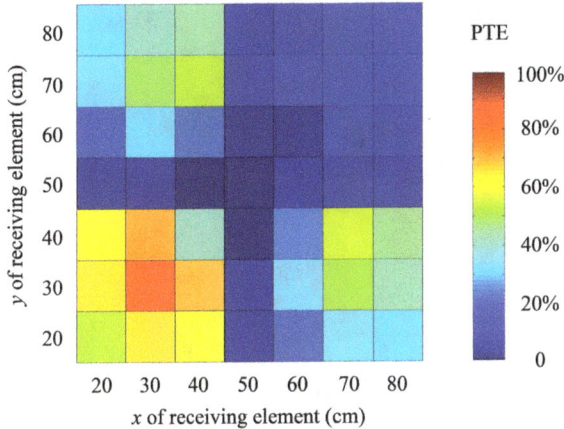

Figure 6.30   Power transmission efficiency (PTE) at various receiver locations, when the transmitter includes only one element in Figure 6.29

power transmitter are fixed at ($x = 30$ cm, $y = 30$ cm), ($x = 30$ cm, $y = 70$ cm), ($x = 70$ cm, $y = 30$ cm), and ($x = 70$ cm, $z = 70$ cm), respectively. The receiving element changes its location in the "$z = -100$ cm" plane. As a z-oriented electric probe, when the receiving element is close to "$x = 0$," "$x = 100$ cm," "$y = 0$," or "$y = 100$ cm" wall, it would be "shorted" by the wall and thus it would be impossible to achieve high power transmission efficiency. Thus in the experiments, the receiving element's location is sampled in the region of (20 cm $\leq x \leq$ 80 cm, 20 cm $\leq y \leq$ 80 cm), with 10 cm as the spacing between two adjacent samples along either x or y.

With the absence of the three elements at Port 2, Port 3, and Port 4 (in other words, when the wireless power transmitter includes only one element at Port 1), the power transmission efficiency between Port 1 and Port 5 is measured and plotted in Figure 6.30. The power transmission efficiency is highly position-sensitive without an antenna array being incorporated into the wireless power transmitter.

With the presence of all the five elements, the 5×5 scattering parameters are measured with a given receiver location. Suppose the four elements at Port 1, Port 2, Port 3, and Port 4 form a phased array. The theoretical maximum power transmission efficiency is calculated from (6.31) and plotted in Figure 6.31 when the receiver location changes in the "$z = -100$ cm" plane. Though better than the 1-element transmitter generally, the power transmission efficiency associated with the 4-element phased array is below 80% throughout the region.

Next, suppose the four elements at Port 1, Port 2, Port 3, and Port 4 construct a parasitic array with Port 1 as the driver element and the other three as parasitic elements. The three parasitic elements are assumed to be terminated by purely reactive loads whose reactance could be adjusted continuously. By making use of the 5×5 scattering parameters obtained from measurements, the maximum power transmission efficiency from the parasitic array to the receiver can be found via (6.25) by sweeping the reflection coefficients' phase between 0 and 360° at the

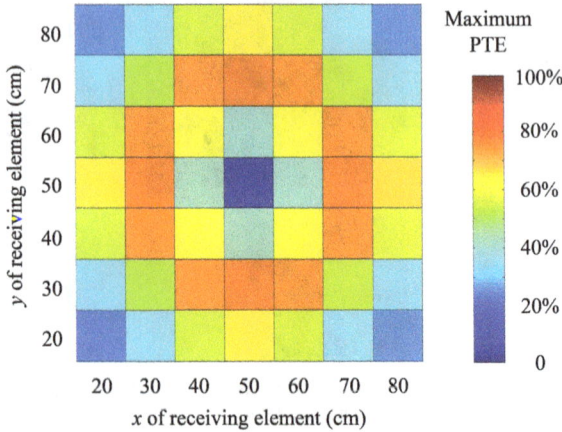

Figure 6.31   Maximum power transmission efficiency (PTE) at various receiver locations, when the four elements in the transmitter form a phased array in Figure 6.29

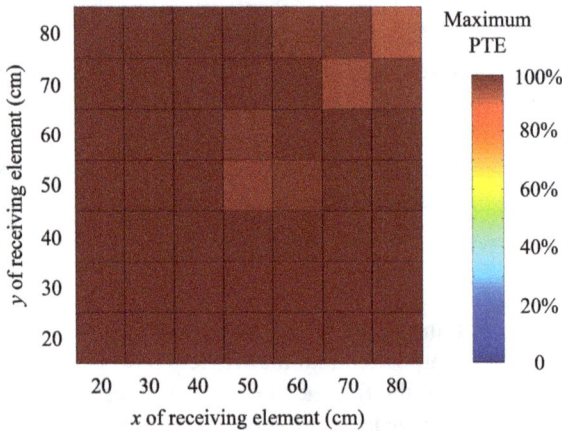

Figure 6.32   Theoretical maximum power transmission efficiency (PTE) at various receiver locations, when the four elements in the transmitter form a parasitic array in Figure 6.29

three parasitic ports. As displayed in Figure 6.32, the maximum power transmission efficiency of the 4-element parasitic array is above 90% when the receiver's location changes in the region of $(20\text{ cm} \leq x \leq 80\text{ cm}, 20\text{ cm} \leq y \leq 80\text{ cm})$. When the receiver's location is at $(x = 50\text{ cm}, y = 50\text{ cm}, z = -100\text{ cm})$, several measured scattering parameters are articulated: $|S_{21}| = -5.99$ dB, $|S_{31}| = -6.76$ dB, and $|S_{41}| = -6.19$ dB. In other words, the coupling between the driver element and each parasitic element is within the optimal range found at the end of Section 6.5

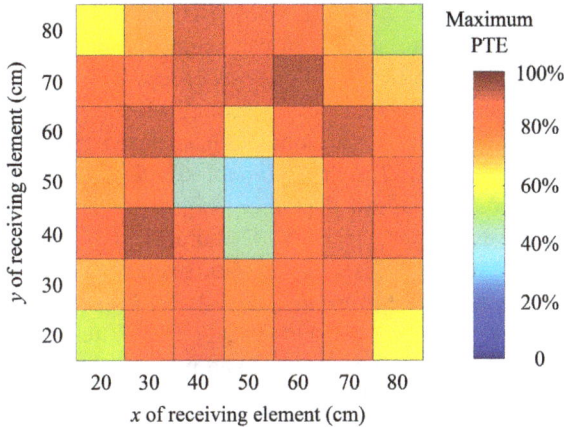

Figure 6.33   Measured maximum power transmission efficiency (PTE) at various receiver locations, when the four elements in the transmitter form a parasitic array in Figure 6.29

(that is, between −7 dB and −3 dB). Though the optimal range was found by assuming there is only one parasitic element, it seems that it also roughly holds true for a parasitic array with multiple parasitic elements.

Finally, at each receiver location, the power transmission efficiency between Port 1 and Port 5 is measured when the terminations of Port 2, Port 3, and Port 4 are switched among the four discrete reactive loads; the largest measured power transmission efficiency among the $4^3 = 64$ combinations is recorded and plotted in Figure 6.33. At most of the receiver locations, the measured maximum power transmission efficiency is close to the values in Figure 6.32. However, at several locations, the measured maximum power transmission efficiency is as poor as 40%, obviously because the four discrete loads are too few or too sparse to approach the optimal power transmission efficiency predicted by the theory. The availability of circuits to tune reactance values with a wide tuning range, low loss, and low cost would definitely facilitate the development of efficient microwave power transmission based on parasitic arrays in fully-enclosed space.

References

[1]   Chen C., Wang X., Lu M. 'Wireless charging to multiple electronic devices simultaneously in enclosed box'. *Presented at Asia-Pacific International Symposium on Electromagnetic Compatibility*; Shenzhen, China. 2016.

[2]   Garnica J., Chinga R.A., Lin J. 'Wireless power transmission: From far field to near field'. *Proceedings of the IEEE*. 2013;101(6):1321–31.

[3]   Wu J., Liang J., Wang X., Chen C., Zhang X., Lu M. 'Feasibility study of efficient wireless power transmission in satellite interior'. *Microwave and Optical Technology Letters*. 2016;58(10):2518–22.

[4]   Goto H., Shinohara N., Mitani T., Dosho H., Mizuno M. 'Study on microwave power transfer to sensors in car engine compartment'. *Presented at IEEE Wireless Power Transfer Conference*; Jeju, Korea. 2014.

[5]   Takano I., Furusu D., Watanabe Y., Tamura M. 'Study on cavity resonator wireless power transfer to sensors in an enclosed space with scatterers'. *Presented at IEEE MTT-S International Conference on Microwaves for Intelligent Mobility*; Nagoya, Japan. 2017.

[6]   Takano I., Furusu D., Watanabe Y., Tamura M. 'Cavity resonator wireless power transfer in an enclosed space with scatterers utilizing metal mesh'. *IEICE Transactions on Electronics*. 2017;E100-C(10):841–9.

[7]   Montgomery K.L., Yeh A.J., Ho J.S., *et al.* 'Wirelessly powered, fully internal optogenetics for brain, spinal and peripheral circuits in mice'. *Nature Methods*. 2015;12(10):969–74.

[8]   Mei H., Thackston K.A., Bercich R.A., Jefferys J.G.R., Irazoqui P.P. 'Cavity resonator wireless power transfer system for freely moving animal experiments'. *IEEE Transactions on Biomedical Engineering*. 2017;64(4):775–85.

[9]   Shin G., Gomez A.M., Al-Hasani R., *et al.* 'Flexible near-field wireless optoelectronics as subdermal implants for broad applications in optogenetics'. *Neuron*. 2017;93(3):509–21.

[10] Umenei A.E. Understanding low-frequency non-radiative power transfer. 2011 [updated 2011; cited 2023]; Available from: https://www.wirelesspowerconsortium.com/data/downloadables/6/8/9/understanding-low-frequency-non-radiative-power-transfer-8_8_11.pdf.

[11] Chabalko M.J., Shahmohammadi M., Sample A.P. 'Quasistatic cavity resonance for ubiquitous wireless power transfer'. *PLoS One*. 2017;12(2): e0169045.

[12] Sasatani T., Yang J., Chabalko M.J., Kawahara Y., Sample A.P. 'Room-wide wireless charging and load-modulation communication via quasistatic cavity resonance'. *Proceedings of the ACM on Interactive, Mobile, Wearable and Ubiquitous Technologies*. 2018;2(4):1–23.

[13] Chabalko M.J., Sample A.P. 'Resonant cavity mode enabled wireless power transfer'. *Applied Physics Letters*. 2014;105(24):243902.

[14] Tsai L.L. 'A numerical solution for the near and far fields of an annular ring of magnetic current'. *IEEE Transactions on Antennas and Propagation*. 1972;20(5):569–76.

[15] Tai C.-T., Rozenfeld P. 'Different representations of dyadic Green's functions for a rectangular cavity'. *IEEE Transactions on Microwave Theory and Techniques*. 1976;24(9):597–601.

[16] Lu M., Bredow J.W., Jung S., Tjuatja S. 'Evaluation of Green's functions of rectangular cavities around resonant frequencies in the method of moments'. *IEEE Antennas and Wireless Propagation Letters*. 2009;8:204–8.

[17] Lu M., Jung S. 'On the well-posedness of integral equations associated with cavity Green's functions around resonant frequencies'. *Microwave and Optical Technology Letters*. 2009;51(6):1476–81.

[18] Wang X., Cheng H., Cao X., Chen C., Lu M. 'Modal analysis based on an integral equation method for characterizing wireless channels in a fully-enclosed environment'. *Progress In Electromagnetics Research B.* 2020;86:59–76.

[19] Panitz M., Hope D.C. 'Characteristics of wireless systems in resonant environments'. *IEEE Electromagnetic Compatibility Magazine.* 2014;3 (3):64–75.

[20] Hope D.C. *Towards a Wireless Aircraft*: University of York, United Kingdom; 2011.

[21] Hill D.A. *Electromagnetic Fields in Cavities: Deterministic and Statistical Theories.* Hoboken, NJ: Wiley-IEEE Press; 2009.

[22] Chabalko M.J., Sample A.P. 'Three-dimensional charging via multi-mode resonant cavity enabled wireless power transfer'. *IEEE Transactions on Power Electronics.* 2015;30(11):6163–73.

[23] Mei H., Huang Y.-W., Thackston K.A., Irazoqui P.P. 'Optimal wireless power transfer to systems in an enclosed resonant cavity'. *IEEE Antennas and Wireless Propagation Letters.* 2015;15:1036–9.

[24] Korhummel S., Rosen A., Popovic Z., 'Over-moded cavity for multiple-electronic-device wireless charging'. *IEEE Transactions on Microwave Theory and Techniques.* 2014;62(4):1074–9.

[25] Sasatani T., Chabalko M.J., Kawahara Y., Sample A.P. 'Multimode quasi-static cavity resonators for wireless power transfer'. *IEEE Antennas and Wireless Propagation Letters.* 2017;16:2746–9.

[26] Balanis C.A. *Antenna Theory: Analysis and Design.* 3rd ed. New York: Wiley-Interscience; 2005.

[27] Nimura S., Furusu D., Tamura M. 'Improvement in power transmission efficiency for cavity resonance-enabled wireless power transfer by utilizing probes with variable reactance'. *IEEE Transactions on Microwave Theory and Techniques.* 2020;68(7):2734–44.

[28] Murata K., Takiya K., Kondo S., Honma N. 'Performance improvement in implicit cavity resonant wireless power transfer using dual parasitic antennas'. *IEEE Antennas and Wireless Propagation Letters.* 2022;21(6):1273–7.

[29] Wang X., Chen C., Wong H., Lu M. 'A reconfigurable scheme of wireless power transmission in fully enclosed environments'. *IEEE Antennas and Wireless Propagation Letters.* 2017;16:2959–62.

[30] Wang X., Wang X., Li M., Lu M. 'Reconfigurable wireless power transmission in fully-enclosed space using antenna array'. *IEEE Access.* 2019;7 (1):173098–110.

Appendices

Appendix A: Analytical and numerical evaluation of a two-fold integral in Chapter 2

The following two-fold integral is derived in Section 2.6.

$$\int_{-L/2}^{+L/2} dx' \int_{-W/2}^{+W/2} dz' \frac{(x_o - x')^2 + (y_o)^2}{\left[(x_o - x')^2 + (y_o)^2 + (z_o - z')^2\right]^{\frac{3}{2}}} \tag{A.1}$$

The integral in (A.1) is evaluated analytically and numerically in this appendix. Because

$$\frac{d}{dz'}\left\{ \frac{z' - z_o}{\beta^2 \sqrt{(z_o - z')^2 + \beta^2}} \right\} = \frac{1}{\left[(z_o - z')^2 + \beta^2\right]^{\frac{3}{2}}} \tag{A.2}$$

the integration with respect to dz' in (A.1) is

$$\int_{-W/2}^{W/2} dz' \frac{(x_o - x')^2 + (y_o)^2}{\left[(z_o - z')^2 + (x_o - x')^2 + (y_o)^2\right]^{\frac{3}{2}}}$$

$$= \frac{\dfrac{W}{2} - z_o}{\sqrt{\left(\dfrac{W}{2} - z_o\right)^2 + (x_o - x')^2 + (y_o)^2}} - \frac{-\dfrac{W}{2} - z_o}{\sqrt{\left(-\dfrac{W}{2} - z_o\right)^2 + (x_o - x')^2 + (y_o)^2}}$$

$$\tag{A.3}$$

The integral in (A.1) therefore becomes

$$
\int_{-L/2}^{+L/2} dx' \int_{-W/2}^{+W/2} dz' \frac{(x_o - x')^2 + (y_o)^2}{\left[(x_o - x')^2 + (y_o)^2 + (z_o - z')^2\right]^{\frac{3}{2}}}
$$

$$
= \int_{-L/2}^{+L/2} dx' \frac{\dfrac{W}{2} - z_o}{\sqrt{\left(\dfrac{W}{2} - z_o\right)^2 + (x_o - x')^2 + (y_o)^2}} \tag{A.4}
$$

$$
- \int_{-L/2}^{+L/2} dx' \frac{-\dfrac{W}{2} - z_o}{\sqrt{\left(-\dfrac{W}{2} - z_o\right)^2 + (x_o - x')^2 + (y_o)^2}}
$$

Because

$$
\frac{d}{dx'}\left\{\ln\left[(x' - x_o) + \sqrt{(x' - x_o)^2 + a^2}\right]\right\} = \frac{1}{\sqrt{(x' - x_o)^2 + a^2}} \tag{A.5}
$$

the two integrals on the right-hand side of (A.4) are

$$
\int_{-L/2}^{+L/2} dx' \frac{\dfrac{W}{2} - z_o}{\sqrt{\left(\dfrac{W}{2} - z_o\right)^2 + (x_o - x')^2 + (y_o)^2}}
$$

$$
= \left(\frac{W}{2} - z_o\right)\ln\left[\left(\frac{L}{2} - x_o\right) + \sqrt{\left(\frac{L}{2} - x_o\right)^2 + \left(\frac{W}{2} - z_o\right)^2 + (y_o)^2}\right]
$$

$$
- \left(\frac{W}{2} - z_o\right)\ln\left[\left(-\frac{L}{2} - x_o\right) + \sqrt{\left(-\frac{L}{2} - x_o\right)^2 + \left(\frac{W}{2} - z_o\right)^2 + (y_o)^2}\right] \tag{A.6}
$$

and

$$
\int_{-L/2}^{+L/2} dx' \frac{-\dfrac{W}{2} - z_o}{\sqrt{\left(-\dfrac{W}{2} - z_o\right)^2 + (x_o - x')^2 + (y_o)^2}}
$$

$$
= \left(-\frac{W}{2} - z_o\right)\ln\left[\left(\frac{L}{2} - x_o\right) + \sqrt{\left(\frac{L}{2} - x_o\right)^2 + \left(\frac{W}{2} + z_o\right)^2 + (y_o)^2}\right]
$$

$$
- \left(-\frac{W}{2} - z_o\right)\ln\left[\left(-\frac{L}{2} - x_o\right) + \sqrt{\left(-\frac{L}{2} - x_o\right)^2 + \left(\frac{W}{2} + z_o\right)^2 + (y_o)^2}\right] \tag{A.7}
$$

Finally, the two-fold integral in (A.1) can be evaluated analytically as

$$
\int_{-L/2}^{+L/2} dx' \int_{-W/2}^{+W/2} dz' \, \frac{(x_o - x')^2 + (y_o)^2}{\left[(x_o - x')^2 + (y_o)^2 + (z_o - z')^2\right]^{\frac{3}{2}}}
$$

$$
= \left(\frac{W}{2} - z_o\right) \ln\left[\left(\frac{L}{2} - x_o\right) + \sqrt{\left(\frac{L}{2} - x_o\right)^2 + \left(\frac{W}{2} - z_o\right)^2 + (y_o)^2}\right]
$$

$$
- \left(\frac{W}{2} - z_o\right) \ln\left[\left(-\frac{L}{2} - x_o\right) + \sqrt{\left(\frac{L}{2} + x_o\right)^2 + \left(\frac{W}{2} - z_o\right)^2 + (y_o)^2}\right]
$$

$$
+ \left(\frac{W}{2} + z_o\right) \ln\left[\left(\frac{L}{2} - x_o\right) + \sqrt{\left(\frac{L}{2} - x_o\right)^2 + \left(\frac{W}{2} + z_o\right)^2 + (y_o)^2}\right]
$$

$$
- \left(\frac{W}{2} + z_o\right) \ln\left[\left(-\frac{L}{2} - x_o\right) + \sqrt{\left(\frac{L}{2} + x_o\right)^2 + \left(\frac{W}{2} + z_o\right)^2 + (y_o)^2}\right]
$$

$$(A.8)$$

The two-fold integral in (A.1) can also be approximated by a two-fold summation:

$$
\int_{-L/2}^{+L/2} dx' \int_{-W/2}^{+W/2} dz' \, \frac{(x_o - x')^2 + (y_o)^2}{\left[(x_o - x')^2 + (y_o)^2 + (z_o - z')^2\right]^{\frac{3}{2}}}
$$

$$
\cong s_x s_z \sum_{n=1}^{N} \sum_{m=1}^{M} \frac{(x_o - x'_{nm})^2 + (y_o)^2}{\left[(x_o - x'_{nm})^2 + (y_o)^2 + (z_o - z'_{nm})^2\right]^{\frac{3}{2}}}
$$

$$(A.9)$$

where

$$
s_x = \frac{L}{N}
$$

$$
s_z = \frac{W}{M}
$$

$$
x'_{nm} = \left(n - \frac{N+1}{2}\right) s_x
$$

and

$$
z'_{nm} = \left(m - \frac{M+1}{2}\right) s_z.
$$

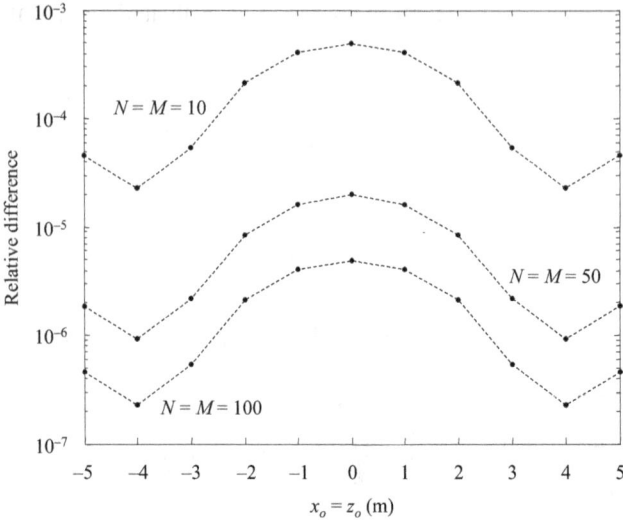

Figure A.1   Numerical results pertinent to evaluating the integral in (A.1)

Some numerical results pertinent to the integral of (A.1) are plotted in Figure A.1. One set of parameters from Section 2.6 is used in Figure A.1: $L = 3$ m, $W = 3$ m, and $y_o = 5$ m. In Figure A.1, the relative difference is calculated as

$$\text{Relative difference} = \frac{|I_{analytical} - I_{numerical}|}{|I_{analytical}|}, \tag{A.10}$$

where $I_{analytical}$ is evaluated via the analytical expression of (A.8) and $I_{numerical}$ is obtained by the two-fold summation of (A.9). In Figure A.1, the relative difference is plotted when x_o and z_o have the same value and their value varies between -5 m and 5 m. The three curves in Figure A.1 are calculated with $N = M = 10$, $N = M = 50$, and $N = M = 100$, respectively. When N and M are as small as 10, the relative difference is below 10^{-3}. When N and M increase, the relative difference becomes smaller, as expected. The numerical results in Figure A.1 indicate that the integrand in (A.1) is very smooth with the parameters of Section 2.6, and as a result, a small number of quadrature points are sufficient to evaluate the integral precisely.

Appendix B: Evaluation of a two-fold integral in Chapter 3

The following two-fold integral is derived in Section 3.5.

$$\int_{-L/2}^{+L/2} dx' \int_{-W/2}^{+W/2} dz' \frac{\left[(x_o - x')^2 + (y_o)^2\right]^2}{\left[(x_o - x')^2 + (y_o)^2 + (z_o - z')^2\right]^3}. \tag{B.1}$$

It is quite similar to (A.1) in Appendix A. The integral in (B.1) can be evaluated with high precision, although it cannot be evaluated analytically.

Because

$$\frac{d}{dz'}\left\{\frac{5\beta^3(z'-z_o)+3\left[\beta^2+(z'-z_o)^2\right]^2\tan^{-1}\left[\frac{(z'-z_o)}{\beta}\right]+3\beta(z'-z_o)^3}{8\beta^5\left[\beta^2+(z'-z_o)^2\right]^2}\right\}$$

$$=\frac{1}{\left[(z'-z_o)^2+\beta^2\right]^3}$$

(B.2)

the integration over z' in (B.1) can be conducted analytically:

$$\int_{-W/2}^{+W/2}dz'\frac{\left[(x_o-x')^2+(y_o)^2\right]^2}{\left[(x_o-x')^2+(y_o)^2+(z_o-z')^2\right]^3}$$

$$=\frac{5\left[(x_o-x')^2+(y_o)^2\right]\left(\frac{W}{2}-z_o\right)}{8\left[(x_o-x')^2+(y_o)^2+\left(\frac{W}{2}-z_o\right)^2\right]^2}-\frac{5\left[(x_o-x')^2+(y_o)^2\right]\left(-\frac{W}{2}-z_o\right)}{8\left[(x_o-x')^2+(y_o)^2+\left(-\frac{W}{2}-z_o\right)^2\right]^2}$$

$$+\frac{3\left[(x_o-x')^2+(y_o)^2+\left(\frac{W}{2}-z_o\right)^2\right]\tan^{-1}\left[\frac{\left(\frac{W}{2}-z_o\right)}{\sqrt{(x_o-x')^2+(y_o)^2}}\right]}{8\sqrt{(x_o-x')^2+(y_o)^2}\left[(x_o-x')^2+(y_o)^2+\left(\frac{W}{2}-z_o\right)^2\right]^2}$$

$$-\frac{3\left[(x_o-x')^2+(y_o)^2+\left(-\frac{W}{2}-z_o\right)^2\right]\tan^{-1}\left[\frac{\left(-\frac{W}{2}-z_o\right)}{\sqrt{(x_o-x')^2+(y_o)^2}}\right]}{8\sqrt{(x_o-x')^2+(y_o)^2}\left[(x_o-x')^2+(y_o)^2+\left(-\frac{W}{2}-z_o\right)^2\right]^2}$$

$$+\frac{3\left(\frac{W}{2}-z_o\right)^3}{8\left[(x_o-x')^2+(y_o)^2+\left(\frac{W}{2}-z_o\right)^2\right]^2}-\frac{3\left(-\frac{W}{2}-z_o\right)^3}{8\left[(x_o-x')^2+(y_o)^2+\left(-\frac{W}{2}-z_o\right)^2\right]^2}$$

(B.3)

After (B.3) is substituted into (B.1), the remaining one-dimensional integration over x' can be approximated by a discrete summation. Since the integrand is smooth, a small number of quadrature points yields high precision. One set of parameters at the end of Appendix A are used as an example: $L = 3$ m, $W = 3$ m, $x_o = 0$, $z_o = 0$, and $y_o = 5$ m. With 50 quadrature points sampled uniformly in the range of $[-L/2, L/2]$ and with uniform sampling weights, the discrete sum achieves at least five digits of precision.

Appendix C: Formulations of linear regression

Given a discrete data set of $(x_1, a_1), (x_2, a_2), \ldots, (x_M, a_M)$, the mathematical operation of linear regression intends to find a linear function $a = c_0 + c_x x$ that fits the discrete dataset. The difference between the linear function and the dataset is

$$c_0 + c_x x_1 - a_1, \ c_0 + c_x x_2 - a_2, \ \ldots, \ c_0 + c_x x_M - a_M. \tag{C.1}$$

Based on the terms in (C.1), the L2 error is defined as

$$E = \sum_{m=1}^{M} (c_0 + c_x x_m - a_m)^2. \tag{C.2}$$

The linear function $a = c_0 + c_x x$ is considered to best fit the discrete dataset when the L2 error is minimized. When the L2 error is minimized, the derivatives of L2 error with respect to c_0 and c_x are zero.

$$\frac{\partial E}{\partial c_0} = 2\sum_{m=1}^{M} (c_0 + c_x x_m - a_m) = 0$$

$$\frac{\partial E}{\partial c_x} = 2\sum_{m=1}^{M} (c_0 + c_x x_m - a_m)x_m = 0 \tag{C.3}$$

From the two equations in (C.3), it is easy to obtain

$$c_x = \frac{\sum_{m=1}^{M} (x_m - \bar{x})(a_m - \bar{a})}{\sum_{m=1}^{M} (x_m - \bar{x})^2} \tag{C.4}$$

and

$$c_0 = \bar{a} - c_x \bar{x}, \tag{C.5}$$

where

$$\bar{x} = \frac{1}{M}\sum_{m=1}^{M} x_m$$

and

$$\bar{a} = \frac{1}{M}\sum_{m=1}^{M} a_m.$$

Another variable "*z*" can be incorporated into the formulations above straightforwardly. To be specific, given a discrete dataset of (x_1, z_1, α_1), (x_2, z_2, α_2), $\ldots$, (x_M, z_M, α_M), a linear function $\alpha = c_0 + c_x x + c_z z$ fits the discrete dataset with

$$c_x = \frac{S_{zz} S_{\alpha x} - S_{xz} S_{\alpha z}}{S_{xx} S_{zz} - (S_{xz})^2} \tag{C.6}$$

$$c_z = \frac{S_{xx} S_{\alpha z} - S_{xz} S_{\alpha x}}{S_{xx} S_{zz} - (S_{xz})^2} \tag{C.7}$$

and

$$c_0 = \bar{\alpha} - c_x \bar{x} - c_z \bar{z}, \tag{C.8}$$

where

$$S_{xx} = \sum_{m=1}^{M} (x_m - \bar{x})^2$$

$$S_{zz} = \sum_{m=1}^{M} (z_m - \bar{z})^2$$

$$S_{xz} = \sum_{m=1}^{M} (x_m - \bar{x})(z_m - \bar{z})$$

$$S_{\alpha x} = \sum_{m=1}^{M} (\alpha_m - \bar{\alpha})(x_m - \bar{x})$$

$$S_{\alpha z} = \sum_{m=1}^{M} (\alpha_m - \bar{\alpha})(z_m - \bar{z})$$

$$\bar{x} = \frac{1}{M} \sum_{m=1}^{M} x_m$$

$$\bar{z} = \frac{1}{M} \sum_{m=1}^{M} z_m$$

and

$$\bar{\alpha} = \frac{1}{M} \sum_{m=1}^{M} \alpha_m.$$

The formulations in (C.6), (C.7), and (C.8) are made use of in Section 5.3.

Index

acoustic power transmission
technology 5
analog-to-digital converter (ADC) 125,
142, 168, 217
antenna array 11, 83
antenna efficiency 36, 212
antenna gain 10, 27
antenna loss 27, 36, 212, 215
array factor 34, 189
artificial satellites 179

band-pass-filter (BPF) 143, 217
beam collection efficiency 210
beam steering 8
beamforming 8

capacitive coupling technology for
wireless power transmission 6
circulators 127
code-division multiple access 155, 164,
175
constructive coefficient 55, 189

digital retro-reflective beamformer 168
digital-to-analog converter (DAC) 125,
144, 168
direction of arrival (DOA) 17, 41, 117
duplexing 122
dyadic Green's functions 234

Earth's geostationary orbit 179, 181
effective aperture 11, 27, 39, 212

electric field integral equation 232
electrical far zone 42
electrical near zone 43, 47
electricity grid 2
electromagnetic absorber 212
electromagnetic compatibility 3
electromagnetic plane
wave 10, 25

Fourier series 80
Fourier transformation 80, 109, 168
discrete 142
frequency-division duplexing 128–30,
138
frequency-division multiple access
155, 164, 175
Friis transmission equation 11, 27
full conjugation antenna array
technique 88, 109, 111, 115
fully-enclosed space 229
microwave power transmission in
231
retro-reflective beamforming
technique in 232

geometrical far zone 42, 65, 111
geometrical near zone 43, 65, 111

Hertzian dipoles 43, 66–7, 75, 90, 112,
152
heterodyne method for retro-reflective
beamforming 121

inductive coupling technology for wireless power transmission 6, 43, 230

Industrial, Scientific, and Medical (ISM) frequency bands 182

Internet of Things 4
microwave power transmission in 133
retro-reflective beamforming in 136

isotropic power density 10, 26

laser propagation 7

light-emitting diode (LED) 139

linear regression 196, 280–1

log-periodic antennas 81

low-frequency electric fields 6

low-frequency magnetic fields 6, 231

matched filter 109

Maxwell's equations 2

Microstrip patch antennas 12, 36, 39, 81, 97, 139, 183, 214

microwave oscillator 141

microwave power transmission technology 1, 5
in fully-enclosed space 231
in Internet of Things 4, 134–5
in space solar power applications 181–8
to mobile targets 9–17

microwave remote sensing 92

microwave-to-DC conversion efficiency 145

microwave-to-DC converter 137

microwave-induced hyperthermia 92

millimeter wave 6, 7, 8

modal analysis 232

optical power transmission technology 6, 7

parasitic arrays 229, 246, 247

Personal Area Networks 4

phase conjugation antenna array technique 85, 87, 109, 111, 115

phase shifter 41, 123, 144, 172, 217

phased array technique 9–17
as receiving antenna 38–42
as transmitting antenna 25
one-dimensional linear phased array 33–6
two-dimensional phased array 36–8
inter-element spacing 13–14, 35, 40, 44, 77
mutual coupling among antenna elements 73–81
power transmission efficiency 27
in electrical far zone 42–7
in electrical near zone 47–53
in geometrical far zone and geometrical near zone 65–73

pilot signal 19

planar antenna elements 44

plane wave region 16, 51, 62

power amplifiers 119, 144

power constructive coefficient 70, 114

power transmission efficiency 3, 11, 27, 108, 149, 183, 211, 230

Poynting vector 1, 26, 32, 38, 52, 200, 209

pulse as pilot signal 141

radio-frequency identification (RFID) tags 15, 134
active RFID tags 134
passive RFID tags 134
tag readers 133

rectenna 223

rectifier 140, 144, 145, 149, 223

reflection coefficient 236, 237, 250, 253, 260, 265, 267, 269

retro-reflective beamforming technique 1, 5, 17
  circuit model of 105–11
  for microwave power transmission 18–19
  for wireless power transmission to multiple targets 151–77
  in electrical far zone 83–6
  in electrical near zone 86–90
  in fully-enclosed space 229, 246
  in geometrical near zone and geometrical far zone 111–17
  in Internet of Things applications 136
  in space solar power applications 182
  practical implementations of 117–30

retro-reflectivity 17

rotating-element electric-field vector method 125

scattering parameters 105, 241, 249

Schottky diodes 145

slot antennas 12

solar power harvesting 179

space solar power satellites (SSPS) 179
  retro-reflective beamforming in 182
  wireless power reception on Earth 220–6
  wireless power transmission in 181

space-division multiple access 164

Terahertz wave 6, 7, 8

Thevenin's theorem 74

three dimensional spatial division multiplexing 92

time-division duplexing 126–7, 137, 138, 155

time-division multiple access 155, 164, 175

time-reversal 18, 109–10, 125

Van Atta method for retro-reflective beamforming 119–20

wireless communication 5, 20, 43, 120, 124, 127, 136, 164, 238

wireless power transmission 1–5
  based on acoustic waves 5
  based on capacitive coupling 6
  based on inductive coupling 6
  based on microwave 5–9
  based on optical waves 7
  based on phased array 9–17
  based on retro-reflective beamforming 17–20
  in fully-enclosed space 229
  in Internet of Things 4, 134
  in space solar power applications 181–8
  to mobile targets 1, 4–5
  to stationary targets 4

wireless sensors 15, 231

Yagi-Uda antennas 81, 248